Asymptotic Analysis *of* Differential Equations

Revised Edition

Asymptotic Analysis *of* Differential Equations

Revised Edition

Roscoe B White
Princeton University, USA

ICP Imperial College Press

Published by

Imperial College Press
57 Shelton Street
Covent Garden
London WC2H 9HE

Distributed by

World Scientific Publishing Co. Pte. Ltd.
5 Toh Tuck Link, Singapore 596224
USA office: 27 Warren Street, Suite 401-402, Hackensack, NJ 07601
UK office: 57 Shelton Street, Covent Garden, London WC2H 9HE

British Library Cataloguing-in-Publication Data
A catalogue record for this book is available from the British Library.

ASYMPTOTIC ANALYSIS OF DIFFERENTIAL EQUATIONS
(Revised Edition)

ISBN-13 978-1-84816-607-3
ISBN-10 1-84816-607-9
ISBN-13 978-1-84816-608-0 (pbk)
ISBN-10 1-84816-608-7 (pbk)

Printed in Singapore by World Scientific Printers.

Preface

The material for this book has resulted from teaching a graduate course in ordinary differential equations at Princeton for several years. Methods of asymptotic analysis covered include dominant balance, the use of divergent asymptotic series, phase-integral methods, asymptotic evaluation of integrals, and boundary layer analysis. The construction of integral solutions and the use of analytic continuation are used in conjunction with the asymptotic analysis, to show the interrelatedness of these methods. Some of the functions of classical analysis are used as examples, to provide an introduction to their analytic and asymptotic properties, and to give derivations of some of the important identities satisfied by them, since it is my experience that students are insufficiently familiar with Whittaker and Watson. There is no attempt to give a complete presentation of all these functions. The emphasis is on the various techniques of analysis: obtaining asymptotic limits, connecting different asymptotic solutions, and obtaining integral representations. Less attention is paid to strict mathematical rigor. A sufficient number of different examples are chosen to demonstrate the various techniques and approaches. Prerequisite material consists of elementary calculus and a basic knowledge of the theory of functions of a complex variable.

In Chapters 1–9 the material in each chapter depends on the results obtained in the previous chapters. These chapters cover: the use of dominant balance for solving equations with a small parameter; a review of the techniques for obtaining exact solutions of soluble differential equations; complex variable theory and analytic continuation; classification of singular points of differential equations and the construction of local expansions to solutions, including nonconvergent asymptotic expansions; phase-integral methods; perturbation theory; asymptotic evaluation of integrals;

basic properties of the Euler gamma function; and methods for finding integral solutions of differential equations. Some basic results concerning the gamma function are in fact used in earlier chapters, so some parts of this chapter should be previewed when necessary. Most of the relations proved concerning $\Gamma(z)$, however, depend on results in Chapters 1–7, so it was impractical to move it to an earlier point in the book. Some of the material in the first chapters may be already known, in which case it can be quickly reviewed or skipped.

The remaining chapters are independent of one another, depending only on the material in the first nine, and can be covered in any order depending on personal interest. All chapters cannot be covered in one semester, and I normally choose Chapters 11, 13, 17, and 18 after the first nine. Chapter 10 introduces the theory of expansions in series of orthogonal polynomials and derives some of their properties, and also introduces the theory of wavelets. Chapter 11 gives the theory of the Airy function, an understanding of which is essential for a good grasp of the WKB treatment of scattering and bound state problems. Chapter 12 continues the development of the theory of phase-integral methods. Chapter 13 further illustrates the use of integral representations and analytic continuation through an examination of the Bessel function. Chapter 14 does the same for the parabolic cylinder function, and Chapter 15 for the Whittaker function. Chapter 16 examines inhomogeneous equations, some arising from the theory of resistive reconnection of a magnetized plasma, and further explores the role of causality using the Green's function technique for the examination of the equations for a parametric instability. Chapter 17 on the Riemann zeta function provides a brief introduction to the distribution of prime numbers. Chapter 18 treats differential equations containing a small parameter giving rise to boundary layers.

Aside from corrections of typos and other errors, this revised edition differs by the addition of Chapter 12, an extension of the techniques of WKB analysis, a rewriting and extension of Chapter 18 using the methods of Kruskal–Newton diagrams, providing a much more powerful way to determine boundary layer scaling, and the addition of several new sections expanding material in different chapters.

Roscoe B. White
Princeton 2009

Acknowledgments

This text is the outcome of interactions with many students and research colleagues over many years. Some of the material presented here is also present in other texts which treat asymptotic methods. Normally not covered are the use of Kruskal–Newton graphs, the derivation of Heading's rules for analytic continuation of phase-integral solutions, the associated derivation of the Stokes constants, and the methods for obtaining integral solutions. These approaches have been integrated to show how they are interrelated.

The book was written using Latex, by Leslie Lamport (Addison Wesley, 1994). The figures were made using Super Mongo, written by Robert Lupton, Princeton University (rhl@astro.princeton.edu), and Patricia Monger (monger@mcmaster.ca). I am grateful to Bedros Afeyan for discussions concerning wavelets, Greg Rewoldt for help with Latex, Leonid Zakharov for help with computing, Thomas Fischaleck for valuable suggestions regarding boundary layer problems, and to Tim Stoltzfus-Dueck and Jeffrey Parker for proofreading several chapters.

Finally, I am grateful to my wife Laura and my daughter Veronica for their patience and support during my many years of teaching and research, and to my brother Bob for encouragement and stimulation.

Contents

Preface v

Acknowledgments vii

List of Figures xvii

1. Dominant Balance 1
   - 1.1 Introduction 2
   - 1.2 Solutions Using Kruskal–Newton Diagrams 3
     - 1.2.1 Third order 3
     - 1.2.2 Non polynomial form 6
     - 1.2.3 Higher order 8
     - 1.2.4 Hidden points 10
   - 1.3 Problems 12
2. Exact Solutions 15
   - 2.1 Introduction 16
   - 2.2 Constant Coefficients 17
   - 2.3 Inhomogeneous Linear Equations 18
   - 2.4 The Fredholm Alternative 20
   - 2.5 The Diffusion Equation 22
   - 2.6 Exact Solutions to Nonlinear Equations 23
   - 2.7 Phase Plane Analysis 24
   - 2.8 Problems 28
3. Complex Variables 31
   - 3.1 Analyticity 32
   - 3.2 Cauchy Integral Theorem 33
   - 3.3 Series Representation 34

3.4 The Residue Theorem 35
3.5 Analytic Continuation 37
3.6 Inverse Functions 39
3.7 Problems 41
4. Local Approximate Solutions 43
4.1 Introduction 44
4.2 Classification 44
4.2.1 Ordinary point 45
4.2.2 Regular singular point 45
4.2.3 Irregular singular point 48
4.3 Asymptotic Series 51
4.3.1 Properties 52
4.3.2 Truncation: A series about $x = 0$. 54
4.3.3 Truncation: A series about $x = \infty$. 56
4.3.4 Truncation: A series about $x = 0$ 57
4.3.5 Asymptotic oscillation 61
4.3.6 Construction of asymptotic series 62
4.3.7 The error function 62
4.4 Origin of the Divergence 63
4.5 Improving Series Convergence 66
4.5.1 Shanks transformation 67
4.5.2 Euler summation 67
4.5.3 Borel summation 68
4.6 The Special Functions of Mathematical Physics 68
4.7 Problems 70
5. Phase-Integral Methods I 73
5.1 Introduction 74
5.2 Solutions far from Singularities 77
5.3 Connection Formulae: Isolated Zero 78
5.4 Derivation of Stokes Constants 81
5.5 Rules for Continuation 83
5.6 Causality 85
5.7 Bound States and Instabilities 86
5.8 Scattering, Overdense Barrier 87
5.9 Scattering, Underdense Barrier 89
5.10 The Budden Problem 93
5.11 The Error Function 96
5.12 Eigenvalue Problems 97
5.13 Problems 101

6. Perturbation Theory 105
6.1 Introduction 106
6.2 Eigenvalues of a Hermitian Matrix 107
6.3 Broken Symmetry Due to Tunneling 110
6.4 Problems 115
7. Asymptotic Evaluation of Integrals 117
7.1 Introduction 118
7.2 End Point 119
7.3 Saddle Point 122
7.4 Finding an Integration Path 126
7.5 Problems 129
8. The Euler Gamma Function 133
8.1 Introduction 134
8.2 The Stirling Approximation 135
8.3 The Euler–Mascheroni Constant 137
8.4 Sine Product Identity 138
8.5 Continuation of $\Gamma(z)$ 140
8.6 Asymptotic $\Gamma(z)$ 140
8.7 Euler Product for Γ 143
8.8 Integral Representation for $1/\Gamma(z)$ 144
8.9 $\Gamma(nx)$ 146
8.10 The Euler Beta Function 146
8.11 Problems 148
9. Integral Solutions 151
9.1 Constructing Integral Solutions 152
9.1.1 Integration by parts 152
9.1.2 Asymptotic value of a series 154
9.1.3 Finding a discrete difference equation 157
9.1.4 Construction from a series 158
9.2 Causal Solutions 159
9.3 Comments 161
9.4 Problems 162
10. Expansion in Basis Functions 165
10.1 Legendre Functions 166
10.1.1 Local analysis 166
10.1.2 Euler integral representation 167
10.1.3 Recurrence relations 169
10.1.4 Laplace integral representation 171
10.1.5 Generating function 173

10.1.6 Legendre polynomials 174
10.2 Orthogonal Polynomials 175
10.3 Wavelets 177
10.3.1 Introduction 177
10.3.2 Scaling function 178
10.3.3 Wavelet basis construction 182
10.3.4 Determining the expansion coefficients 184
10.3.5 Examples 186
10.3.6 Time–frequency analysis 190
10.4 Problems 193
11. Airy 195
11.1 WKB Analysis 196
11.2 Argand Plot of the Airy Function 198
11.3 Fourier–Laplace Integral Representation 200
11.4 Asymptotic Limits 201
11.4.1 Large negative z 201
11.4.2 Large positive z 204
11.4.3 Small $|z|$ 206
11.5 Mellin Integral Representation 207
11.6 Matching Local Solutions 208
11.7 The Wronskian 210
11.8 Problems 211
12. Phase-Integral Methods II 213
12.1 Introduction 213
12.2 Stokes Phenomena and Integral Representations 214
12.3 Scattering, Overdense Barrier 221
12.3.1 WKB analysis 222
12.3.2 Integral representation 223
12.4 Scattering, Underdense Barrier 227
12.4.1 Integral representation 228
12.5 The Budden Problem 230
12.5.1 WKB analysis 230
12.5.2 Integral representation 232
12.6 Bound state 236
12.6.1 WKB analysis 236
12.6.2 Integral representation 238
12.7 Concerning Accuracy 241
12.8 Conclusion 243
12.9 Problems 246

13. Bessel 247
13.1 Local Analysis 248
13.1.1 Local at zero 248
13.1.2 Analytic continuation in ν 249
13.1.3 Local at infinity 250
13.2 WKB Analysis 250
13.3 Integral Representations 252
13.3.1 Fourier–Laplace representation 252
13.3.2 The Hankel functions 252
13.3.3 Asymptotic limits 253
13.3.4 Relation to Bessel and Neumann functions 255
13.3.5 The Wronskian 257
13.3.6 Sommerfeld integral representation 258
13.3.7 Mellin integral representation 261
13.4 Generating Function 262
13.5 Matching Local Solutions 262
13.6 Imaginary Argument 263
13.7 Gaussian–Bessel integrals 264
13.8 Problems 267
14. Weber and Hermite 271
14.1 Local Analysis at Infinity 272
14.2 Local Analysis at Zero 273
14.3 Euler Integral Representation 274
14.4 Fourier–Laplace Integral Representations 275
14.5 Orthogonality 279
14.6 The Wronskian 280
14.7 Mellin Integral Representation 281
14.8 Problems 283
15. Whittaker and Watson 285
15.1 Local Analysis at Infinity 286
15.2 Local Analysis at Zero 287
15.3 Euler Integral Representation 287
15.4 Relation Between $M_{\lambda,\mu}(z)$ and $W_{\lambda,\mu}(z)$ 290
15.5 Fourier–Laplace Integral Representation 291
15.6 Mellin Integral Representation 292
15.7 Special Cases 294
15.7.1 The error function 294
15.7.2 The logarithmic integral function 295
15.8 Problems 296

16. Inhomogeneous Differential Equations 299
16.1 The Driven Oscillator 300
16.2 The Driven Weber Equation 301
16.3 The Struve Equation 303
16.3.1 Local analysis at zero 303
16.3.2 Local analysis at infinity 304
16.4 Resistive Reconnection 305
16.5 Resistive Internal Kink 306
16.6 A Causal Inhomogeneous Problem 307
16.7 Driven Oscillator 312
16.8 A Driven Overdamped Oscillator 313
16.9 Stokes Modification of Boundary Conditions 316
16.10 Inhomogeneous Weber equation 318
16.11 Problems 324
17. The Riemann Zeta Function 327
17.1 Introduction 328
17.2 $\zeta(s)$ and $\zeta(1-s)$ 329
17.3 The Euler Product for $\zeta(s)$ 331
17.4 Distribution of Prime Numbers 332
17.5 Public Key Codes 336
17.6 Stirling Revisited 338
17.7 Problems 341
18. Boundary Layer Problems 343
18.1 Introduction 344
18.2 Layer Location 346
18.3 Layer at Left Boundary 349
18.4 Layer in Domain Center 352
18.5 Layer at Right Boundary 354
18.6 Nested Boundary Layers 356
18.7 Inhomogeneous Equations 360
18.8 Simplification through Expansion 364
18.8.1 A corner layer 364
18.8.2 Nested layers 365
18.9 Coupled Equations 367
18.10 The Nonlinear Cole Equation 372
18.11 Failure of Asymptotic Matching 379
18.12 Boundary Layers in the Complex Plane 381
18.12.1 Crystal growth 381
18.12.2 Viscous fingering 385

18.12.3 Model equation with $a = b = 0$ 386
18.12.4 Model equation with $a \neq 0$ and $b = 0$ 387
18.12.5 Model equation with $a \neq 0$ and $b \neq 0$ 388
18.13 Problems 390
Appendix A. Lagrange's Theorem 393

Appendix B. Integral Solution for Eq. 18.129 395

Bibliography 397

Index 401

List of Figures

1.1 Kruskal–Newton diagram. 3

1.2 Original Kruskal–Newton diagram, from *Methods of Series and Fluxions*, by Isaac Newton. 4

1.3 Kruskal–Newton diagram showing lines of dominant balance. 5

1.4 Convergence of iteration of Eq. 1.4, $\epsilon = 0.3$. 6

1.5 Kruskal–Newton diagram showing lines of dominant balance for Eq. 1.8. 7

1.6 Kruskal–Newton diagram showing lines of dominant balance for Eq. 1.16. 9

1.7 Kruskal–Newton diagram with inaccessible points, Eq. 1.20. 10

2.1 Phase plot for Eq. 2.48. 25

2.2 Numerical integration of Eq. 2.47. 26

3.1 Analytic continuation of the series $\sum_k z^k$ successively from $z = 0$ to $z = a, b, c$. 38

3.2 Analytic continuation of $f = \sqrt{1 - z}$. 39

4.1 Number of terms kept in optimal asymptotic expansion of Eq. 4.48. 55

4.2 Numerical $y(x)$ and asymptotic approximation to Eq. 4.48. . 55

4.3 Number of terms kept in optimal asymptotic expansion to Eq. 4.53. 58

4.4 Numerical $y(x)$ and asymptotic approximation to Eq. 4.53. . 58

4.5 Magnitude of terms in asymptotic series vs n for Eq. 4.56. . . 59

4.6 The number of terms retained in the asymptotic sum, N, and the analytic expression $4/\sqrt{x}$ for Eq. 4.56. 59
4.7 Numerical integration of $W(x)$ for both solutions (smooth), and the optimal asymptotic series approximations (jagged) for Eq. 4.56. 60
4.8 The error function, showing optimal asymptotic expressions, which do not exist for $|x| < 0.7$. 64
4.9 Numerical integration of $G(x)$ (smooth), and the optimal asymptotic series approximations (jagged). 65

5.1 Stokes diagram for $Q = (z - z_l)(z - z_2)(z - z_3)^2/(z - z_4)$ with $z_1 = l + i$, $z_2 = -1 - i$, $z_3 = 1 - i$, and $z_4 = -1 + i$. . . 76
5.2 Stokes diagram for a first order turning point. 79
5.3 Large scale Stokes diagram for the Bessel function. 82
5.4 Stokes plot for the bound state problem. 86
5.5 Stokes plot for the overdense barrier. 88
5.6 $Q = z^2$, Stokes plot. 90
5.7 Stokes plot for the underdense barrier. 91
5.8 Underdense barrier, rotated Stokes plot. 92
5.9 $Q = z^2 + b^2$, transmission and reflection vs b. 93
5.10 $Q = z^2$, Stokes plot. 94
5.11 Stokes diagram for Budden problem. 95
5.12 Stokes diagram during search for eigenvalue. 98
5.13 Stokes structure for $K < K_c$. 99
5.14 Stokes structure for $K > K_c$. 100

6.1 Eigenvalues of a Hermitian matrix vs perturbation. 111
6.2 Probability of close eigenvalues in a random Hermitian matrix. 111
6.3 Q function for the double potential. 112
6.4 Stokes plot for the double potential. 112
6.5 Even and odd parity solutions. 113

7.1 Integration path for Eq. 7.8. 121
7.2 Integrand for a saddle point of width w. 123
7.3 Contours of equal magnitude of $e^{ixt-xt^2/2}$, showing the integration contour and the saddle. 126
7.4 Integration path for Eq. 7.24. 127

8.1 Plot of $\Gamma(x)$. 135

8.2 Integration path for evaluation of I. 138
8.3 Integration path for continuation of Γ. 141
8.4 Integration path for evaluation of Γ, $Re z \to +\infty$. 141
8.5 Integration path for evaluation of Γ, $Re z \to -\infty$. 142
8.6 Integration path for 1/Γ. 145

9.1 Integration contour in the t and s planes. 155
9.2 Asymptotic expression and a numerical summation of Eq. 9.11. 156

10.1 Integration paths for the Legendre function. 168
10.2 Integration path for the Laplace representation of P_ν. 172
10.3 Integration path for the Laplace representation of Q_ν. 173
10.4 Nested function spaces, with $V_{k+1} = W_k \bigoplus V_k$. 183
10.5 The functions given by the length two filter, Eqs. 10.104, 10.105. 186
10.6 A wavelet approximation using the Haar wavelet system with $0 \le k \le 4$. 187
10.7 A wavelet approximation using the Haar wavelet system with $0 \le k \le 8$. 188
10.8 Distribution of the wavelet amplitudes for the expansion given in Fig. 10.7. 189
10.9 The functions given by the length four filter, Eq. 10.81. . . . 189
10.10 An example of a Malvar–Wilson basis function. 191

11.1 Stokes diagram for the Airy function. 197
11.2 Argand diagram of the Airy function Ai(z) and its WKBJ approximation ψ_- continued in the complex plane. 199
11.3 Argand diagram of the Airy function Ai(z) and its uniformly valid approximation $\mathrm{Ai}_{\mathrm{unif}}(z)$. 199
11.4 Integration paths for the Airy function. 201
11.5 Integration paths showing saddle points for $Re z < 0$. 202
11.6 Integration paths showing saddle points for $Re z > 0$. 205
11.7 Airy function by matching local expansions. 209
11.8 Optimal number of terms in asymptotic expansions for Airy. 210

12.1 Integration path showing contours of $Re\phi$ just before z arrives at the Stokes line. Positive values of $Re\phi$ are black, and negative values green. Following the line of steepest descent from the upper saddle point leads to the asymptote at $t \sim e^{-i2\pi/3}$ without encountering the lower saddle point. . . . 214
12.2 Integration path showing contours of $Re\phi$ with z at the Stokes line. Positive values of $Re\phi$ are black, and negative values green. Following the line of steepest descent from the upper saddle point leads to the the lower saddle point. 215
12.3 Integration path showing contours of $Re\phi$ just after z passes the Stokes line. Positive values of $Re\phi$ are black, and negative values green. Now to complete the integration one must pass also through the lower saddle point. 216
12.4 Integration path for $Q = z^n$ and positive real z, showing some level contours, the saddle points, the cut, and the integration path. Negative levels of $Re\phi$ are shown in green. . . 218
12.5 Integration path for $Q = z^n$ with z above the Stokes line at $z \sim e^{-\pi i/(n+2)}$ showing some level contours, the saddle points, the cut, and the integration path. Negative levels of $Re\phi$ are shown in green. 219
12.6 Integration path for $Q = z^n$ and z on the Stokes line at $z \sim e^{-\pi i/(n+2)}$ showing some level contours, the saddle points, the cut, and the integration path. Negative levels of $Re\phi$ are shown in green. 220
12.7 Integration path for $Q = z^n$ and z rotated past the Stokes line at $z \sim e^{-\pi i/(n+2)}$ showing some level contours, the saddle points, the cut, and the integration path. Negative levels of $Re\phi$ are shown in green. 221
12.8 Stokes plot for the overdense barrier. 222
12.9 Integration contour for the overdense barrier, z real positive. 224
12.10 Integration path for the overdense barrier, z real negative. . . 225
12.11 Magnitude of the Stokes constant for the overdense barrier, and absolute squares of reflection and transmission amplitudes vs b. 226
12.12 Underdense barrier, rotated Stokes plot. 227
12.13 Underdense barrier, reflection, transmission and Stokes constant. 229
12.14 Stokes diagram for Budden problem. 231

12.15 Contour for the Budden problem, $z = r$. 232
12.16 Contour for the Budden problem, $z = re^{-i\pi}$. 233
12.17 Power absorption at the Budden singularity. The dashed line shows the result using the Stokes constants for isolated singularities. 234
12.18 Contour for the Budden problem, $z = re^{-2i\pi}$. 235
12.19 Stokes diagram for the parabolic cylinder function. 237
12.20 Saddle point for large positive z evaluation of the parabolic cylinder function. 239
12.21 Saddle point for large imaginary z evaluation of the parabolic cylinder function. 240
12.22 Saddle point for large negative z evaluation of the parabolic cylinder function. 241
12.23 Stokes plot for $Q = E - z^4$, showing anti-Stokes lines. . . . 242

13.1 Stokes plot for the Bessel equation. 251
13.2 Integration paths for the Hankel functions. 253
13.3 Deformed integration paths for the Hankel functions. 256
13.4 Sommerfeld contours for the Bessel functions. 259
13.5 Sommerfeld contour for the Bessel function for ν integer. . . 260
13.6 Local approximations to $J_0(x)$. 263

14.1 Saddle point for large positive z evaluation using the Euler integral representation. 275
14.2 Saddle point for large negative z evaluation using the Euler integral representation. 276
14.3 Integration path for a Fourier–Laplace representation of the parabolic cylinder function. 277

15.1 Integration path for the Euler representation of the Whittaker function. 289
15.2 Integration path for the Mellin representation of the Whittaker function. 294

16.1 Integration path for the inhomogeneous Weber equation for $z > 0$. 302
16.2 Integration path for the inhomogeneous Weber equation for $z < 0$. 303
16.3 Integration path for $Q > 1/4$. 310
16.4 Integration path for $Q < 1/4$. 311

16.5 Stokes diagram for Eq. 16.64. 315
16.6 Analytic continuation of Eq. 16.69, showing the integration contour for negative real z, and the phase of $(t^2-1)^{-1/4}$. 316
16.7 Integration paths for $z > 0$ (lower) and $z < 0$ (upper) showing saddle points. 318
16.8 Integration path for $z < 0$. 321
16.9 Integration path for the continuation of the subdominant term, for $z \sim e^{-i\pi/2}$. 322

17.1 Integration contour for $\zeta(s)$. 329
17.2 Contour for integral representation of the zeta function. . . 330
17.3 Distribution of prime numbers. 333
17.4 Difference between the distribution of primes and the analytic expression. 334

18.1 Kruskal–Newton diagram for Eq. 18.3. 346
18.2 Layer on the left. 347
18.3 Layer on the right. 348
18.4 Internal layer. 349
18.5 Kruskal–Newton diagram for Eq. 18.6. 350
18.6 Lowest order uniform solution (red) and numerical integration for Eq. 18.6 with $\epsilon = .1$. 351
18.7 Kruskal–Newton diagram for Eq. 18.13. 352
18.8 Lowest order uniform solution (red) and numerical integration for Eq. 18.13 with $\epsilon = 0.2$. 353
18.9 Kruskal–Newton diagram for Eq. 18.19. 355
18.10 Lowest order uniform solution (red) and numerical integration for Eq. 18.19 with $\epsilon = .1$. 356
18.11 Kruskal–Newton diagram for Eq. 18.19. 357
18.12 Lowest order uniform solution (red) and numerical integration for Eq. 18.27 with $\epsilon = .1$. 360
18.13 The Kruskal–Newton plot for Eq. 18.49. The inhomogeneity is represented by the dotted vertical line at $p = 1$. The empty circles mark intersection points with support lines 1, 2 and 3. 361
18.14 Kruskal–Newton diagram for Eq. 18.59. 364
18.15 Kruskal–Newton diagram for Eq. 18.65. 366
18.16 Integration contours for Eq. 18.68. 367

18.17 Kruskal–Newton diagram for the $m \geq 2$ tearing mode. 368
18.18 Kruskal–Newton diagram for the $m = 1$ tearing mode. 370
18.19 Kruskal–Newton diagram for the $m = 1$ tearing mode. 371
18.20 Kruskal–Newton diagram for Eq. 18.85. 372
18.21 Numerical and first order uniform solutions to the Cole equation, $\epsilon = 0.02$, $A = 2$, $B = 2$, $\beta = 1/3$ and $A = -1$, $B = 1.5$, $\beta = -1$. 376
18.22 Numerical and lowest order uniform solutions to the Cole equation, for $\epsilon = 0.002$, in domain c, $A = .25$, $B = .25$, and in the lower right quadrant, $\epsilon = 0.01$ $A = .5$, $B = -.5$. 377
18.23 Numerical and lowest order uniform solutions to the Cole equation, for $\epsilon = 0.002$, in domain a, $A = -.5$, $B = 2$, $\beta = -3$, and in domain b, $A = 2$, $B = .5$. 378
18.24 Solution domains for Eq. 18.85. Series solutions converge in all domains but a, b and c. 379
18.25 Kruskal–Newton diagram for Eq. 18.109. 380
18.26 Numerical solutions of Eq. 18.109 with $y(0) = 1$, $y(1) = e$, $\epsilon = 0.001$ and $\epsilon = 0.01$. 380
18.27 The Kruskal–Newton diagram for Eq. 18.113. The dotted line represents the inhomogeneity and empty circles mark intersection points with the support lines. 382
18.28 The steepest decent path for the integral Eq. 18.115 (left), showing the contribution from the pole (right). 383
18.29 Integration paths for $z > 0$ (lower) and $z < 0$ (upper) showing saddle points. 385
18.30 The Kruskal–Newton diagram for Eq. 18.124. 386
18.31 Kruskal–Newton diagram for Eq. 18.128. The shaded dot represents the term $a\phi$ with a scaled to lie on support line 2. 388
18.32 Kruskal–Newton diagram for Eq. 18.130. The term multiplied by b is represented by a series of shaded dots. The parameter b is scaled, so no dot lies below support line 2. . . 389

B.1 Left: The steepest descent contour for Eq. B.4 if $Im(w) < 0$. Right: Same but with $Im(w) > 0$. 396

Chapter 1

Dominant Balance

Isaac Newton was born in Woolstorpe, Lincolnshire, England, in 1642.[1] An intense and lonely boy, he began early to keep notebooks of his ideas and calculations, a practice he continued all his life. He also had mechanical talents and while in school constructed many models in wood, including water clocks, sun dials, and a watermill. Early influences included Aristotle and Euclid, but also Descartes and Galileo. He graduated from Cambridge in 1665. While at Cambridge he developed the general binomial theorem and his work with this and other power series led to the formulation of differential and integral calculus (Turnbull [1951]); (Boyer [1959]), necessary tools for his investigation of planetary motion. Intensely secretive and easily hurt by criticism, he kept these results to himself, only revealing what he knew when he believed that someone else was close to discovering it and taking the credit away from him. His submission to the newly formed Royal Society of his discovery that white light was composed of all colors, which could be separated by a prism, led to a bitter dispute with Hooke, resolving him to further secrecy. Finally the astronomer Edmond Halley asked him in 1680 what path a comet would take moving in the gravitational field of the sun if the force of attraction varied inversely as the square of the distance. Newton gave the answer, an ellipse with the sun at one focus, and remarked that he had worked it out twenty years earlier. Halley convinced him to publish these results, and Newton then wrote the *Principia*, published in 1687. In this work appears for the first time the expression $e^x = \sum x^n/(1 \cdot 2 \cdot 3 \ldots n)$ and geometric methods of solving equations with small

[1] Sources for biographical material for this and other chapters include Franceschetti [1999], Feingold [2004], Turnbull [1951] [1961], Bell [1965], Gleick [1992] [2003], Airy [1896] and Watson [1922], along with the MacTutor History of Mathematics Archive [2003].

parameters, which he used to find approximate solutions to algebraic equations. These methods have been generalized and further developed by Kruskal [1963].

1.1 Introduction

Asymptotic analysis often involves the evaluation of mathematical expressions which contain a small parameter $0 < \epsilon \ll 1$. To simplify the solution of equations involving such expressions it is often possible to find two or more large terms which dominate the solution, other terms giving only small corrections to the value obtained by neglecting them entirely. In this case an iteration procedure can often be constructed which will give the solution to any degree of accuracy. The simplest such expressions are polynomials of the form

$$\sum_{p,q} C_{p,q}\epsilon^q x^p = 0, \tag{1.1}$$

with $C_{p,q}$ constants of order unity with respect to ϵ, and the sum over an index set for p, q. To find solutions of this equation using the idea of dominant balance we assume a particular ordering of a solution, i.e. take $x \sim \epsilon^\alpha$. The polynomial then becomes

$$\sum_{p,q} C_{p,q}\epsilon^{q+\alpha p} = 0. \tag{1.2}$$

Note that terms of the same order in ϵ in this equation are represented by a line $q + \alpha p = constant$. Each term in the polynomial can be represented as a point in the p, q plane, as shown in Fig. 1.1. All points above the line represent terms smaller than points on the line, and all points below the line represent terms larger than those on the line. A shift to a position with $q + \alpha p$ differing by an integer represents a difference in magnitude by one order of ϵ.

To find all possible combinations of dominant balance for a given polynomial, find all possible placements of a line so that it includes two or more terms of the polynomial, with all other points lying above the line. Each such placement represents a potential solution to the equation whereby the dominant balance of the solution is given by the points on the line, and all points above the line are associated with terms which are small corrections to this solution. Graphically this may be understood as bringing the line up from below until it makes contact with a point, and then rotating it

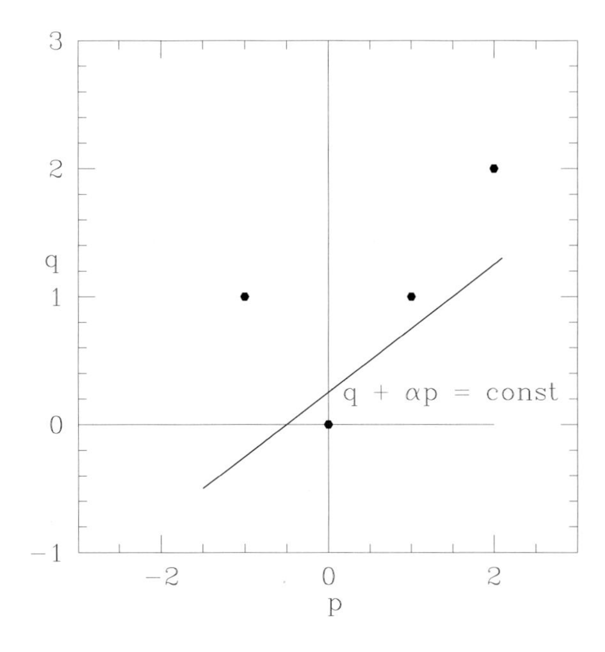

Fig. 1.1 Kruskal–Newton diagram.

one way or the other until it makes contact with a second point. Such a plot is known as a Kruskal–Newton diagram, and is best understood with a few examples. Shown in Fig. 1.2 is a page from Newton's book on the method of fluxions, showing the first such diagram used by Newton to find the dominant terms in a problem illustrating the development of differential calculus. Figure 1 shows the placement for terms of the form x^q, y^p, arranged linearly in the variables p, q. The actual terms in the calculation are shown in Fig. 2, with the line DE passing through dominant terms, with higher order terms indicated by stars above the line.

1.2 Solutions Using Kruskal–Newton Diagrams

1.2.1 *Third order*

Consider the polynomial

$$1 - x - \epsilon x^2 - 2\epsilon^3 x^3 = 0, \tag{1.3}$$

50 *Methods of series and fluxions* [1, 2, §2

ad quem casum cæteri duo casus sunt reducibiles. E terminis in quibus specie radicalis (*y*, *p*, *q* vel *r* &c) non reperitur selige depressissimum respectu dimen sionum indefinitæ speciei (*x* vel *z* &c), dein alium terminum in quo sit ill: species radicalis selige, talem nempe ut progressio dimensionum utriusq præfatæ speciei a termino priùs assumpto ad hunc terminum continuata, quàn maximè potest descendat vel minimè ascendat.(27) Et siqui sint alij termin quorum dimensiones cum hâc progressione ad arbitrium continuatâ con veniant, eos etiam selige. Deniq; ex his selectis terminis tanquam nihilo æquali bus quære valorem dictæ speciei radicalis,(28) et quotienti appone.

Cæterùm ut hæc regula magis elucescat, placuit insuper ope sequentis diagram matis(29) exponere. Descripto angulo recto *BAC*, latera ejus *BA*, *AC* divido in partes æquales, et inde normales erigo distribuentes angulare spatium in æquali: quadrata vel parallelogramma, quæ concipio denominata esse a dimensionibu specierum *x* et *y*, prout vides in fig 1 inscriptas.(30) Deinde cùm æquatio aliqu:

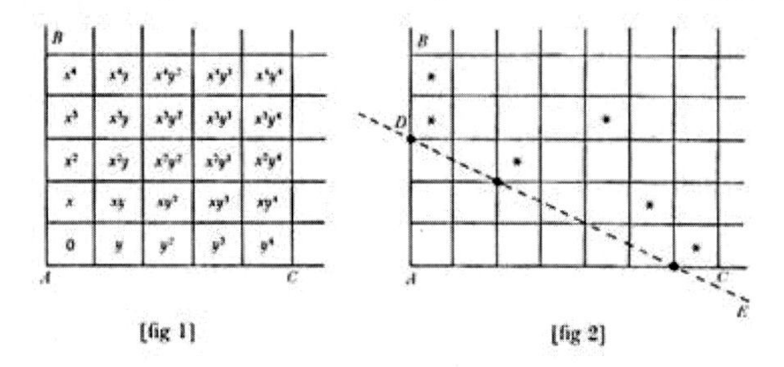

proponitur, parallelogramma singulis ejus terminis correspondentia insigni notâ aliquâ, et Regulâ ad duo vel forte plura ex insignitis parallelogrammi

Fig. 1.2 Original Kruskal–Newton diagram, from *Methods of Series and Fluxions*, by Isaac Newton.

with the associated Kruskal–Newton diagram shown in Fig. 1.3. Terms in the polynomial are represented by dots. There are three possible placements of straight lines passing through two or more points with all remaining points above the line, as shown in the figure. Now for each line, treat the points above the line as a small perturbation, giving for line 1, with dominant terms $x, 1$,

$$x = 1 - [\epsilon x^2 + 2\epsilon^3 x^3], \tag{1.4}$$

with the terms in brackets associated with points above line 1, and thus small, since $x \simeq 1$ and $\epsilon \ll 1$. Now treat this equation iteratively.

In general an equation of the form

$$x = f(x), \tag{1.5}$$

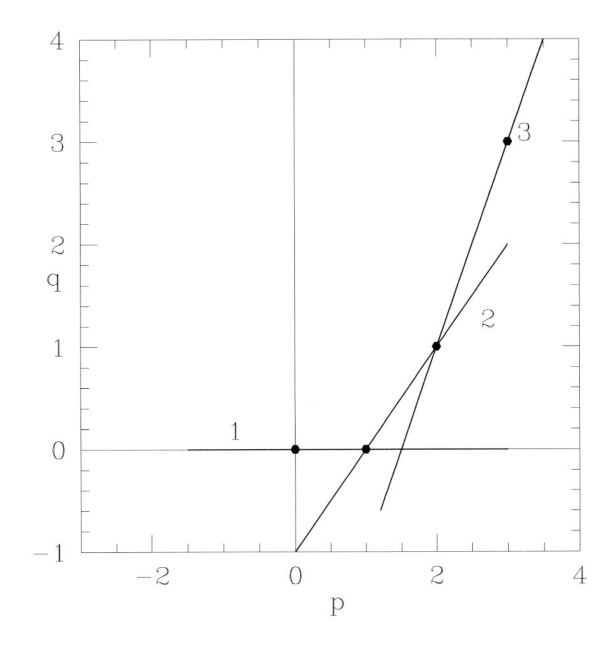

Fig. 1.3 Kruskal–Newton diagram showing lines of dominant balance.

is easily seen to converge to a fixed point x_f under the iteration scheme $x_{n+1} = f(x_n)$ provided the initial guess is sufficiently close to the solution and that $|f'(x_f)| < 1$. In this case $x_0 = 1$, and the fixed point cannot be far from this. We have $|f'| = |2\epsilon x + 6\epsilon^3 x^2| \simeq 2\epsilon$ for $x \simeq 1$, and thus the iteration converges. Convergence for $\epsilon = 0.3$ is shown in Fig. 1.4, with the function value shown as a downward sloping arc in red. Successive iterations are shown by the straight line segments starting at (1,0). After ten iterations the iterated value is correct to 15 places and converges much more rapidly if ϵ is smaller.

For line 2 the dominant terms are $x, \epsilon x^2$. By inspection $x \simeq 0$ is not a solution, so dividing by x we obtain

$$x = -\frac{1}{\epsilon} + \left[\frac{1 - 2\epsilon^3 x^3}{x\epsilon}\right], \tag{1.6}$$

giving the solution $x_0 \simeq -1/\epsilon$ and once again $|f'(x_0)| < 1$. For line 3 the dominant terms are $\epsilon x^2, 2\epsilon^3 x^3$ and we find

$$x = -\frac{1}{2\epsilon^2} + \left[\frac{1 - x}{2\epsilon^3 x^2}\right], \tag{1.7}$$

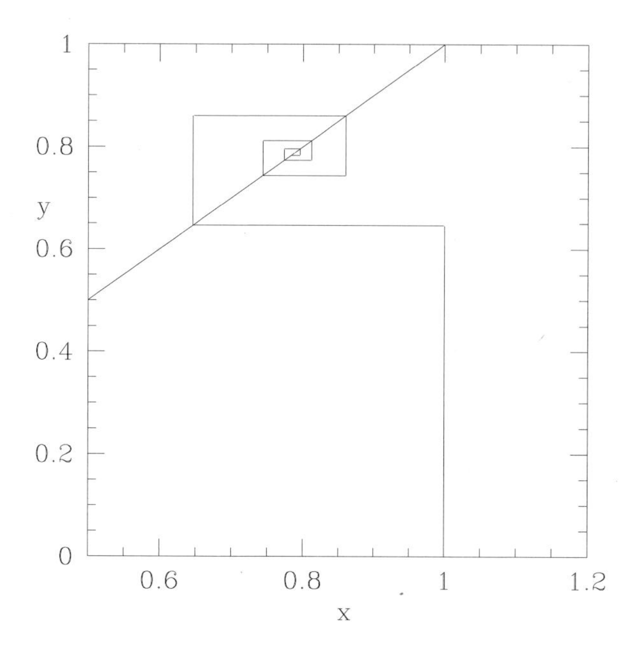

Fig. 1.4 Convergence of iteration of Eq. 1.4, $\epsilon = 0.3$.

giving a solution $x_0 \simeq -1/2\epsilon^2$ and once again $|f'(x_0)| < 1$. The original equation is a cubic polynomial and these three solutions give the complete set.

In general, to obtain a solution once a dominant balance is determined, simply treat the small terms as constant, and solve the resulting equation for x. The initial approximation x_0 is obtained by setting the small terms to zero, and evaluating them with x_n gives the iteration for x_{n+1}.

1.2.2 *Non polynomial form*

Consider the equation

$$2e^{-x} + \epsilon x^2 - 1 = 0. \tag{1.8}$$

The second derivative of this function is $2e^{-x} + 2\epsilon > 0$, and the function tends to $+\infty$ for $x \to \pm\infty$ and thus there are either two solutions or none. An associated Kruskal–Newton diagram is shown in Fig. 1.5. The points corresponding to the second and third terms are obvious. The placement of a point representing $2e^{-x}$ depends on the magnitude of x, since this term

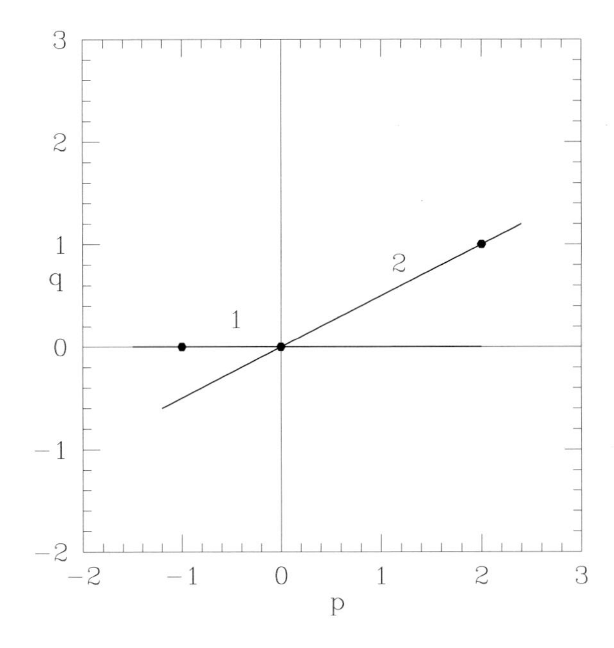

Fig. 1.5 Kruskal–Newton diagram showing lines of dominant balance for Eq. 1.8.

is not a simple power. In fact for various values of x it can assume the magnitude of x^p with p taking on any real value. The point must be along the line $q = 0$, and if it is placed to the right of the origin the resulting dominant balance would be $2e^{-x} + \epsilon x^2 \simeq 0$, which is impossible for x real. Thus place it to the left of the origin, giving the two dominant balances shown by the two lines in the figure. Line 1 gives the balance $2e^{-x} \simeq 1$ with $x_0 = ln(2)$. Treating the small term ϵx^2 perturbatively we find the iteration scheme

$$x_{n+1} = ln\left(\frac{2}{1-\epsilon x_n^2}\right), \tag{1.9}$$

which can easily be seen to converge. For the second root line 2 gives the balance $\epsilon x^2 \simeq 1$ with $x_0 = 1/\sqrt{\epsilon}$. We then find the iteration scheme

$$x_{n+1} = \frac{1}{\sqrt{\epsilon}} - \frac{2e^{-x_n}}{\sqrt{\epsilon}(\sqrt{\epsilon}x_n + 1)}, \tag{1.10}$$

and these two iteration schemes give all possible roots.

Consider the equation

$$\frac{2cosx}{1+x} = 1 + 2\epsilon x^2 \tag{1.11}$$

where we are interested only in roots with $x > 0$. Multiplying by the denominator we have $1 + x - 2cosx + 2\epsilon x^2 + 2\epsilon x^3 = 0$. For $x > 0$ the only possible dominant balance is given by $2cosx = 1 + x$. It is important to realize that the initial guess x_0 need not be perfect, so write $cosx \simeq 1 - x^2/2$ giving $x_0 = (\sqrt{5} - 1)/2 \simeq 0.6$. The iteration $x_{n+1} = f(x_n)$ of the form

$$x_{n+1} = 2cosx_n - 1 - 2\epsilon x_n^2 - 2\epsilon x_n^3 \tag{1.12}$$

has $|f'(x_0)| \simeq 2sin(x_0) > 1$ so does not converge. If $f'(x_f)$ is not small, the convergence can often be improved by using the modified iteration

$$x_{n+1} = \frac{x_n}{2} + \frac{f(x_n)}{2} \tag{1.13}$$

and in this case produces a rapidly convergent sequence. Another way of finding the zeros of a function $F(x)$, used by Newton but discovered independently and first published by Joseph Raphson, known as the Newton–Raphson method, is given by using the first term of the Taylor series $F(x) = F(x_0) + F'(x_0)(x - x_0)$ and setting $F(x) = 0$, giving the iteration

$$x_{n+1} = x_n - \frac{F(x_n)}{F'(x_n)}. \tag{1.14}$$

Similarly, by choosing a second point near to x_0 and approximating the function as being linear, one finds the iteration

$$x_{n+1} = \frac{F(x_n)x_{n-1} - F(x_{n-1})x_n}{F(x_n) - F(x_{n-1})}. \tag{1.15}$$

Obviously, if root finding occupies a central part of a large numerical code it is important to find a reliable rapidly convergent scheme. Whatever iteration scheme is used, the Kruskal–Newton procedure gives a good initial guess x_0.

1.2.3 *Higher order*

Consider the equation

$$\epsilon x^5 + x^4 - 2\epsilon x^2 + \epsilon^2 = 0. \tag{1.16}$$

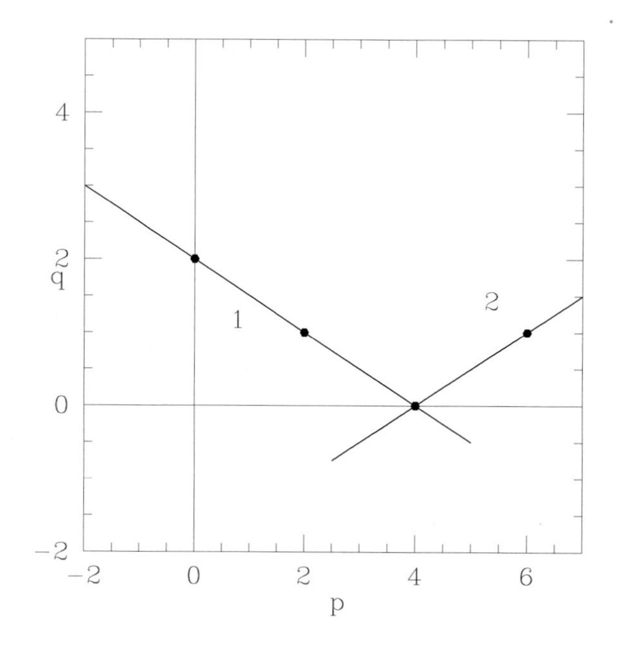

Fig. 1.6 Kruskal–Newton diagram showing lines of dominant balance for Eq. 1.16.

An associated Kruskal–Newton diagram is shown in Fig. 1.6, with the two dominant balances shown by the two lines. Line 1 gives the balance

$$x^4 - 2\epsilon x^2 + \epsilon^2 \simeq 0 \tag{1.17}$$

which is a quadratic equation in x^2, giving four iteration schemes with $x_0 = \pm\sqrt{\epsilon}$

$$x_{n+1} = \pm\sqrt{\epsilon \pm i\sqrt{\epsilon x_n^5}}. \tag{1.18}$$

Line 2 gives the dominant balance $\epsilon x^5 + x^4 \simeq 0$. Since $x \simeq 0$ is not a solution of the original equation, we find the iteration

$$x_{n+1} = -\frac{1}{\epsilon} + \frac{2\epsilon x_n^2 - \epsilon^2}{\epsilon x_n^4}, \tag{1.19}$$

with $x_0 = -1/\epsilon$, and these five iteration schemes give all possible roots.

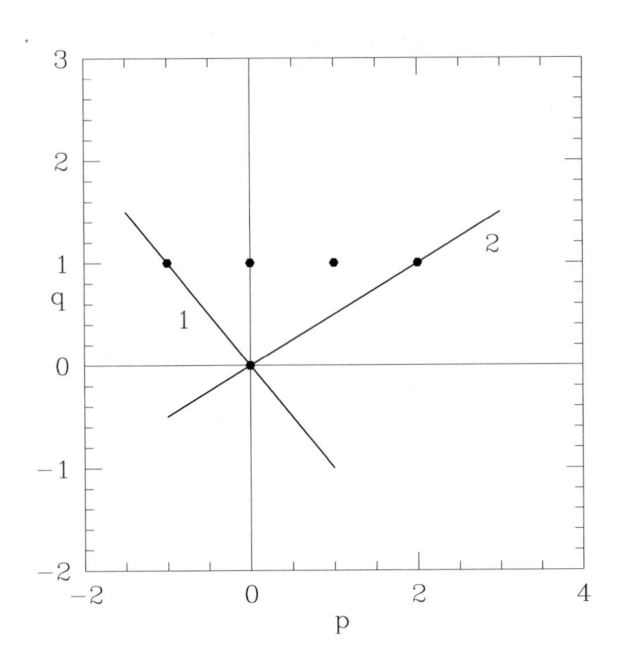

Fig. 1.7 Kruskal–Newton diagram with inaccessible points, Eq. 1.20.

1.2.4 *Hidden points*

It may often happen that some points in the Kruskal–Newton diagram are not accessible by any line giving dominant balance. An example is given in Fig. 1.7 by the equation

$$2\epsilon x^{-1} - 6\epsilon + \epsilon x + \epsilon x^2 - 1 = 0. \tag{1.20}$$

The only possible lines giving dominant balance are lines 1 and 2, and the points at $q = 1, p = 0, 1$ cannot be reached. Line 1 gives the iteration

$$x_{n+1} = 2\epsilon + \epsilon x_n^3 + \epsilon x_n^2 - 6\epsilon x_n, \tag{1.21}$$

with $x_0 = 2\epsilon$, and line 2 gives the iterations

$$x_{n+1} = \pm\sqrt{\left[\frac{1}{\epsilon} + 6 - x_n - \frac{2}{x_n}\right]} \tag{1.22}$$

with $x_0 = \pm 1/\sqrt{\epsilon}$, and these three iteration schemes give all possible roots.

As has been seen, these methods often reduce the order and complexity of a problem to a tractable level. It is easy to see that it is impossible to lose solutions, i.e. to come up with fewer iteration schemes than the order of the equation. Because it regularly finds all solutions this method is more powerful than perturbation theory, which loses some solutions if they happen not to be analytic functions of ϵ. The worst thing that can happen is that all points lie on a single line in the Kruskal–Newton diagram, in which case nothing at all is achieved by the method.

Note that in cases in which a solution behaves as $1/\epsilon$ or even some higher power, in the limit of $\epsilon \to 0$ the solution does not exist. Attempting a perturbation theory solution to the problem does not recover such a root, because it is not an analytic function of ϵ. The recovery of such non-analytic roots using perturbation methods is called singular perturbation theory. The Kruskal–Newton approach has no problem with such roots since it does not depend on analyticity of the root in ϵ in any way. Inverse and fractional powers regularly occur.

1.3 Problems

1. Use a Kruskal–Newton diagram to find dominant balances. Set up iteration schemes and show they converge.

$$\epsilon^2 x^4 - \epsilon x^2 - x - 1 = 0, \qquad \epsilon^2 x^3 + \epsilon x^2 - 2x + 4 = 0.$$

2. Use a Kruskal–Newton diagram to find dominant balances. Set up iteration schemes and show they converge.

$$\frac{\epsilon}{x^2} - \frac{2}{x} + x + 3\epsilon x^2 = 0, \qquad \epsilon^2 + 3\epsilon x - 4x^2 + \epsilon x^3 = 0.$$

3. Find the roots of the following equation to seven significant figures.

$$.01x^3 - x^2 + x + 6 = 0.$$

4. Find the real roots of $2/(1 - \epsilon x^2) = e^x$ to order ϵ.

5. For $\epsilon << 1$ how many roots does the following equation have for $x > -1/\epsilon$, $k = O(1)$, $k > 1$?

$$(1 + \epsilon x)^x = k.$$

Set up an iteration scheme, and show it converges.

6. Use a Kruskal–Newton diagram to find dominant balances. Set up an iteration scheme for each root and show it converges.

$$\epsilon x^3 + x^2 - 2x + 1 = 0.$$

7. Find the real roots of

$$\frac{sin(x)}{1 - 2x^2} = 1 + \epsilon x^2.$$

8. Find the real roots of

$$\frac{2cos(x)}{1 + x^2} = 1 + \epsilon x^2.$$

9. Find the leading behavior and iteration schemes for all roots.

$$\epsilon x^8 - \epsilon^2 x^6 + x - 2 = 0, \qquad \epsilon^2 x^8 - \epsilon x^6 + x - 2 = 0.$$

10. Find the roots of the following equation to seven significant figures.

$$-\epsilon + x + x^2\epsilon^2 - \epsilon^2 x^3 = 0, \qquad \epsilon = .01.$$

11. Find the roots of the following equation to seven significant figures.

$$20 + x - \frac{1}{x} - x^2 = 0.$$

12. Zeros of the Wilkinson polynomial are given by

$$(x-1)(x-2)...(x-20) + \epsilon x^{19} = 0.$$

Is a Kruskal–Newton diagram of any help? Write a simple iteration scheme for the root approaching $x = n$ for $\epsilon \to 0$. To lowest order, show that the roots at 15, 16 collide for small ϵ. Estimate the value of ϵ and x for this collision. Do the iteration schemes converge at this x? Sometimes numerical evaluation of df/dx using $(f(x+d) - f(x))/d$ is easier than analytic differentiation!

13. Use a Kruskal–Newton diagram to find dominant balances. Set up iteration schemes and show they converge.

$$\epsilon^2 x^3 - x^2 + \epsilon x + 5 = 0$$
$$\epsilon + 3\epsilon x^2 - x - \epsilon^3 x^3 = 0$$

14. Find the real roots of the following equation to seven significant figures for $\epsilon = .01$.

$$\frac{1}{x} - \epsilon - 2x^2 + \epsilon x^3 = 0.$$

15. Find iteration schemes for solutions for the real roots of

$$2sin(x) = x + \epsilon x^3.$$

Find bounds on x_0 so that the iteration will converge, i.e. how accurate does x_0 have to be?

16. For $\epsilon << 1$ draw the Kruskal–Newton diagram for

$$-\epsilon^3 + 10\epsilon^2 x - 20\epsilon x^2 + x^4 = 0.$$

What are the orders of the solutions p, ie $x \sim \epsilon^p$? Can you find an iteration scheme?

How is the problem changed if $\epsilon = 0.1$, or $\epsilon = 10^{-2}$?

Chapter 2

Exact Solutions

The Bernoulli family contributed to calculus, probability theory, the calculus of variations, hydrodynamics, and mathematical physics. The family left Belgium for Basel, Switzerland in the late seventeenth century. Johann I (born in 1667) and Jakob I (born in 1654) were brothers, among the first to recognize the power of calculus and to extend it to the calculus of variations, allowing them to discover the catenary, the shape of a rope hanging between two points. Johann visited Paris and gave a series of lectures on calculus for l'Hôpital, who then published a textbook using Johann's notes and letters, claiming the work to be his own. Jakob publicly posed the problem of finding the trajectory of minimum time of descent between two points, receiving correct solutions from Newton, Leibnitz, and Johann. Daniel, the son of Johann I, wrote a book on the kinetic theory of gases, and Bernoulli's law, relating pressure and velocity of a flowing fluid, describing the aerodynamic lift of a wing and the force on a sail. He also gave the first solution to a nonlinear Riccati equation, named after the man who had discussed the problem but never solved it. Daniel's major work is *Hydrodynamique*, published in 1738, and written in such a way so that all results are a consequence of the conservation of energy.

Daniel Bernoulli and Euler were together in St. Petersburg for six years, and developed together the theory of the mechanics of flexible and elastic bodies. In 1773 he returned to the University of Basel, where he successively held chairs in medicine, metaphysics, and natural philosophy.

2.1 Introduction

Ordinary differential equations of order n are equations of the form

$$y^{(n)} = F(x, y(x), y^{(1)}(x), \ldots, y^{(n-1)}(x)), \tag{2.1}$$

where $y^{(k)} = d^k y(x)/dx^k$, and x is a real variable, y a real function. This last requirement is, however, often relaxed in the course of finding solutions: x can be extended to the complex plane z and also y can be allowed to be complex. Equation 2.1 is called linear and homogeneous if $F(x, ay(x), ay^{(1)}(x), \ldots, ay^{(n-1)}(x)) = aF$. It is linear and inhomogeneous if there is an additional term which is only a function of x. A solution to a linear equation possesses n constants of integration. Write the nth order linear homogeneous equation as

$$Ly = y^{(n)} + p_{n-1}(x)y^{(n-1)} \ldots + p_0(x)y = 0. \tag{2.2}$$

Solutions $y_j(x)$ are said to be linearly dependent if there exist nonzero constants c_j such that

$$\sum_j c_j y_j(x) = 0 \qquad all \quad x. \tag{2.3}$$

Since this is true for all x we can differentiate, giving

$$\begin{vmatrix} y_1 & y_2 & \cdots & y_n \\ y_1^{(1)} & y_2^{(1)} & \cdots & y_n^{(1)} \\ \cdots & \cdots & \cdots & \cdots \\ y_1^{(n-1)} & y_2^{(n-1)} & \cdots & y_n^{(n-1)} \end{vmatrix} \begin{vmatrix} c_1 \\ c_2 \\ \cdots \\ c_n \end{vmatrix} = 0. \tag{2.4}$$

This set of equations has a solution only if the determinant $W(x)$, known as the Wronskian, is zero for all x. Consider an initial value problem, with y and the first $n-1$ derivatives of y given at $x = 0$. Write the solution in the form

$$y(x) = \sum_j c_j y_j(x) = 0. \tag{2.5}$$

Then the boundary conditions $y^{(k)}(0) = a_k, k \in [0, n-1]$ give $\sum_j c_j y_j^{(k)}(0) = a_k, k \in [0, n-1]$, and this set of equations has a nontrivial solution only if the Wronskian is nonzero.

Write the Wronskian as

$$W(x) = \epsilon_{ijk\ldots p} y_i y_j^{(1)} y_k^{(2)} \ldots y_p^{(n-1)}, \tag{2.6}$$

with $\epsilon_{ijk\ldots}$ the *nth* order Levi-Civita tensor, defined through setting $\epsilon_{1234\ldots} = 1$, $\epsilon_{ijk\ldots}$ odd under odd permutations of these indices, and zero otherwise, and differentiate. Differentiating the first factor produces

$$\epsilon_{ijk\ldots} y_i^{(1)} y_j^{(1)} y_k^{(2)} \ldots, \tag{2.7}$$

which is zero since $\epsilon_{ijk\ldots} = -\epsilon_{jik\ldots}$. Similarly when the derivative acts on every factor but the last the result is zero. Differentiating the last factor gives

$$\frac{dW(x)}{dx} = \epsilon_{ijk\ldots lp} y_i y_j^{(1)} y_k^{(2)} \ldots y_l^{(n-2)} y_p^{(n)}. \tag{2.8}$$

Use Eq. 2.2 to substitute $y_p^{(n)}$, and note that again all terms give zero except the $-p_{n-1} y_p^{(n-1)}$ term, giving

$$\frac{dW(x)}{dx} = -p_{n-1} \epsilon_{ijk\ldots p} y_i y_j^{(1)} y_k^{(2)} \ldots y_p^{(n-1)} = -p_{n-1} W. \tag{2.9}$$

Integrating we find

$$W(x) = c e^{-\int p_{n-1}(x) dx}, \tag{2.10}$$

with c a constant. This is called Abel's formula, and often allows one to evaluate the Wronskian for all x by calculating it at a single point, determining the constant c. Note that this equation also implies that $W(x)$ cannot vanish at a point unless $p_{n-1} = \infty$, which would be a singular point of the differential equation.

2.2 Constant Coefficients

If the functions $p_k(x)$ are all constants the differential equation 2.2 is easily solved by the substitution

$$y(x) = e^{rx}, \tag{2.11}$$

giving the polynomial in r, $P(r) = r^n + \sum_0^{n-1} p_j r^j$, with $Le^{rx} = P(r)e^{rx} = 0$. If the roots of this polynomial r_k are all distinct the solutions are simply $y_k(x) = e^{r_k x}$. Otherwise, suppose that there is an m-fold degeneracy, so that $Le^{rx} = e^{rx}(r - r_1)^m Q(r)$. Then examine

$$L\left(\frac{d}{dr} e^{rx}\right) = (r - r_1)^m \frac{d}{dr}(e^{rx} Q(r)) + m(r - r_1)^{m-1} e^{rx} Q(r). \tag{2.12}$$

This expression is zero for $r = r_1$, so $\frac{d}{dr}e^{rx} = xe^{rx}$ is also a solution for $r = r_1$. This process can be continued for $m - 1$ derivatives, giving as solutions

$$e^{r_1 x}, xe^{r_1 x}, \ldots, x^{m-1}e^{r_1 x}. \tag{2.13}$$

Thus no matter what the degeneracy, n independent solutions of the differential equation can be constructed.

2.3 Inhomogeneous Linear Equations

Consider an inhomogeneous differential equation of the form

$$Ly(x) = f(x). \tag{2.14}$$

If independent solutions of the homogeneous equation $Ly(x) = 0$ are known, a solution of Eq. 2.14 is readily found using the method of Green's functions. The Green's function for Eq. 2.14 is defined to be the solution to

$$LG(x, a) = \delta(x - a), \tag{2.15}$$

with $\delta(x)$ the Dirac delta function, which is actually not a function but a distribution, and can be defined in many different ways. Some of the possible definitions of δ and the related Heaviside function θ are

$$\begin{aligned}
\delta(x) &= lim_{a\to 0}\begin{bmatrix} \frac{1}{a} & |x| < a/2 \\ 0 & |x| > a/2 \end{bmatrix}, \\
\delta(x) &= lim_{a\to 0}\frac{1}{\sqrt{\pi a}}e^{-x^2/a}, \\
\delta(x) &= lim_{a\to 0}\frac{1}{\pi}\frac{a}{x^2 + a^2}, \\
\delta(x) &= lim_{a\to 0}\frac{sin(ax)}{\pi x}, \\
\delta(x) &= lim_{a\to\infty}\frac{1}{2\pi}\int_{-a}^{a} e^{ixt}dt, \\
\delta(x) &= lim_{N\to\infty}\frac{1}{2\pi}\sum_{m=-N}^{N} e^{imx}, \\
\theta(x) &= \int^{x} \delta(t)dt,
\end{aligned} \tag{2.16}$$

with the limits all understood to be taken after the delta function is multiplied by a sufficiently smooth function and integrated over x. Some of the properties of these functions, again understood to hold only upon being integrated over x, are

$$\begin{aligned}
\delta(x) &= \delta(-x), \\
\delta'(x) &= -\delta'(-x), \\
x\delta(x) &= 0, \\
x\delta'(x) &= -\delta(x), \\
\delta(ax) &= \frac{\delta(x)}{a}, \\
f(x)\delta'(x) &= -\delta(x)f'(x), \\
\delta(f(x)) = \sum_j \frac{1}{f'(x_j)}\delta(x - x_j) \quad \textit{with} \quad f(x_j) &= 0, \\
\theta'(x) &= \delta(x).
\end{aligned} \tag{2.17}$$

We illustrate the use of these functions with the derivation of the Green's function for a general linear second order differential equation with constant coefficients

$$Ly(x) = \left[\frac{d^2}{dx^2} + p_1\frac{d}{dx} + p_0\right] y = f(x). \tag{2.18}$$

To solve Eq. 2.15 note that everywhere but $x = a$ the solution for $G(x-a)$ is a sum of two independent solutions of the homogeneous equation. Thus write

$$G(x,a) = [A_1y_1 + A_2y_2][1 - \theta(x-a)] + [B_1y_1 + B_2y_2]\theta(x-a), \tag{2.19}$$

where we have used $1 - \theta(x-a)$ rather than $\theta(a-x)$ so that all θ and δ functions have the same argument, and it can be suppressed during the calculation. Now calculate the first and second derivatives of $G(x,a)$, substitute into $LG(x,a)$ and make use of the fact that $Ly_1 = Ly_2 = 0$. Using also $f(x)\delta'(x) = -\delta(x)f'(x)$ we find

$$\begin{aligned}
LG(x,a) = \delta(x-a)[&-(A_1y_1' + A_2y_2') + (B_1y_1' + B_2y_2') \\
&-p_1(A_1y_1 + A_2y_2) + p_1(B_1y_1 + B_2y_2)] = \delta(x-a).
\end{aligned} \tag{2.20}$$

This is a second order differential equation so we should be left with two constants of integration. Our solution has four arbitrary constants, and we must fix two of them. One condition is given by Eq. 2.20. It is convenient to choose the second by setting the term proportional to p_1 in this equation equal to zero, a condition equivalent to demanding that $G(x,a)$ be continuous at $x = a$, although with a discontinuous derivative. We thus have the two conditions

$$\begin{vmatrix} y_1(a) & y_2(a) \\ y_1'(a) & y_2'(a) \end{vmatrix} \begin{vmatrix} B_1 - A_1 \\ B_2 - A_2 \end{vmatrix} = \begin{vmatrix} 0 \\ 1 \end{vmatrix}, \tag{2.21}$$

and the solution

$$\begin{aligned} G(x,a) = {} & [A_1 y_1(x) + A_2 y_2(x)][1 - \theta(x-a)] \\ & + [B_1 y_1(x) + B_2 y_2(x)]\theta(x-a), \end{aligned} \tag{2.22}$$

with $B_1 = A_1 - y_2(a)/W(a)$ and $B_2 = A_2 + y_1(a)/W(a)$, and $W(a) = y_1(a)y_2'(a) - y_1'(a)y_2(a)$ is the Wronskian.

The solution to the inhomogeneous equation is then

$$y(x) = \int G(x,a)f(a)da, \tag{2.23}$$

with the condition that the range of integration in a includes the entire range of x of interest. The constants A_1, A_2 are still free, and can be used to match the boundary conditions for $y(x)$.

2.4 The Fredholm Alternative

Consider the differential equation

$$\frac{d^2y}{dx^2} + y = f(x), \tag{2.24}$$

with periodic boundary conditions, $y(0) = y(2\pi)$ and $y'(0) = y'(2\pi)$, and use the Green's function method of the last section. Two solutions of the homogeneous equation are $y_1 = sin(x)$, and $y_2 = cos(x)$. The Wronskian is $W(x) = -1$ and the Green's function is

$$G(x,a) = \begin{cases} A_1 sin(x) + A_2 cos(x) & x < a \\ B_1 sin(x) + B_2 cos(x) & x > a \end{cases}, \tag{2.25}$$

with the conditions

$$B_1 = A_1 + cos(a), \qquad B_2 = A_2 - sin(a), \tag{2.26}$$

and the solution

$$y(x) = \int G(x,a) f(a) da, \tag{2.27}$$

or

$$y(x) = \int_0^x [B_1 sin(x) + B_2 cos(x)] f(a) da$$
$$+ \int_x^{2\pi} [A_1 sin(x) + A_2 cos(x)] f(a) da. \tag{2.28}$$

Now attempt to determine the constants A_1 and A_2 by matching the boundary conditions. We find

$$y(0) = \int_0^{2\pi} A_2 f(a) da, \qquad y(2\pi) = \int_0^{2\pi} B_2 f(a) da. \tag{2.29}$$

Equating these expressions and using Eq. 2.26 we find

$$0 = \int_0^{2\pi} sin(a) f(a) da. \tag{2.30}$$

For the second boundary condition we have, using $B_2(0) = B_2(2\pi) = A_2$

$$y'(0) = \int_0^{2\pi} A_1 f(a) da, \qquad y'(2\pi) = \int_0^{2\pi} B_1 f(a) da, \tag{2.31}$$

and using again Eq. 2.26 we find

$$0 = \int_0^{2\pi} cos(a) f(a) da. \tag{2.32}$$

These conditions on $f(x)$, known as solubility conditions, have a very simple interpretation. The integrals above are the projections of the function $f(x)$ onto all functions belonging to the null space of the operator $d^2/dx^2 + 1$ and satisfying the given boundary conditions. Since all such functions satisfy the differential equation and the boundary conditions and also map into zero with this operator, they can never appear on the right-hand side of Eq. 2.24, and thus if $f(x)$ contains any parts of this kind the equation cannot be solved. Also note that A_1 and A_2 are not determined by these boundary conditions, since the homogeneous solutions satisfy them. Thus even with the boundary conditions satisfied, there are two arbitrary constants in the solution.

This is a particular example of the Fredholm alternative theorem, which states:

Suppose L is an nth order differentiable operator with n boundary conditions, $B_1 = 0, B_2 = 0 \ldots B_n = 0$. We are invited to solve $Ly = f$ with $B_k(y) = 0$, $k = 1, 2 \ldots n$. Then either

(1) there is one unique solution provided $f(x)$ is continuous
or
(2) $Ly = 0$ has at least one nontrivial solution, $y = w$. In this case $Ly = f$ has a solution only if $(w, f) = 0$ for each w satisfying $Lw = 0$, where (w, f) is an inner product defined on the space,[1] and as seen above, appears naturally with the use of the Green's function. These conditions are known as the solubility conditions.

2.5 The Diffusion Equation

As an example of a simple partial differential equation which can be solved exactly, consider the diffusion equation

$$\frac{\partial}{\partial t} f = D \frac{\partial^2}{\partial x^2} f, \tag{2.33}$$

with the additional requirement that the normalization $\int f dx = 1$ for all time. First note that this equation is invariant under $x \to cx$ and $t \to c^2 t$. Thus one would guess that $f = f(x^2/t)$. Introduce a new variable $z = x^2/t$. But we find that the normalization

$$\int_{-\infty}^{\infty} f(x, t) dx = 1, \tag{2.34}$$

upon substituting the variable z becomes

$$\frac{\sqrt{t}}{2} \int_{-\infty}^{\infty} f(z) dz/z = 1, \tag{2.35}$$

and thus $f(x, t)$ cannot be simply a function of z but rather

$$f(x, t) = kg(z)/\sqrt{t}. \tag{2.36}$$

Substitute into the differential equation using $\partial_t z = -z/t$, $\partial_x z = 2x/t$, $\partial_x^2 z = 2/t$, and find a second order differential equation for $g(z)$, with one

[1] In the example given the inner product is particularly simple, since we are dealing with real functions defined on the real interval $(0, 2\pi)$. In general the functions w belong to the null space of the adjoint operator L^*.

solution $g(z) = e^{-z/(4D)}$. By looking at dominant balance of the two solutions to the differential equation for large z, show that the second solution is not normalizable. Evaluating k we find

$$f(x,t) = \frac{e^{-\frac{x^2}{4Dt}}}{2\sqrt{\pi Dt}}. \tag{2.37}$$

Note from Eq. 2.16 that this in fact gives a representation of the delta function as $t \to 0$.

It was shown by Lord Raleigh that with $D = 2$ and $t = n$ the solution to Eq. 2.33, $f(n,x)$ gives the probability that the sum of n random variables in the interval [0,1] equals x, giving the interpretation of diffusion as a sequence of random walks, although this expression was only introduced by Karl Pearson in 1905. These ideas were essential for Einstein to develop the theory of Brownian motion.

2.6 Exact Solutions to Nonlinear Equations

There are very few cases allowing exact integration. The simplest is the case of separation of variables, i.e. the differential equation can be put in the form

$$M(x)dx + N(y)dy = 0, \tag{2.38}$$

which has the immediate solution $\int^x M(x)dx + \int^y N(y)dy = C$.

Another is the Bernoulli equation

$$y'(x) = a(x)y(x) + b(x)y^p(x), \tag{2.39}$$

which is solved by substituting $u = y^{1-p}$, after which the equation becomes linear in $u(x)$.

The Riccati equation is

$$y'(x) = a(x)y^2(x) + b(x)y(x) + c(x) \tag{2.40}$$

for which there is no general method, but if one solution $y_1(x)$ can be found, then the substitution $y(x) = y_1(x) + u(x)$ leads to a Bernoulli equation for $u(x)$.

Occasionally a nonlinear equation can be linearized by the proper substitution. Consider

$$y''(x) + c(y')^2 = 0. \tag{2.41}$$

The substitution $u = e^{cy}$ gives the trivial equation $u'' = 0$.

The differential equation

$$M(x, y(x)) + N(x, y(x))y'(x) = 0 \tag{2.42}$$

reduces to $df(x, y(x))/dx = 0$ provided that $\partial_y M = \partial_x N$.

2.7 Phase Plane Analysis

Often insight can be obtained concerning the behavior of solutions of coupled nonlinear equations by examining the phase space of the individual dependent variables. The trajectory in the phase space may move off to infinity, approach some limiting set consisting of a point, a circle, or some higher dimensional object, and it is sometimes possible to ascertain that one of these cases holds without fully solving the differential equations. We illustrate this technique with a well known problem (Bender-Orszag [1978]), namely:

At $t = 0$ a pig initially at (1,0) starts to run around the unit circle with constant speed v. At $t = 0$ a farmer initially at the origin runs with constant speed v and instantaneous velocity directed toward the pig. Does the farmer catch the pig?

Without loss of generality take $v = 1$. The position of the pig is given by $\vec{p} = (cos(t), sin(t))$. Let the position of the farmer be given by $\vec{f} = (x, y)$. We are interested in the separation between them so define $s^2 = (cos(t) - x)^2 + (sin(t) - y)^2$. Then define q through

$$cos(t) - x = scos(q), \qquad sin(t) - y = ssin(q), \tag{2.43}$$

or $\vec{s} = s(cos(q), sin(q))$. At $t = 0$ we have $x = y = 0$, $s = 1$, and thus $q = 0$. The instantaneous direction of the velocity of the farmer is given by

$$\frac{dy}{dx} = \frac{sin(q)}{cos(q)}. \tag{2.44}$$

Writing $(dx/dt)^2 + (dy/dt)^2 = 1$ and substituting we find

$$\frac{dx}{dt} = cos(q), \qquad \frac{dy}{dt} = sin(q). \tag{2.45}$$

Now differentiate Eq. 2.43 and substitute Eq. 2.45, multiply by $cos(q)$ and by $sin(q)$, and add and subtract, giving

$$\frac{ds}{dt} = sin(q-t) - 1, \qquad s\frac{dq}{dt} = cos(q-t). \tag{2.46}$$

Let $\theta = q - t$, giving the simple coupled set of nonlinear equations

$$\dot{s} = sin(\theta) - 1, \qquad s(\dot{\theta} + 1) = cos(\theta), \tag{2.47}$$

with initial conditions $t = 0, s = 1, \theta = 0$. The angle θ is simply seen to be the angle between $\vec{s}$ and $\vec{p}$. We wish to discover whether at any time $s = 0$.

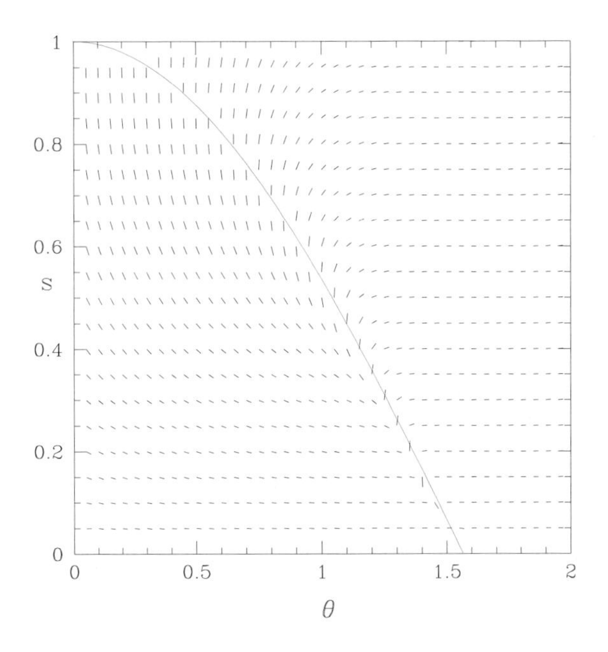

Fig. 2.1 Phase plot for Eq. 2.48.

Examine the θ, s plane, and use the fact that

$$\frac{ds}{d\theta} = \frac{s(sin(\theta) - 1)}{cos(\theta) - s}, \tag{2.48}$$

and note that for $\theta = 0$, $ds/d\theta = -s/(1-s)$, so $ds/d\theta < 0$ for $0 < s < 1$. Thus as the trajectory starts moving down from $s = 1$, $\theta = 0$, the variable θ must increase. Furthermore for $s = 0$ we have $ds/d\theta = 0$ except at the point $\theta = \pi/2$, and thus this is the only possible point the trajectory could arrive at $s = 0$. The phase plot is shown in Fig. 2.1. Also shown is the function $s = cos(\theta)$, along which the phase is vertical, except for the point $s = 0$. As this curve is approached $ds/d\theta$ rotates toward the vertical, and thus it acts as a barrier, forcing the solution downward.

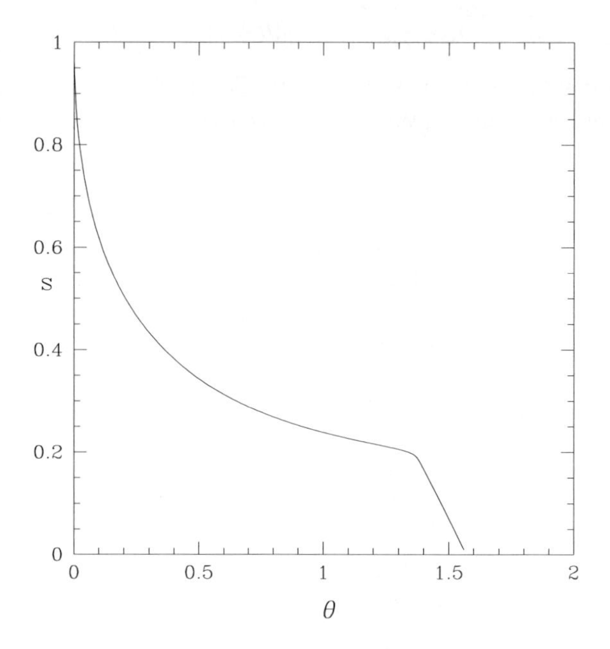

Fig. 2.2 Numerical integration of Eq. 2.47.

We cannot solve for $s(t)$, $\theta(t)$, but we can look at the asymptotic solution near $s = 0$. Write

$$\begin{aligned} \theta(s) &= \theta_0 + \theta_0' s + \theta_0'' s^2/2 + \ldots \\ \dot{\theta} &= \theta_0' \dot{s} + \theta_0'' s\dot{s} + \ldots . \end{aligned} \tag{2.49}$$

Expanding we find

$$\begin{aligned} cos(\theta) &= cos(\theta_0) - (\theta_0' s + \theta_0'' s^2/2) sin(\theta_0) + \frac{(\theta_0' s)^2}{2} cos(\theta_0) + \ldots \\ sin(\theta) &= sin(\theta_0) + (\theta_0' s + \theta_0'' s^2/2) cos(\theta_0) - \frac{(\theta_0' s)^2}{2} sin(\theta_0) + \ldots . \end{aligned} \tag{2.50}$$

Now substitute these equations into Eq. 2.47 giving

$$cos(\theta_0) - (\theta_0's + \theta_0''s^2/2)sin(\theta_0) + \frac{(\theta_0's)^2}{2}cos(\theta_0) = s(\theta_0'\dot{s} + \theta_0''s\dot{s} + 1),$$

$$sin(\theta_0) + (\theta_0's + \theta_0''s^2/2)cos(\theta_0) - \frac{(\theta_0's)^2}{2}sin(\theta_0) - 1 = \dot{s}. \quad (2.51)$$

Now solve order by order in s. Order zero gives $cos(\theta_0) = 0$ so $\theta_0 = \pi/2$, as we knew it had to be from the phase plot. The above equations then reduce to

$$-(\theta_0's + \theta_0''s^2/2) = s(\theta_0'\dot{s} + \theta_0''s\dot{s} + 1),$$

$$-\frac{(\theta_0's)^2}{2} = \dot{s}. \quad (2.52)$$

From the second equation we see that $\dot{s}$ is second order in s and thus from the terms in the first equation of first order in s we have $\theta_0' = -1$, and second order gives $\theta_0'' = 0$.

But now near $s = 0$ we find $\dot{s} = -s^2/2$ and thus

$$\frac{1}{s} + c = \frac{t}{2}. \quad (2.53)$$

The constant of integration comes from the initial part of the trajectory and we cannot evaluate it, but now we note that no matter what this constant is, the only solution for $s \to 0$ is $t \to \infty$, and thus the farmer never catches the pig. Numerical integration up to $t = 200$ is shown in Fig. 2.2. Note that the asymptotic solution $\theta = \pi/2 - s$ is very good for $s < 0.2$, and that this part of the solution is along the barrier $s = cos(\theta)$ shown in Fig. 2.1.

2.8 Problems

1. Find the Green's function for the third order inhomogeneous linear differential equation $y''' = f(x)$.

2. Given one solution to a general second order linear differential equation $y_1(x)$, use Abel's formula to find a second solution.

3. Use the method of Green's functions to find the solution to

$$x\frac{d^2y}{dx^2} - 6\frac{dy}{dx} + \frac{6y}{x} = x^4.$$

4. Solve

$$\frac{d^2y}{dx^2} + 2\frac{dy}{dx} - 2y = e^x.$$

5. Solve $y'' + (x-2)y' - xy = 0$.

6. Solve $x^3y' = 2y^2 + 3x^2 = 0$.

7. Solve $y' = 2xy + y = 0$.

8. Find a solution with $y = 1$, $dy/dt = 0$ at $t = 0$ using the Green's function:

$$\frac{d^2y}{dt^2} + \nu\frac{dy}{dt} + \omega_0^2 y = sin(\omega t).$$

9. If $u = 1 + x^3/3! + x^6/6! + ...$
$v = x/1! + x^4/4! + x^7/7! + ...$
$w = x^2/2! + x^5/5! + x^8/8! + ...,$
prove that $u^3 + v^3 + w^3 - 3uvw = 1$. (Putnam exam, 1939)

10. Consider the matrix equation

$$\begin{bmatrix} 0 & 1 \\ 0 & 1 \end{bmatrix} \begin{vmatrix} x_1 \\ x_2 \end{vmatrix} = \begin{vmatrix} y_1 \\ y_2 \end{vmatrix}.$$

Find conditions on y such that a solution exists. Relate this to the Fredholm alternative theorem.

11. Consider the differential equation

$$\frac{d^2y}{dx^2} + y = sin^n(x),$$

with periodic boundary conditions on $x = [0, \pi]$. For what values of n does this equation have a solution? Is the solution unique? Explain.

12. Consider the differential equation

$$\frac{d^2y}{dx^2} + y = cos^n(x),$$

with boundary conditions $y(0) = 0$, $y(1) = 0$ on $x = [0, \pi]$. For what values of n does this equation have a solution? Is the solution unique? Explain.

13. Find the approximate distance between the line $s = cos(\theta)$ and the path of the solution in Fig. 2.2, by setting the slope $ds/d\theta = -1$ in Eq. 2.48.

14. Solve

$$\frac{d^2y}{dx^2} + \frac{dy}{dx} - 2y = e^x$$

for $y(0) = 1$, $y(1) = e$ using the Green's function.

Chapter 3

Complex Variables

Augustin-Louis Cauchy was born in Paris in 1789. He was educated by his father, and Laplace and Lagrange were regular visitors to their home. He entered the École Polytechnique in 1805, and then went to the engineering school École des Ponts et Chaussees. He was an engineer in the army of Napoleon, working on the port facilities at Cherbourg, and in 1815, after failing for some time to find a position, became a professor at the École Polytechnique. In 1847 he announced that he had a proof of Fermat's last theorem, but Ernst Kummer showed that his proof depended on unique factorization, a property true for real integers, but not for the complex numbers used in his proof. He was responsible for introducing rigorous concepts of function and limit, including both differentiation and integration. He defined integration, which previously had been simply the inverse of differentiation, as a limit of a sum using small intervals. He provided precise definitions of convergence for series and made rigorous the comparison, ratio, and integral tests. He was responsible for developing the theory of functions of a complex variable, and his major achievement was the Cauchy integral theorem for analytic functions. He also developed techniques for the diagonalization of matrices and the use of Fourier transforms. He worked on Fresnel's wave theory and on the dispersion and polarization of light. He worked on waves in elastic media, and developed the theory of stress, introducing the 3x3 stress tensor, and wrote a large number of papers on various mechanical questions, especially problems concerning vibrating strings. He also wrote a treatise on the economics of risk aversion, based on the solution to the St. Petersburg paradox. He developed the kinetic theory of gases and derived Boyle's law. He wrote over eight hundred research articles and published five textbooks.

3.1 Analyticity

This chapter reviews only a few basic concepts of the theory of complex variables to establish notation for future analysis. Normally we will be interested in real solutions of differential equations on the real line, but we will see that properties in the complex plane determine the nature of the solutions and provide means of finding asymptotic limits. Consider a function of the complex variable $z = x + iy = re^{i\theta}$, and write it in terms of its real and imaginary parts

$$f(z) = u(x, y) + iv(x, y). \tag{3.1}$$

A function is said to be analytic at $z = z_0$ if

$$\frac{df}{dz} = lim_{z\to z_0} \frac{f(z) - f(z_0)}{z - z_0} \tag{3.2}$$

exists and is independent of the path along which $z \to z_0$. Write this in terms of the real variables

$$\frac{df}{dz} = lim_{x\to x_0, y\to y_0} \frac{u(x, y) + iv(x, y) - u(x_0, y_0) - iv(x_0, y_0)}{x + iy - x_0 - iy_0}. \tag{3.3}$$

Taking the limit from above ($x = x_0$) and from the side ($y = y_0$) we find

$$\frac{df}{dz} = -i\frac{\partial u}{\partial y} + \frac{\partial v}{\partial y} = \frac{\partial u}{\partial x} + i\frac{\partial v}{\partial x}, \tag{3.4}$$

giving the Cauchy–Riemann conditions

$$\frac{\partial v}{\partial x} = -\frac{\partial u}{\partial y}, \qquad \frac{\partial v}{\partial y} = \frac{\partial u}{\partial x}, \tag{3.5}$$

from which it follows also that both of the functions u, v satisfy Laplace's equation, i.e.

$$\frac{\partial^2 v}{\partial x^2} + \frac{\partial^2 v}{\partial y^2} = 0, \tag{3.6}$$

and similarly for u. If $f(z)$ is not analytic at some point it is said to be singular there.

Also note that $\nabla u \cdot \nabla v = 0$, i.e. surfaces of constant u are orthogonal to surfaces of constant v.

3.2 Cauchy Integral Theorem

From the Cauchy–Riemann conditions we can prove the Cauchy integral theorem, which states that if $f(z)$ is analytic in a simply connected domain bounded by the contour C then

$$\int_C f(z)dz = 0. \tag{3.7}$$

To prove this write everything in terms of the real functions and variables

$$\int_C f(z)dz = \int_C (u + iv)(dx + idy) = \int_C (udx - vdy) + i \int_C (vdx + udy). \tag{3.8}$$

Now recall the Stokes theorem

$$\oint \vec{A} \cdot d\vec{s} = \int (\nabla \times \vec{A}) \cdot d\vec{a}, \tag{3.9}$$

where the domain defining the integration must be simply connected. For the first term of Eq. 3.8 use $\vec{A} = (u, -v)$ and for the second term use $\vec{A} = (v, u)$. Using the Cauchy–Riemann conditions we find

$$\int_C f(z)dz = 0. \tag{3.10}$$

From this we can prove that an integral $\int_{z_1}^{z_2} f(z)dz$ is independent of the integration path provided that $f(z)$ is analytic in the domain defined by the two paths. To prove this let C_1 and C_2 be the two paths connecting the points z_1 and z_2 and simply write

$$\int_{C_1} f(z)dz - \int_{C_2} f(z)dz = \oint f(z)dz = 0. \tag{3.11}$$

We then can prove the Cauchy integral formula, which states that if $f(z)$ is analytic inside the closed contour C and u a point inside the contour, then

$$f(z) = \frac{1}{2\pi i} \int_C \frac{f(u)du}{u - z}. \tag{3.12}$$

To prove this note that $f(u)/(u - z)$ is analytic everywhere inside C except for the point $u = z$ and deform the contour into a small circle of radius ϵ around z. (For rigor, there are a number of intermediate steps, e.g. that the sum of two analytic functions is analytic, the product of two analytic functions is analytic, etc., which are left to the reader.) Write $u - z = \epsilon e^{i\theta}$, $du = i\epsilon e^{i\theta} d\theta$, use the continuity of $f(u)$ to write $f(u) = f(z) + O(\epsilon)$, and take the limit as $\epsilon \to 0$.

3.3 Series Representation

Suppose that a power series in $(z - z_0)$ converges to a function of the complex variable z,

$$f(z) = \sum_{n=0}^{\infty} a_n (z - z_0)^n. \tag{3.13}$$

Then by differentiating this equation we find

$$f^{(k)}(z_0) = k! a_k, \tag{3.14}$$

and thus

$$f(z) = \sum_{n=0}^{\infty} \frac{f^{(n)}(z_0)}{n!} (z - z_0)^n, \tag{3.15}$$

the Taylor series (Taylor [1715]) for the function $f(z)$ about the point z_0.

A complex series $\sum c_n$ is said to be absolutely convergent if the real series $\sum |c_n|$ is convergent. Two tests for absolute convergence are the ratio test (D'Alembert)

$$lim_{n\to\infty} \left| \frac{c_{n+1}}{c_n} \right| = \lambda < 1, \tag{3.16}$$

and the radical test (Cauchy)

$$lim_{n\to\infty} |c_n|^{1/n} = \lambda < 1. \tag{3.17}$$

From the ratio test the radius of convergence for a power series is

$$R = lim_{n\to\infty} \left| \frac{a_n}{a_{n+1}} \right|. \tag{3.18}$$

The Taylor series in Eq. 3.15 converges in a circle about the point z_0 with a radius given by the distance to the nearest singularity of $f(z)$. Clearly the convergence test in Eq. 3.18 gives a condition on $|z - z_0|$, so the domain of convergence is a circle with center z_0. By hypothesis the function is analytic within the circle defined by the first singularity, and by the construction above, the series must converge. Conversely, if the series converged in a larger circle the function defined by it would be analytic at the point of singularity, a contradiction.

If $f(z)$ has a pole of order p at $z = z_0$ then in a similar manner it can be written in the form $f(z) = g(z)/(z - z_0)^p$ and $g(z)$ is analytic with a Taylor series about z_0. Equivalently $f(z)$ has a series of the form

$$f(z) = \sum_{n=0}^{\infty} \frac{g^{(n)}(z_0)}{n!}(z - z_0)^{n-p}. \tag{3.19}$$

3.4 The Residue Theorem

Now apply the Cauchy integral theorem to a function which is analytic except for possessing isolated poles in a given simply connected domain D. Consider the integral

$$I = \int_C f(z)dz, \tag{3.20}$$

where the contour C encloses the domain D in a counterclockwise sense. Then use the integral theorem to contract the contour to a series of separate contours each encircling one of the singularities z_k of $f(z)$. At each singularity construct the series expansion of $f(z)$ in the neighborhood of z_k,

$$f(z) = \sum_n a_n^k (z - z_k)^n. \tag{3.21}$$

Then using

$$\oint z^m dz = 2\pi i \delta_{m,-1}, \tag{3.22}$$

we find that

$$I = 2\pi i \sum_k a_{-1}^k \tag{3.23}$$

which is known as the Cauchy residue theorem. The coefficient of $(z-z_k)^{-1}$ in the expansion about the point z_k is called the residue of the function $f(z)$ at the point z_k. This expression is very useful in the evaluation of integrals.

The residue theorem can also be used to evaluate series. For example consider the series

$$\sum_{n=1}^{\infty} \frac{1}{n^2}. \tag{3.24}$$

Construct an integral which contains this series as a part of the sum over the residues,

$$I = \oint_C \frac{dz}{z^2} \frac{cos(\pi z)}{sin(\pi z)} \tag{3.25}$$

where C is a large square contour including the integers $-M, M$, and the numerator is chosen so that the residue is positive at each integer. Evaluate this integral in two ways. First bound it for large M. Along the vertical part of the contour at the right take $z = M + 1/2 + iy$ so that along the right side we have

$$\frac{cos(\pi z)}{sin(\pi z)} = \frac{e^{-\pi y} - e^{\pi y}}{e^{-\pi y} + e^{\pi y}} < 1. \tag{3.26}$$

A similar bound exists along the left vertical path, taking $z = -M-1/2+iy$. Along the top the terms in $e^{\pi y}$ dominate, and the ratio tends to 1 for large M, and similarly along the bottom with the terms $e^{-\pi y}$. We then have

$$\oint_C \frac{dz}{z^2} \frac{cos(\pi z)}{sin(\pi z)} \to 0 \tag{3.27}$$

for $M \to \infty$. Now evaluate the sum of the residues. From the positive and negative integers the contribution to the integral is

$$4\pi i \sum_{n=1}^{\infty} \frac{1}{n^2}. \tag{3.28}$$

The singularity at $z = 0$ is third order. Write

$$\frac{cos(\pi z)}{sin(\pi z)} = \frac{g(z)}{z} \tag{3.29}$$

and find the Taylor expansion for $g(z)$. We readily find

$$g(z) = 1/\pi - \pi^2 z^2/3 + O(z^3). \tag{3.30}$$

Adding the residue at $z = 0$ to those from the nonzero integers then gives for the sum

$$\sum_{n=1}^{\infty} \frac{1}{n^2} = \frac{\pi^2}{6}. \tag{3.31}$$

3.5 Analytic Continuation

If a function is defined and analytic on a bounded set which contains a limit point the Taylor series provides a unique analytic continuation to the rest of the plane, with the exception of points where the function is singular. A simple example which illustrates this is the series

$$\frac{1}{1-z} = 1 + z + z^2 + \ldots \tag{3.32}$$

which converges for $|z| < 1$. It is important to realize that although the series does not converge for $|z| > 1$, the function $f(z)$ defined by the analytic continuation of the series still exists and is analytic everywhere in the complex plane except at the point $z = 1$, given by the expression $1/(1-z)$. The series provides a unique analytic continuation of the function from the domain $|z| < 1$ to the rest of the plane. Choose another point a within the circle $|z| < 1$ and construct the Taylor series about it, by equating

$$\sum_l z^l = \sum_k a_k (z-a)^k, \tag{3.33}$$

and by differentiating and setting $z = a$ find

$$a_l = \sum_k \frac{\Gamma(l+1)}{\Gamma(l-k+1)\Gamma(k+1)} a^{k-l}. \tag{3.34}$$

In this case we also know the closed form of the function, so we also have $a_l = f^{(l)}(a)/l!$. Since $f^l(a) = l!/(1-a)^{l-1}$ we recognize this relation as a simple binomial series. This new series is again convergent in a circle up to the singularity at $z = 1$. The radius of convergence is given by Eq. 3.18, giving

$$R = |a - 1|, \tag{3.35}$$

and thus is always a circle making contact with the point $z = 1$. By a successive choice of points a, b, c it is possible to continue into the entire complex plane minus the point $z = 1$ as shown in Fig. 3.1. This is not a

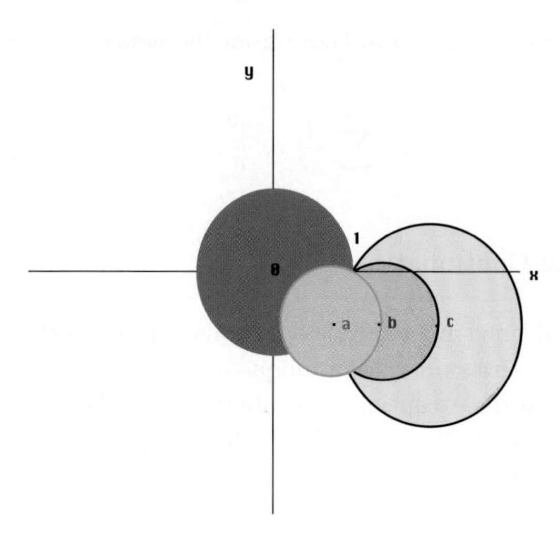

Fig. 3.1 Analytic continuation of the series $\sum_k z^k$ successively from $z = 0$ to $z = a, b, c$.

convenient way to carry out analytic continuation, it merely demonstrates the possibility. A much more powerful method is through the use of integral representations.

Analytic continuation in some cases is only to the cut plane, that is continuation in an arbitrary manner throughout the plane results in a multi-valued function. For example, the function $f = \sqrt{1-z}$, with $f(0) = +1$ is analytic at $z = 0$, but there is a cut emanating from the point $z = 1$, and continuation from the origin to a point on the real line with $x > 1$ depends on whether the line is approached from above or from below. Near $z = 1$ write $z = 1 - \epsilon e^{i\theta}$ so that $\theta = 0$ on the real axis between zero and one. Going above the cut, θ goes from zero to $-\pi$ and f changes phase from zero to $-\pi/2$, giving $f = -i\sqrt{\epsilon}$. Going below the cut, θ goes from zero to π and f changes phase from zero to $\pi/2$, giving $f = i\sqrt{\epsilon}$. This continuation is shown in Fig. 3.2. This method of keeping track of the sheet to which the continuation is made is essential, and will appear in many problems.

Continuation through a cut defines a multi-valued extension of the function, which is single valued on a topological covering space called the Riemann sheet. In the case of the square root the Riemann sheet has only two surfaces, continuation around the singularity twice returns f to the value on the first sheet.

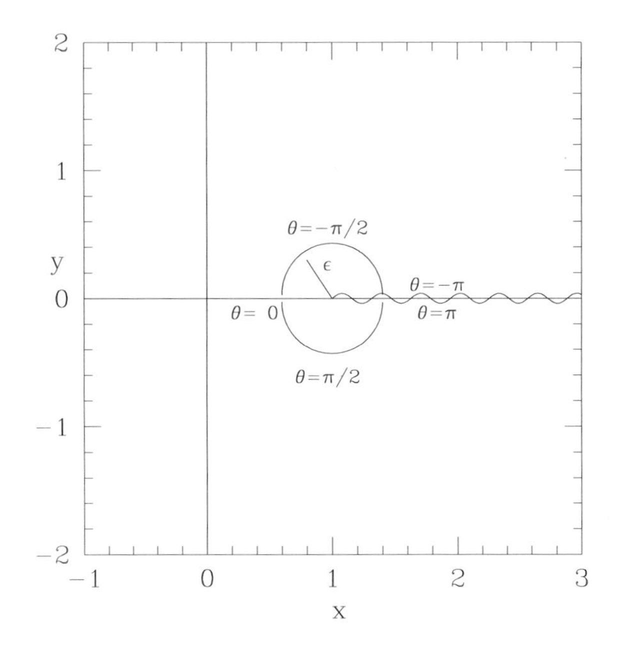

Fig. 3.2 Analytic continuation of $f = \sqrt{1-z}$.

3.6 Inverse Functions

The residue theorem gives a means of inverting an analytic function. Suppose that $f(z)$ is analytic in a circle $|z| < R$ with $f(0) = 0$, $f'(0) \neq 0$ and that $f(z) \neq 0$ for $0 < |z| < R$. Choose a contour C given by $s = re^{i\theta}$ with $r < R$. Then

$$g(w) = \frac{1}{2\pi i} \int_C \frac{sf'(s)}{f(s) - w} ds \tag{3.36}$$

defines an analytic function of w with $z = g(w)$ the only local inverse of

$$f(z) = w, \tag{3.37}$$

which tends to zero at $w = 0$. First note that by direct substitution into Eq. 3.36 we find that $g(f(z)) = z$, provided that there is only one zero of the denominator within the contour. The analyticity of the function g is restricted to $w < m = min_\theta |f(s)|$. To prove that g is analytic in this domain note that for any point w with $|w| < m$ with z on C we have $|f(z)| > |w|$, and thus the two functions $f(z)$ and $f(z) - w$ have the same number of zeros for $|z| < r$. Since $f(z)$ has a single zero at $z = 0$ we conclude that Eq. 3.37 has a single root, which we name $g(w)$, inside C. Then for $|w| < m$ note that

$$\frac{1}{f(s) - w} = \frac{1}{f(s)} + \frac{w}{f^2(s)} + \frac{w^2}{f^3(s)} + \dots, \tag{3.38}$$

which converges uniformly as long as s is inside C and $w < m$. Multiply this series by the function $sf'(s)$ and integrate term by term. Then we find

$$g(w) = \sum_0^\infty w^n \frac{1}{2\pi i} \int_C \frac{sf'(s)}{[f(s)]^{n+1}} ds, \tag{3.39}$$

which proves Eq. 3.36.

3.7 Problems

1. Show that the function

$$\frac{1}{1+\sqrt{2-z}}$$

has two series about the point $z = 0$, one with a radius of convergence of 1, the other with a radius of convergence of 2.

2. Evaluate the integral

$$\int_{-\infty}^{\infty} \frac{dx}{x^2 - x + 1}.$$

3. Evaluate the integral

$$\int_{-\infty}^{\infty} \frac{(x^2+3)dx}{(x^2 - x + 1)(x^2 - 2x + 2)}.$$

4. Evaluate the integral

$$\int_{-\infty}^{\infty} \frac{(x^2-1)dx}{(x^2+1)^3}.$$

5. Evaluate the integral

$$\int_{0}^{\infty} \frac{xdx}{(x^3+1)}.$$

6. Evaluate the integral

$$\int_{0}^{\infty} \frac{cos(ux)du}{(a^2+u^2)}.$$

7. Find the sum

$$\sum_{0}^{\infty} \frac{1}{1+n^2}.$$

8. Show that

$$\sum_{0}^{\infty} \frac{(-1)^n}{(2n+1)^3} = \frac{\pi^3}{32}.$$

9. Let

$$g(w) = \frac{1}{\pi i} \int_C \frac{s(s+1)}{s^2 + 2s - w} ds,$$

with C given by $|s| = 1$. Find the Taylor series of $g(w)$ at $w = 0$ and evaluate it.

10. Let

$$g(w) = \frac{1}{2\pi i} \int_C \frac{s^2(3s+2)}{s^3 + s^2 - w} ds,$$

with C given by $|s| = 1/2$. Find the first two terms of the Taylor series of $g(w)$ at $w = 0$.

11. Show that

$$\int_0^\infty \frac{\cos(mt)}{t^2 + 1} dt = \frac{\pi e^{-m}}{2}.$$

Chapter 4

Local Approximate Solutions

Henri Poincaré was born in Nancy, France in 1854. He graduated from the Lyceé in Nancy in 1871. He served in the Franco-Prussian war of 1870 in the Ambulance Corps. He then studied at the École Polytechnique and the École des Mines and spent most of his career at the University of Paris. He came close to developing the theory of relativity. He published, at the same time as Einstein's 1905 publication, a paper which was mathematically equivalent but maintained the concept of the fixed universal ether. He discovered that the relativistic Lorentz transformations were those that left invariant Maxwell's equations. He studied the three-body problem and found that the differential equations had chaotic solutions, this discovery being the origin of chaos theory. He invented the procedure for plotting the points of a solution in a particular section corresponding to a given time period or a particular value of one variable in order to understand the nature of the chaos generated. The development of these ideas was limited by the lack of rapid numerical computing at the time, and they were fully implemented only in recent decades. He developed the field of algebraic topology, and the Poincaré conjecture, that every simply connected three-dimensional manifold is homeomorphic to the three-sphere, has only been proven recently by G. Perelman, at least everyone expects that the proof is correct. Poincaré gave an incorrect proof himself, but later found a counter example for one step of the proof. He was also the originator of the theory of analytic functions of several complex variables. In the process of studying differential equations he also discovered asymptotic series, and formulated the methods to make them useful.

4.1 Introduction

It is often difficult or even impossible to obtain exact global solutions to a given differential equation. In this case it is extremely useful to have a local approximation, either in the form of a power series convergent within some range of the variable or in the form of an asymptotic series valid only to a limited accuracy. In particular if certain boundary conditions are imposed at say $z = \pm\infty$ or at $z = 0$ and $z = \infty$ the local approximations can be used to determine that combination of independent solutions satisfying the boundary conditions and to find the nearby behavior of the solution. The nature of the local approximation which exists is determined by the local properties of the differential equation. There are three main types of local approximations: series which converge with infinite radius of convergence, with terms in the series taking the form $z^n/(n+a)!$ for large n; series which converge conditionally, i.e. for a restricted range of z, with terms taking the form $z^n(n+a)!/(n+b)!$; and asymptotic divergent expansions which take the form $z^n(n+a)!$. Even though it may be more reassuring to use the convergent series, they have the defect that as z increases, more and more terms are needed in order to obtain a given accuracy. The asymptotic expansions have the problem of limited accuracy, but the benefit of producing that accuracy with typically very few terms. The conditionally convergent expansions are midway between these extremes, and commonly appear in theories in which the limited range of convergence is associated with a transition in the nature of the solution.

However, even if local solutions can be obtained in the vicinity of the points of interest, the problem remains to connect these solutions to each other. One may know which combination of solutions gives the desired boundary conditions, for example, at $z = \infty$, but knowing how this translates into the correct combination of the solutions obtained locally near zero or at $z = -\infty$ is nontrivial. There are two principal methods of achieving this global connection: WKB theory and the construction of integral solutions. These methods will be examined in Chapters 5 and 9.

4.2 Classification

Consider the nth order linear differential equation

$$Ly = y^{(n)} + p_{n-1}(x)y^{(n-1)} \dots p_0(x)y = 0. \tag{4.1}$$

Local approximations to the solution of the differential equation through series expansions are classified according to the behavior of the functions $p_k(x)$ at the point in question. The classification is due to Fuchs [1866].

4.2.1 *Ordinary point*

A point x_0 is called an ordinary point if all the $p_k(x)$ are analytic at this point, naturally considered as functions of the complex variable $z = x + iy$. In this case the solutions of the differential equation possess Taylor series which converge in a circle with radius at least as large as the distance to the nearest singularity of the p_k.

4.2.2 *Regular singular point*

A point x_0 is called a regular singular point if the functions $(x-x_0)^n p_0, (x-x_0)^{n-1}p_1, \ldots (x-x_0)p_{n-1}$ are analytic at x_0. A simple way to remember this is to think of the solution $y(x)$ as a power of $x - x_0$, and as $y^{(n)} \sim y/(x-x_0)^n$. Then this condition reduces to the requirement that no term in the differential equation be more singular than the highest derivative term. In this case there is at least one solution of the form

$$y = (x - x_0)^\alpha A(x), \tag{4.2}$$

where $A(x)$ is analytic with a Taylor series converging to the nearest singularity of the $p_k(z)$. The power α is called the indicial exponent and the equation which determines it the indicial equation. If $n \geq 2$ there is a second solution of the form

$$y = C(x)(x - x_0)^\beta, \tag{4.3}$$

or possibly, if $\alpha - \beta$ is an integer,

$$y = (x - x_0)^\alpha A(x) ln(x - x_0) + C(x)(x - x_0)^\beta. \tag{4.4}$$

If $n > 2$, for each new linearly independent solution there is a new analytic function and either a new indicial equation or another power of $ln(x - x_0)$.

It is instructive to see how the $ln(x)$ terms appear. Consider the modified Bessel equation

$$y'' + \frac{1}{x}y' - (1 + \frac{\nu^2}{x^2})y = 0. \tag{4.5}$$

The origin is a regular singular point, so write

$$y = \sum a_n x^{n+\alpha}. \tag{4.6}$$

Substituting into the differential equation we find

$$(\alpha^2 - \nu^2)a_0 x^{\alpha-2} + [(1+\alpha)^2 - \nu^2]a_1 x^{\alpha-1} +$$
$$\sum_2^\infty ([(n+\alpha)^2 - \nu^2]a_n - a_{n-2})x^{n+\alpha-2} = 0, \tag{4.7}$$

with solutions $\alpha = \pm\nu$, $a_k = 0$ for k odd, giving (see the first section of Chapter 8)

$$a_{2k} = \frac{a_0\Gamma(1+(\alpha-\nu)/2)\Gamma(1+(\alpha+\nu)/2)}{2^{2k}\Gamma(k+1+(\alpha-\nu)/2)\Gamma(k+1+(\alpha+\nu)/2)}. \tag{4.8}$$

The standard solution is given by setting $\alpha = \nu$ and taking $a_0 = 1/\Gamma(\nu+1)$,

$$I_\nu(x) = \sum_0^\infty \frac{(\frac{x}{2})^{2k+\nu}}{k!\Gamma(\nu+k+1)}. \tag{4.9}$$

If $\nu \neq$ *integer* there are two independent functions, $I_\nu, I_{-\nu}$, but for ν integer it is easy to see that the second solution is the same as the first.

To construct a second solution for $\nu = 0$ proceed in a similar manner as with an equation with constant coefficients. Introduce the operator

$$Ly = y'' + \frac{1}{x}y' - y, \tag{4.10}$$

and apply it to Eq. 4.6, giving

$$Ly = \alpha^2 a_0 x^{\alpha-2}, \tag{4.11}$$

where we have used the fact that the a_k are defined through Eq. 4.8, so the sum in Eq. 4.7 is zero for all α. Now find

$$L\frac{dy}{d\alpha} = 2\alpha a_0 x^{\alpha-2} + \alpha^2 a_0 ln(x) x^{\alpha-2}, \tag{4.12}$$

which is also zero for $\alpha = 0$, and thus $dy/d\alpha|_{\alpha=0}$ is a solution.

For $\nu = 0$ we have

$$y_0(x) = a_0 \sum_0^\infty \frac{(\frac{x}{2})^{2k} x^\alpha}{[(1+\alpha/2)(2+\alpha/2)...(k+\alpha/2)]^2}. \tag{4.13}$$

Differentiating with respect to α we have

$$\frac{dy_0(x)}{d\alpha}|_{\alpha=0} = ln(x)y_0 - a_0 \sum_0^\infty \frac{(\frac{x}{2})^{2k}[\frac{1}{k} + \frac{1}{k-1} + \ldots + 1]}{k!^2}. \tag{4.14}$$

The usual choice for the second solution is given by $a_0 = -1$, and adding $(ln(2) - \gamma)I_0$ with $\gamma = 0.5772...$ the Euler–Mascheroni constant we find

$$K_0(x) = -[ln(x/2) + \gamma]I_0(x) + \sum_0^\infty \frac{(\frac{x}{2})^{2k}[\frac{1}{k} + \frac{1}{k-1} + \ldots + 1]}{k!^2}. \tag{4.15}$$

This choice is determined by requiring simple behavior at $x = \infty$ and cannot be justified here.

It is not always the case that differentiating with respect to α produces immediately a second solution. Consider

$$Ly(x) = \left(\frac{d^2}{dx^2} - \frac{1}{x}\right) y(x) = 0. \tag{4.16}$$

The point $x = 0$ is a regular singular point. Substitute

$$y = \sum a_n x^{n+\alpha}, \tag{4.17}$$

giving

$$Ly(x) = \alpha(\alpha - 1)a_0 x^{\alpha-2} + \sum_1^\infty [(n + \alpha)(n + \alpha - 1)a_n - a_{n-1}]x^{n+\alpha-2}, \tag{4.18}$$

giving the recurrence relation $(n + \alpha)(n + \alpha - 1)a_n = a_{n-1}$, and $Ly = \alpha(\alpha - 1)a_0 x^{\alpha-2}$. The indicial equation gives $\alpha = 0, 1$, but $\alpha = 0$ has a problem in the recurrence relation for $n = 1$. Taking $\alpha = 1$ gives the solution

$$y_1 = \sum a_n x^{n+1}, \qquad a_n = \frac{a_{n-1}}{n(n+1)} = \frac{a_0}{(n+1)!n!}. \tag{4.19}$$

To find a second solution note that

$$L\frac{dy_1(x)}{d\alpha}|_{\alpha=1} = \frac{a_0}{x}, \tag{4.20}$$

and look for another solution to the same inhomogeneous equation. Take

$$g = \sum b_n x^{n+\beta} \tag{4.21}$$

giving

$$Lg(x) = \beta(\beta-1)b_0 x^{\beta-2} + \sum_1^{\infty}[(n+\beta)(n+\beta-1)b_n - b_{n-1}]x^{n+\beta-2}, \tag{4.22}$$

and we set $Lg = a_0/x$. Clearly $\beta = 1$ does not work; the first term is zero and the second begins with a constant. But $\beta = 0$ gives

$$Lg(x) = -\frac{b_0}{x} + \sum_2^{\infty}[n(n-1)b_n - b_{n-1}]x^{n-2}, \tag{4.23}$$

so a solution to the differential equation is given by $f(x) = dy_1/d\alpha|_{\alpha=1} + g$, with $b_0 = a_0$, b_1 free. But then note that the b_1 terms simply reproduce the y_1 solution, so there is no need to include them, and we have the solution

$$f = \frac{dy_1}{d\alpha}|_{\alpha=1} + a_0. \tag{4.24}$$

To evaluate this write

$$y_1 = a_0 \sum \frac{x^{n+\alpha}}{(n+\alpha)(n+\alpha-1)...(1+\alpha)\alpha} \tag{4.25}$$

and differentiate with respect to α, giving

$$\frac{dy_1}{d\alpha}|_{\alpha=1} = ln(x)y_1 + a_0 \sum \frac{x^{n+1} + \left(-2\sum_1^n \frac{1}{k} - \frac{1}{n+1}\right)}{(n+1)!n!}. \tag{4.26}$$

This gives for the second solution

$$y_2(x) = ln(x)y_1 + a_0 + a_0 \sum \frac{x^{n+1}\left(-2\sum_1^n \frac{1}{k} - \frac{1}{n+1}\right)}{(n+1)!n!}. \tag{4.27}$$

4.2.3 *Irregular singular point*

If a point is neither an ordinary nor a regular singular point, it is called an irregular singular point. In this case there is no convergent series representation of the solution around this point. The solution to the differential equation $y(z)$ then has an essential singularity at this point, meaning that neither the limit of $y(z)$ nor of $1/y(z)$ exists. In any neighborhood of an essential singularity the function takes on every complex value, except possibly one (Picard's great theorem). Nevertheless it is often possible to find local asymptotic series, that is series which approach the solution to within some small error and thereafter diverge. In spite of this divergent behavior an asymptotic series, if truncated after the appropriate number of terms, can give a very accurate and useful local representation of the solution.

To order the terms near an irregular singular point, we first must introduce the concept of asymptotic dominance. A function $g(x)$ is said to dominate over the function $f(x)$ at $x = x_0$, written $f(x) \ll g(x)$ if

$$lim_{x \to x_0} \frac{f(x)}{g(x)} = 0. \tag{4.28}$$

Also we say $f(x)$ is asymptotic to $g(x)$, or $f(x) \sim g(x)$ as $x \to x_0$ if $f(x) - g(x) \ll g(x)$ for $x \to x_0$. Note that $x \ll -5$ for $x \to 0$ and $f(x) \sim 0$ is not possible.

It is easily seen that a convergent series solution at an irregular singular point is impossible. Consider the simplest linear homogeneous second order differential equation that one can write with $x = 0$ an irregular singular point

$$x^3 y'' = y, \tag{4.29}$$

and attempt a solution about $x = 0$ of the form $y = \sum a_n x^{n+\alpha}$. Substituting we find the conditions

$$a_0 = 0, \qquad a_{n+1} = (n + \alpha - 1)(n + \alpha)a_n, \tag{4.30}$$

and the attempted series is identically zero. In general one finds that either the series does not exist, or that it is divergent for all x.

In this case, since there is no series, divergent or otherwise, we follow the suggestion of Poincaré and attempt a solution of the form $y = e^S$. Substitute and find

$$x^3[S'' + (S')^2] = 1. \tag{4.31}$$

Now look for the dominant behavior as $x \to 0$, by trial and error. First attempt $S'' \gg (S')^2$. Solving $x^3 S'' \sim 1$, we find $S' \sim -1/(2x^2)$. But then $S'' \ll (S')^2$ and the attempt fails. Trying $S'' \ll (S')^2$ we find $S' \sim \pm x^{-3/2}$ giving $S'' \sim \mp \frac{3}{2} x^{-5/2}$, which is consistent with the assumption. We then find

$$y \sim e^{\mp 2x^{-1/2}} \qquad x \to 0. \tag{4.32}$$

Now improve this estimate. Write $S(x) = \mp 2x^{-1/2} + g(x)$ and substitute into Eq. 4.31 giving

$$\mp \frac{3}{2} x^{-5/2} + g'' + x^{-3} \pm 2x^{-3/2} g' + (g')^2 = x^{-3}. \tag{4.33}$$

Again we must find dominant balance by trial and error. The correct choice is dominance of the $x^{-5/2}$ and the $x^{-3/2}g'$ terms, giving $g' \sim 3/(4x)$. This gives $g \sim (3/4)ln(x)$ plus a constant of integration, which contributes only to the normalization of $y(x)$. If one proceeds one step further by writing $S(x) = \mp 2x^{-1/2} + (3/4)ln(x) + h(x)$ one finds that $h(x) \ll const$. The fact that $h(x) \ll const$ means that further development of $y(x)$ can be achieved using a series.

Instead of continuing to guess dominant balance it is possible at this point to continue these two possible asymptotic forms by writing

$$y = Ax^{3/4} e^{\mp 2x^{-1/2}} W(x) = e^{S(x)} W(x), \tag{4.34}$$

and seek a series solution for $W(x)$, which we know must begin with a constant. The functional form $e^{S(x)}$ is called the leading asymptotic form of the solution. It is always characterized by the fact that the further development of the solution consists of a series begining with a constant. Substituting this expression into Eq. 4.29 we find a second order differential equation for each choice of asymptotic behavior

$$W'' + \left(\frac{3}{2x} \pm \frac{2}{x^{3/2}} \right) W' - \frac{3}{16x^2} W = 0. \tag{4.35}$$

Note that the point $x = 0$ is again an irregular singular point, also for the function $W(x)$. Because $W(x)$ satisfies a second order differential equation there is a second solution with different behavior at $x = 0$, but we are interested only in that solution with $W(0) = 1$. In spite of the fact that $W(x)$ has an irregular singular point at $x = 0$, the fact that it begins with a constant means that a series does exist. Because of the fractional power of x it is necessary to use a series of the form

$$W(x) = \sum_{0}^{\infty} a_n x^{n/2}. \tag{4.36}$$

One can avoid the fractional powers by writing the differential equation in terms of the variable $s = \sqrt{x}$, but $s = 0$ remains an irregular singular point, and little is gained by this change of variable. Substituting the series we find the solution

$$a_{n+1} = \mp \frac{(n-1/2)(n+3/2)a_n}{4(n+1)}. \tag{4.37}$$

This series is clearly divergent, but is nevertheless useful. For large n we have

$$|a_{n+1}| \simeq \frac{n|a_n|}{4} \tag{4.38}$$

or

$$\frac{dln(|a_n|)}{dn} \simeq ln(n) - ln(4). \tag{4.39}$$

Integrating in n we have $ln(|a_n|) \simeq nln(n) - n - nln(4)$, or

$$|a_n| \simeq e^{nln(n)-n-nln(4)}, \tag{4.40}$$

growing faster than any power of n. Thus the terms $a_n x^{\alpha n}$ for any α and any x eventually become very large with large n.

4.3 Asymptotic Series

An asymptotic sequence ϕ_n for $x \to x_0$ is a sequence of functions such that $\phi_{n+1} << \phi_n$ as $x \to x_0$. The sequences $\phi_n = (x - x_0)^n$ for $x \to x_0$ and $\phi_n = x^{-n}$ for $x \to \infty$ are both asymptotic sequences. For any such sequence of functions the series $\sum a_n \phi_n(x)$ is called an asymptotic series. This definition clearly includes the cases of series which converge, either for all x or for a restricted range of x, but normally when we refer to an asymptotic series it is understood that the series in question is divergent.

The study of asymptotic expansions was first made by Poincaré. A series $\sum_0^\infty a_n x^n$ is said to be asymptotic to a function $y(x)$ at $x = 0$, or

$$y(x) \sim \sum_0^\infty a_n x^n, \tag{4.41}$$

if for $x \to 0$ we have $y(x) - \sum_0^N a_n x^n \ll x^N$ for all N. Equivalently this means that

$$y(x) - \sum_0^N a_n x^n \sim a_m x^m, \tag{4.42}$$

with a_m the first nonzero term after N.

These definitions are easily extended to non-integer powers, by writing

$$y(x) \sim \sum_{0}^{\infty} a_n x^{\alpha n}, \tag{4.43}$$

if for $x \to 0$ we have $y(x) - \sum_0^N a_n x^{\alpha n} \ll x^{\alpha N}$ for all N.

4.3.1 *Properties*

If an asymptotic expansion of a function $y(x)$ exists, the a_n are unique, and given by the limiting procedure

$$a_0 = lim_{x\to 0} y(x), \qquad a_1 = lim_{x \to 0} \frac{y(x) - a_0}{x^{\alpha}} \qquad \text{etc.} \tag{4.44}$$

Although it is true that given a function $y(x)$, the asymptotic series representing it is unique, it is not true that an asymptotic series uniquely defines a function. This is because to $y(x)$ one can add any function which decreases faster than any power of x without changing the asymptotic expansion. For example, $y(x) + Ce^{-1/x}$ has the same asymptotic expansion about $x = 0$ for all C, and $y(x) + Ce^{-x^2}$ the same asymptotic expansion about $x = \infty$ for all C. Thus an asymptotic expansion defines an equivalence class of functions, differing from one another by terms which tend to zero faster than any power in the expansion. The class of all functions which are asymptotically equal to $y(x)$ is defined as the sum of the asymptotic series. All functions which are asymptotically equal to $y(x)$ differ by functions which have asymptotic expansions with all coefficients equal to zero. For example, the function e^{-x^2} has an asymptotic expansion for large x, $\sum a_n x^{-n}$ with all terms equal to zero.

It is clear that every asymptotic series possesses a sum. Consider the series $\sum a_n x^n$. Construct a nested sequence of sets U_n, all including the point $x = 0$, with $U_{n+1} \subset U_n$, with U_n the set of all $x \in U_{n-1}$ such that $|a_{n+1}x^{n+1}| \leq |a_n x^n|/2$. This construction is always possible for any sequence a_n since by the definition of an asymptotic sequence $\phi_{n+1} << \phi_n$ so the boundary of U_{n+1} can be shrunk inside of U_n until this condition is satisfied. For each n define a function $\mu(x)$ equal to one for x in U_{n+1} and zero outside U_n. Then for x in U_n we have

$$|a_{n+p}\mu_{n+p}x^{n+p}| \leq 2^{-p}|a_n x^n|, \tag{4.45}$$

by construction of the the $U's$ if x is in U_{n+p}, and because the left side vanishes when x is outside U_{n+p}. Let

$$f(x) = \sum_0^\infty \mu_n(x) a_n x^n. \tag{4.46}$$

This series converges for all x. Actually this series terminates for all x, because for some N the functions $\mu_n = 0, n > N$. Furthermore $f \sim \sum a_n x^n$, and thus this function defines the equivalence class of the sum. To see this fix N and let x be in U_{N+1}. Then $\mu(x) = 1$ for $n = 1, 2, \ldots N$ and

$$|f(x) - \sum_0^N a_n x^n| \le \sum_{N+1}^\infty |\mu_n(x) a_n x^n|$$

$$\le |a_{N+1} x^{N+1}| \sum_{N+1}^\infty 2^{N+1-n} = 2|a_{N+1} x^{N+1}| << x^N. \tag{4.47}$$

Furthermore this result gives a bound on the error given by truncating the asymptotic series for the representation of f, as being less than the first neglected term $|a_{N+1} x^{N+1}|$, provided x is in U_{N+1}. (It is not necessary to use the factor 2 to define the U_n, any factor larger than 1 will do.)

Thus the best approximation to $f(x)$ is given by truncating the asymptotic series at that value of N giving the minimum value of $|a_{N+1} x^{N+1}|$. Note that the truncation point N depends on x, and thus the asymptotic approximation to $f(x)$ is discontinuous at the values of x where N changes.

It is straightforward to see that if $f(x)$ is represented by an asymptotic series, then the integral of $f(x)$ is represented by the series obtained integrating term by term. However, differentiation is not as straightforward, because of the possible presence of terms which tend to zero faster than any power in the expansion, but which upon differentiation can lead to terms which do not. If $f(x) \sim \sum a_n x^n$ is known or assumed not to possess any such "asymptotically zero" terms, then the series can be differentiated term by term giving $f'(x) \sim \sum n a_n x^{n-1}$.

It requires some revision of prejudices to understand that almost any arbitrary series represents a perfectly well-defined function. If the series diverges this can be an indication that the function is not analytic at the point of the expansion, not that the function is infinite or otherwise ill defined. Furthermore the asymptotic expansion, properly truncated, often gives a remarkably good approximation to the function in question. A Taylor series has very different properties. For example, the expansion of

$1/(1-z)$ about $z=0$, if examined as though it were an asymptotic series, indicates that truncation occurs at $n=\infty$, i.e. including all terms, for $|z|<1$, and no terms for $|z|>1$. In the next sections we illustrate these properties with three examples.

4.3.2 *Truncation: A series about* $x = 0$.

Consider first an example of a differential equation for which $x=0$ is an irregular singular point

$$y'' + \frac{y'}{x^2} + \frac{y}{x^2} = 0, \tag{4.48}$$

with boundary condition $y(0)=1$. Substituting the form $y \sim e^S$ we find

$$S'' + (S')^2 + \frac{S'}{x^2} + \frac{1}{x^2} = 0. \tag{4.49}$$

One dominant balance for small x is given by the last two terms, with leading behavior $S'=-1$ so this solution approaches a constant at zero. In this case going to higher order is not necessary. If nevertheless we write $S'=-1+g'$ we find

$$g'' + 1 - 2g' + (g')^2 + \frac{g'}{x^2} = 0, \tag{4.50}$$

with solution $g'=-x^2$. Note that there is another dominant balance of this equation with $g'=-x^{-2}$, but this is more singular than the leading value of S' we have chosen, so is not acceptable. In fact it is the leading order of the second solution. The second solution is found by balancing the second and third terms of Eq. 4.49, giving $S'=-x^{-2}$, producing a solution which becomes infinite at $x=0$ as $y \sim e^{1/x}$, so the requirement that $y(0)$ approach a constant is sufficient to eliminate this solution. Since $y(x)$ itself tends to 1 at $x=0$ there is no need to write $y(x)=e^S W(x)$ and find an equation and expansion for $W(x)$. The boundary condition requires that we choose the finite solution, so we attempt a series of the form $y(x)=\sum a_n x^n$. We find a solution given by a divergent series with

$$a_{n+1} = -\frac{(n^2-n+1)}{n+1} a_n, \tag{4.51}$$

and $a_0=1$. Truncation of the series occurs at a given x for $x=(n+1)/(n^2-n+1)$, which for small x gives $n \simeq 1/x$. Figure 4.2 shows the numerically integrated solution (black) as well as the optimally truncated asymptotic

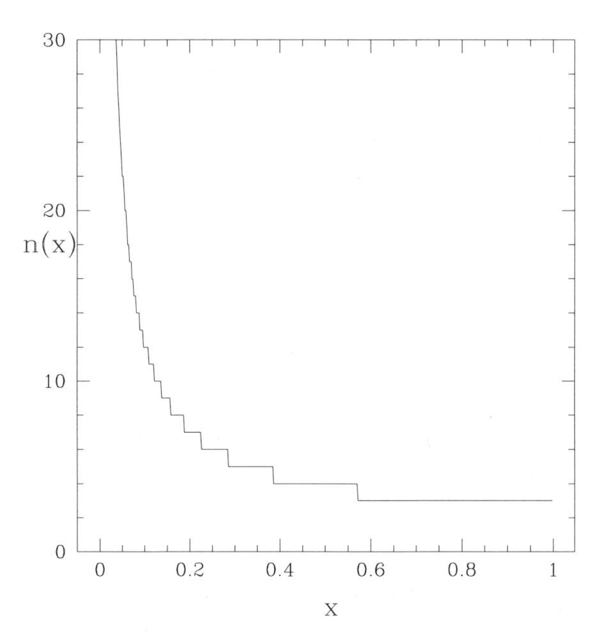

Fig. 4.1 Number of terms kept in optimal asymptotic expansion of Eq. 4.48.

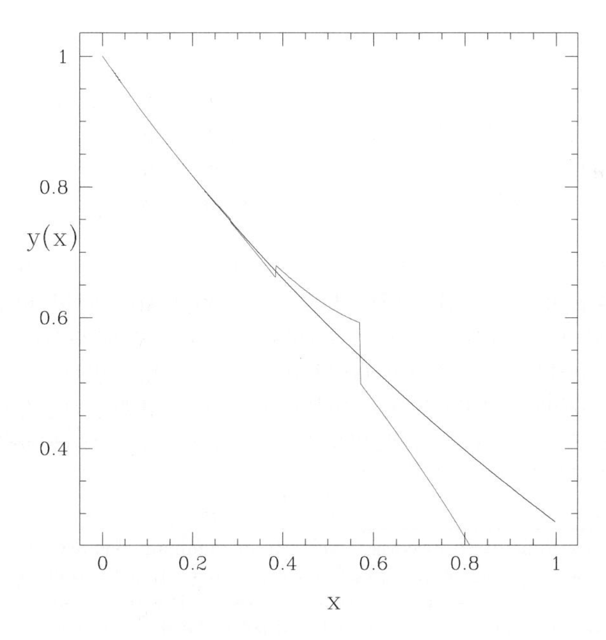

Fig. 4.2 Numerical $y(x)$ and asymptotic approximation to Eq. 4.48.

solution (red). The discontinuities in the asymptotic solution are at points at which the number of terms changes. At $x \simeq 0.35$ the number of terms changes from five to four, and at $x \simeq 0.55$ it changes from four to three. The numerical solution was integrated using the asymptotic value for small x to determine the values of the function and the first derivative. Figure 4.1 shows the number of terms in the optimal expansion as a function of x.

For large x, dominant balance gives $S' = c/x$ with $c = 1/2 \pm i\sqrt{3}/2$, so we have for the two solutions[1]

$$y(x) \sim \sqrt{x} sin\left(\frac{\sqrt{3}}{2} ln(x)\right), \quad \sqrt{x} cos\left(\frac{\sqrt{3}}{2} ln(x)\right). \tag{4.52}$$

However, there is no way at this point of knowing which combination of asymptotic solutions at large x corresponds to each asymptotic solution for small x.

4.3.3 *Truncation: A series about $x = \infty$.*

Consider an example of a differential equation for which $x = \infty$ is an irregular singular point

$$y'' - y' - 2\frac{y'}{x} + \frac{y}{x^2} = 0, \tag{4.53}$$

with boundary condition $y(\infty) = 1$. Substituting the form $y \sim e^S$ we find

$$S'' + (S')^2 - S' - 2\frac{S'}{x} + \frac{1}{x^2} = 0. \tag{4.54}$$

One dominant balance for large x is given by $S' = 1/x^2$ so this solution approaches a constant at infinity. The second solution is obtained by balancing the second and third terms, giving $S' = 1$, a solution which behaves as $y \sim e^x$ at large x. Thus the requirement that the solution approach a constant at infinity is sufficient to determine the solution. Since $y(x)$ itself tends to 1 at $x = \infty$ there is no need to write $y(x) = e^S W(x)$ and find an equation and expansion for $W(x)$.

---

[1] It is not that difficult to discover a three-term dominant balance. Choosing two terms and solving for S, write under each term its value. If the two chosen terms are larger than any others, two-term balance is the result. If instead, three terms are of equal magnitude, multiply S by a constant and solve the resulting three-term relation for the constant.

To match the boundary condition we attempt a series of the form $y(x) = \sum a_n x^{-n}$, giving a solution of a divergent series with

$$a_{n+1} = -\frac{(n^2+3n+1)}{n+1} a_n, \tag{4.55}$$

and $a_0 = 1$. Truncation of the series occurs at a given x for $x = (n^2+3n+1)/(n+1)$, which for large x gives $n \simeq x$. Figure 4.4 shows the numerically integrated solution (black) as well as the optimally truncated asymptotic solution (red). The discontinuities in the asymptotic solution are at points at which the number of terms changes. For $x < 4.8$ there are less than four terms in the series and the asymptotic approximation becomes useless. The numerical solution was integrated using the asymptotic value for large x to determine the values of the function and the first derivative. Figure 4.3 shows the number of terms in the optimal expansion as a function of x.

The point $x = 0$ is a regular singular point, and the indicial equation gives $y \sim x^\alpha$ with $\alpha = 3/2 \pm \sqrt{5}/2$. Thus both solutions approach $y = 0$ at $x = 0$, but each with a different power of x.

4.3.4 *Truncation: A series about $x = 0$*

Return to the case given by Eq. 4.35. In this case there is no asymptotic series for $y(x)$ itself; one must first extract the leading asymptotic behavior, $y(x) = e^S W(x)$, and find a differential equation for $W(x)$

$$W'' + \left(\frac{3}{2x} \pm \frac{2}{x^{3/2}}\right) W' - \frac{3}{16x^2} W = 0. \tag{4.56}$$

Dominant balance of the two solutions for $y(x)$ gives $S = \mp x^{-1/2} + 3ln(x)/4$, and the two signs in the equation for W reflect the choice of the solution. Note that also $W(x)$ has two solutions, but we are interested only in that for which $W(0) = 1$. The function W does possess an asymptotic series, but because of the fractional powers of x in the equation the series must be a series in $\sqrt{x}$. It is given by $W = \sum a_n x^{n/2}$, with solution (see the first section of Chapter 8)

$$a_n = \frac{(\pm)^n \Gamma(n-1/2)\Gamma(n+3/2)}{4^n \Gamma(n+1)(-\pi)} a_0. \tag{4.57}$$

Using the series to approximate $y(x) = Ax^{3/4} e^{\mp x^{-1/2}} W(x)$ for a given x the minimum error is achieved by truncating the series at the point where the terms begin to increase in size. The magnitude of the terms in the series

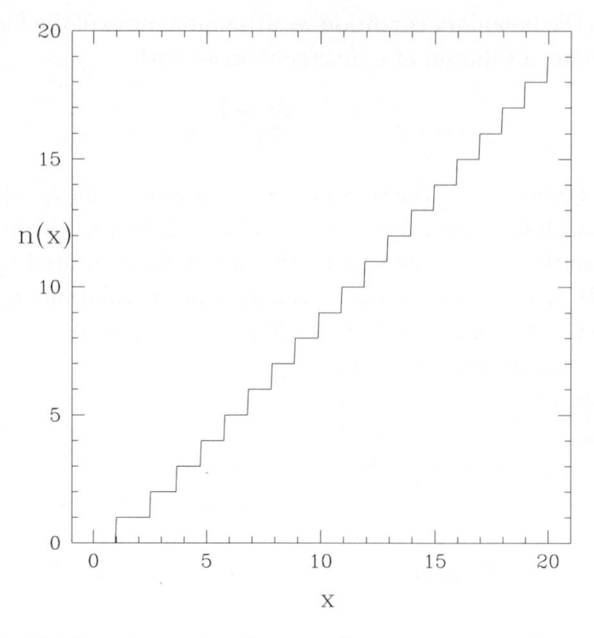

Fig. 4.3 Number of terms kept in optimal asymptotic expansion to Eq. 4.53.

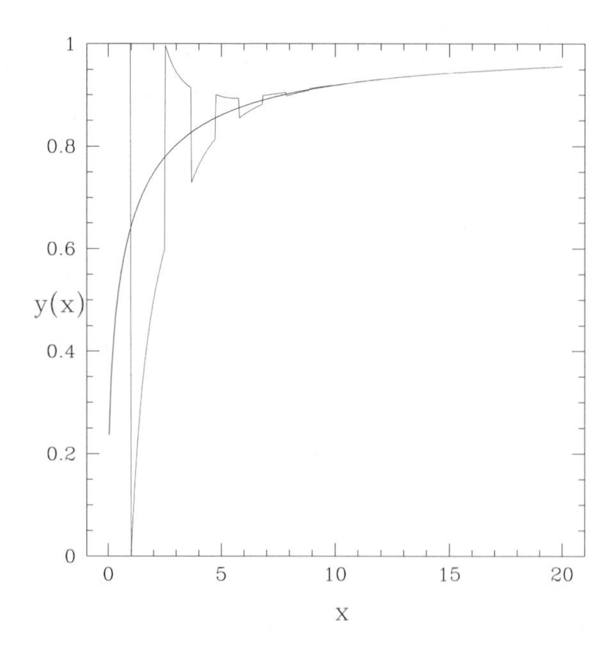

Fig. 4.4 Numerical $y(x)$ and asymptotic approximation to Eq. 4.53.

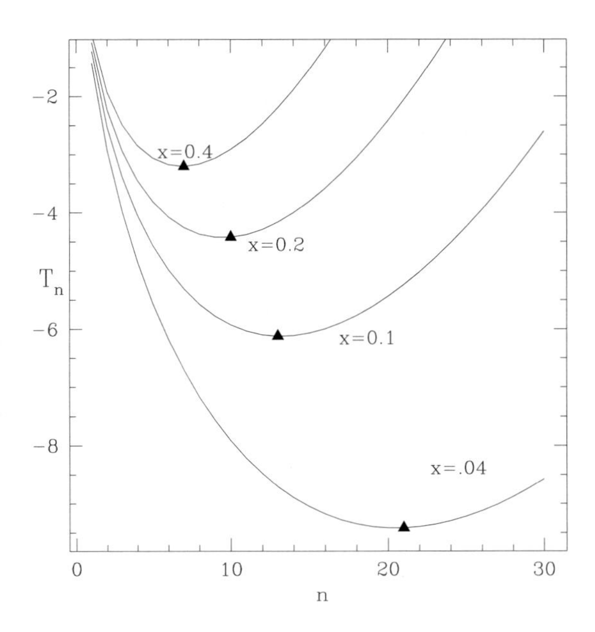

Fig. 4.5 Magnitude of terms in asymptotic series vs n for Eq. 4.56.

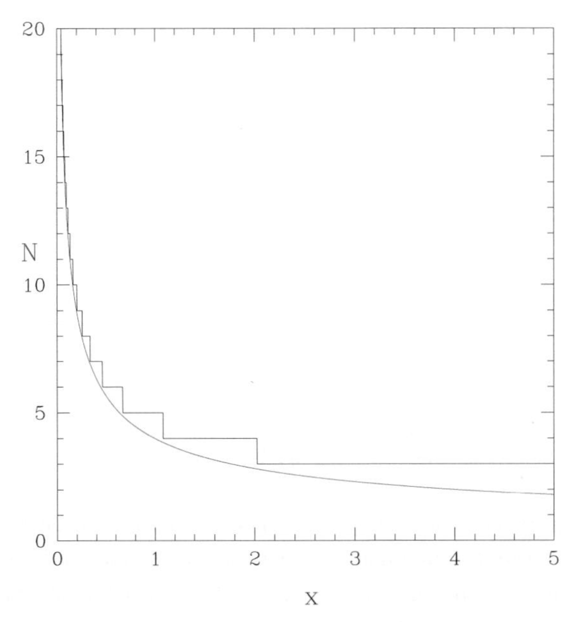

Fig. 4.6 The number of terms retained in the asymptotic sum, N, and the analytic expression $4/\sqrt{x}$ for Eq. 4.56.

is shown versus n for four different values of x in Fig. 4.5. Shown is $T_n = log_{10}(|a_n x^{n/2}|)$. For each value of x, to produce the optimal asymptotic expression, the series must be truncated at the minimum of this curve, marked by a triangular point. The ratio $r_n = a_{n+1}/a_n$ asymptotically for large n tends to $n/4$. The asymptotic series should be truncated leaving off the first term for which $r_n > 1/\sqrt{x}$. Thus for small x the number of terms kept is approximately $N = 4/\sqrt{x}$. Shown in Fig. 4.6 is the truncation as a function of x, along with this asymptotic estimate.

Finally, shown in Fig. 4.7 is a comparison of the optimal asymptotic expansions with numerical integrations of the differential equation given in Eq. 4.56 for $W(x)$. The asymptotic approximation produces discontinuities at the points where the number of terms retained in the expansion changes.

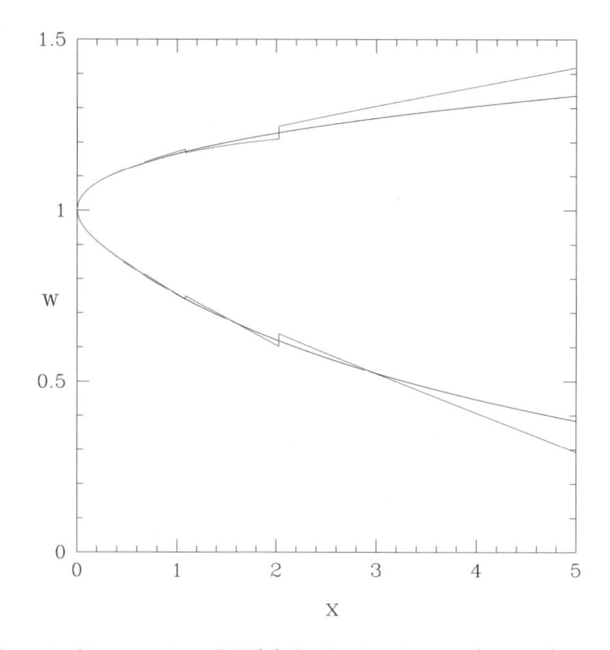

Fig. 4.7 Numerical integration of $W(x)$ for both solutions (smooth), and the optimal asymptotic series approximations (jagged) for Eq. 4.56.

The discontinuity near $x = 2$, at the transition from three terms to four terms is fairly large, the one near $x = 1$ is barely visible, and transitions for smaller x produce discontinuities smaller than the thickness of the line. The boundary conditions for the numerical integration were obtained by using the asymptotic series to evaluate the function and its derivative at the point

$x = 0.04$, where from Fig. 4.5 we see that the error in the representation of the function is approximately 10^{-9}. The two solutions were integrated from this point to $x = 5$. It is extremely difficult to carry out integrations very close to irregular singular points; numerical schemes invariably contain differences of terms which each become infinite as the singularity is approached, producing inaccuracies in the numerical integration. On the other hand, as the singularity is approached the asymptotic approximation becomes very good, and a numerical integration is not necessary.

Finally, as a practical matter, note that it is a waste of time to make the substitution $x = 1/t$ when examining asymptotic behavior at large x. It is not necessary to know whether the solution possesses a Taylor series, has an irregular singular point, or an irregular singular point at ∞ in order to make the substitution $y \sim e^S$. This substitution provides a means of determining the correct dominant balance whatever the nature of the solution.

4.3.5 *Asymptotic oscillation*

If a function is asymptotically oscillatory, of the form

$$y(x) \to a(x)sin(b(x)), \tag{4.58}$$

for large x for example, it is not correct to find asymptotic expressions for the functions $a(x) \to a_0(x)$ and $b(x) \to b_0(x)$ and write $y(x) \sim a_0(x)sin(b_0(x))$. This is because the zeros of the asymptotic solution will not exactly coincide with those of an exact solution, and the asymptotic relation will fail in the vicinity of the zeros. In addition, substituting the form $y(x) = a(x)sin(b(x))$ converts an equation which is linear in $y(x)$ into a nonlinear equation for $a(x), b(x)$ so is of no use. Instead write $y(x) = a_1(x)sin(b_0(x)) + a_2(x)cos(b_0(x))$ with b_0 the leading order term for the oscillation, normally easily found, and find two coupled linear equations for the two functions a_1, a_2. Asymptotic approximation of these functions then produces an expression which correctly approximates $y(x)$, although in terms of our original definition we cannot write $y(x) \sim a_1(x)sin(b_0(x)) + a_2(x)cos(b_0(x))$, because of the non-coincident zeros.

As an example, consider the equation

$$x^2y'' + y' + y = 0 \tag{4.59}$$

and examine asymptotic behavior for $x \to 0$ and for $x \to \infty$. Writing $y = e^S$ we have

$$S'' + (S')^2 + \frac{S'}{x^2} + \frac{1}{x^2} = 0. \tag{4.60}$$

For $x \ll 1$ there are two dominant balances. Balancing the second and third terms gives $S = 1/(3x) + lnx$ and thus $y \sim xe^{1/3x}$. Balancing the third and fourth terms gives $S = constant$, so a series exists. Writing $y = \sum a_n x^n$ we find the divergent series $a_{n+1} = (n^2 - n + 1)a_n/(n+1)$.

Now examine large x. There is a three-term dominant balance involving all but the third term. In this case write $S' = c/x$ giving

$$c^2 - c + 1 = 0, \qquad c = 1/2 \pm i\sqrt{3}/2 \tag{4.61}$$

and the asymptotic forms of the solutions are

$$y(x) \sim \sqrt{x} sin[(\sqrt{3}/2)ln(x)], \qquad y(x) \sim \sqrt{x} cos[(\sqrt{3}/2)ln(x)]. \tag{4.62}$$

4.3.6 *Construction of asymptotic series*

Recognizing that asymptotic series are divergent in a sense gives one a large degree of freedom in their construction. That is, the "proof" of the correctness of the expansion does not exist in the usual sense in which it does for convergent series. One common means of constructing asymptotic series is through the evaluation of integrals, to be discussed in Chapter 7. But it is also possible to simply assume an expansion of the form $\sum a_n x^{\alpha - n}$ and solve for the a_n, as is possible for the error function.

4.3.7 *The error function*

The error function is defined through

$$erf(z) = \frac{2}{\sqrt{\pi}} \int_0^z e^{-t^2} dt. \tag{4.63}$$

Suppose that for large z we have

$$erf(z) = \sum_0^\infty a_n e^{-z^2} z^{\alpha - n}, \tag{4.64}$$

and differentiate this equation, obtaining

$$\frac{2}{\sqrt{\pi}} = \sum_0^\infty \left[(\alpha - n) a_n z^{\alpha - n - 1} - 2a_n z^{\alpha - n + 1} \right]. \tag{4.65}$$

The largest power of z gives $a_0 = -1/\sqrt{\pi}$ and $\alpha = -1$ and the rest of the series gives

$$a_{2k} = -(k - 1/2)a_{2k-2} \tag{4.66}$$

and also $a_1 = 0$, giving (see the first section of Chapter 8) $a_{2k} = (-1)^k\Gamma(k+1/2)a_0/\sqrt{\pi}$. Because of differentiating the initial equation there is a constant of integration in the final asymptotic expression

$$erf(z) \simeq C - \frac{2}{\pi z}e^{-z^2}\sum_0^\infty (-1)^k\Gamma(k+1/2)z^{-2k}, \tag{4.67}$$

and the constant of integration is evaluated by taking the limit $|z| \to \infty$. But the limit of $erf(z)$ depends on the sign of the real part of z. Finally we have the asymptotic expansions

$$erf(z) \simeq \begin{bmatrix} -1 - \frac{1}{\pi z}e^{-z^2}\sum_0^\infty(-1)^k\Gamma(k+1/2)z^{-2k} & Re(z) < 0 \\ -\frac{1}{\pi z}e^{-z^2}\sum_0^\infty(-1)^k\Gamma(k+1/2)z^{-2k} & Re(z) = 0 \\ 1 - \frac{1}{\pi z}e^{-z^2}\sum_0^\infty(-1)^k\Gamma(k+1/2)z^{-2k} & Re(z) > 0 \end{bmatrix} \tag{4.68}$$

and these series clearly have zero radius of convergence.

The change in the form of the asymptotic expansion is due to the Stokes phenomenon, to be studied in Chapter 5, and the asymptotic expression, in spite of appearing to be discontinuous, is in fact the asymptotic approximation for large $|z|$ to a function which is continuous in the complex plane. The asymptotic expressions, as well as the numerical error function, evaluated on the real axis, are shown in Fig. 4.8.

4.4 Origin of the Divergence

An alternate derivation of the asymptotic expansion of the error function reveals the origin of the divergence of the series. Write the error function as

$$erf(z) = 1 - \frac{2}{\sqrt{\pi}}\int_z^\infty e^{-u^2}du, \tag{4.69}$$

so that for large positive z the integral is a small correction, and make the substitution $u = z + t$ giving

$$erf(z) = 1 - \frac{2}{\sqrt{\pi}}e^{-z^2}\int_0^\infty e^{-2zt-t^2}dt. \tag{4.70}$$

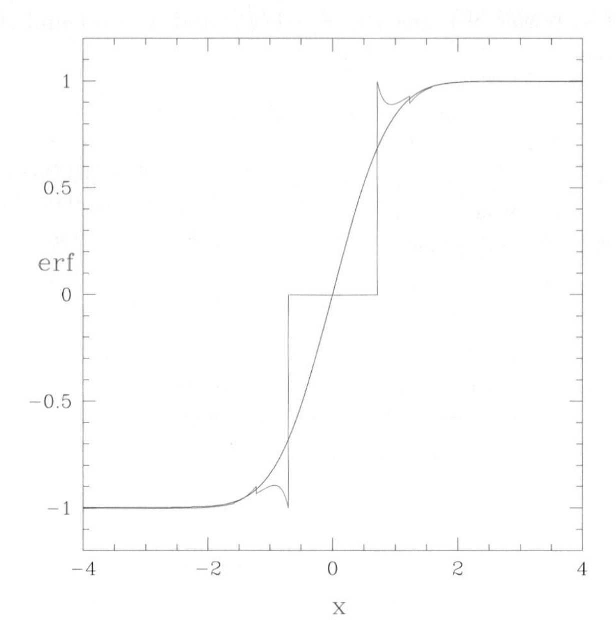

Fig. 4.8 The error function, showing optimal asymptotic expressions, which do not exist for $|x| < 0.7$.

Now expand e^{-t^2} and change variables to $s = 2zt$, giving

$$erf(z) = 1 - \frac{2}{\sqrt{\pi}} e^{-z^2} \int_0^\infty e^{-s} \sum_0^\infty \frac{(-1)^n}{n!} \left(\frac{s}{2z}\right)^{2n} \frac{ds}{2z}. \tag{4.71}$$

Now exchange orders of integration. However, the integral of a convergent series $\int dx[\sum_n f_n(x)]$ converges and is given by the sum of the integrals $\sum_n [\int f_n(x)dx]$ only provided that the series converges uniformly in the range of x under consideration. In this case we have the series

$$\sum_0^\infty \frac{(-1)^n}{n!} x^{2n}, \tag{4.72}$$

with $x = s/(2z)$ and the range of x is $[0, \infty]$. The series diverges for $x = \infty$, so changing orders of summation and integration does not result in a convergent series. Performing the integrals term by term anyway we find

$$erf(z) = 1 - \frac{2}{\sqrt{\pi}} e^{-z^2} \sum_0^\infty \frac{(-1)^n}{n!} \frac{\Gamma(2n+1)}{(2z)^{2n+1}}, \tag{4.73}$$

and using the identity Eq. 8.55 to find

$$\Gamma(2n+1) = \frac{2^{2n}\Gamma(n+1/2)\Gamma(n+1)}{\sqrt{\pi}}, \tag{4.74}$$

we recover Eq. 4.68, and indeed the series is not convergent. However it is readily verified that the series is asymptotic. This calculation demonstrates very clearly how a series which before integration is convergent for all values of the argument becomes, upon integrating term by term, a series which for small argument initially decreases but finally diverges. Derivations of asymptotic series from integrals, which is a very common method of obtaining them, generally involve an improper interchange of orders of integration and summation.

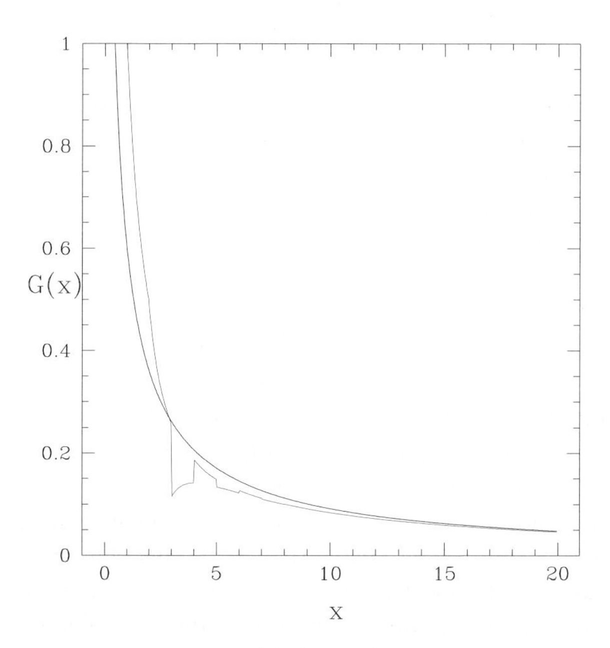

Fig. 4.9 Numerical integration of $G(x)$ (smooth), and the optimal asymptotic series approximations (jagged).

A more blatant example of the disregard of the necessary conditions for exchanging the order of summation and integration is given by the integral (Olver [1974])

$$G(x) = \int_0^\infty \frac{e^{-xt}}{1+t} dt. \tag{4.75}$$

Ignoring the fact that the series converges only for $|t| < 1$, and the integration is from zero to infinity, expand the denominator, giving

$$G(x) \simeq \int_0^\infty e^{-xt} \sum_0^\infty (-1)^n t^n \, dt, \tag{4.76}$$

and upon integrating term by term we have

$$G(x) \simeq \sum_0^\infty \frac{(-1)^n \Gamma(n+1)}{x^{n+1}} \tag{4.77}$$

which clearly diverges for all x. Nevertheless this asymptotic series, if truncated appropriately, provides a reasonably accurate approximation to the integral. The series is to be truncated at the term $n!/x^{n+1}$ for $x = n$. Figure 4.9 shows the optimal asymptotic approximation and a numerical integration for the function $G(x)$. After a little consideration, the success of this improper expansion is not so surprising, because for large x the entire contribution to the integral comes from small t, where the expansion is in fact valid.

4.5 Improving Series Convergence

There are several methods of defining summation procedures which make slowly converging series converge more rapidly, or make divergent series converge, and at the same time reproduce the correct value if the series does converge. As seen in the last chapter a series converges only within a radius given by the nearest singularity of the function expanded. This means that as the limiting value is approached the convergence is slower and slower, making numerical evaluation very difficult. The fact that this happens for very simple functions such as $1/(1-z)$ makes it reasonable to suspect that there must be sufficient information contained in the first few terms to allow evaluation, if only it could be extracted. Some of the methods of improving the evaluation through the construction of auxilliary sequences are due to Euler, Shanks, Padé, and Borel. As examples we give here the Shanks, Euler, and Borel summation methods. Other methods can be found for example in Whittaker and Watson [1962], or Bender-Orszag [1978].

4.5.1 *Shanks transformation*

For a given series $\sum a_n$ let the partial sums be

$$A_n = \sum_0^n a_k, \tag{4.78}$$

and assume this can be written in terms of the final result plus a transient

$$A_n = S + \beta t^n, \tag{4.79}$$

with $|t| < 1$ so that as $n \to \infty$ the transient term tends to zero. Then solving the three equations for A_{n-1}, A_n, A_{n+1} we find

$$S = \frac{A_{n+1}A_{n-1} - A_n^2}{A_{n+1} + A_{n-1} - 2A_n}. \tag{4.80}$$

If the sequence of A_n in fact does possess exactly one transient of this form, S would be a true constant. But even if it does not, the sequence

$$S_n = \frac{A_{n+1}A_{n-1} - A_n^2}{A_{n+1} + A_{n-1} - 2A_n} \tag{4.81}$$

may converge much more rapidly than the sequence A_n, due to the removal of the primary transient. As examples the series

$$a_k = z^k \tag{4.82}$$

gives immediately the correct answer, i.e. S_n is independent of n, even if z is larger than 1. The sequence

$$A_n = 1 - \frac{1}{2} + \frac{1}{3} + \ldots + \frac{(-1)^n}{n} \tag{4.83}$$

is correct to only two places for $n = 100$, whereas the Shanks transformation gives six place accuracy.

4.5.2 *Euler summation*

Given a series

$$\sum_0^\infty a_n, \tag{4.84}$$

the Euler sum is defined as

$$lim_{z\to 1}\sum_0^\infty a_n z^n, \tag{4.85}$$

and in case the initial series is only algebraically divergent the Euler sum still exists.

4.5.3 *Borel summation*

A means of summing divergent series is due to Borel. Given a series

$$\sum_0^\infty a_n z^n, \tag{4.86}$$

define

$$\phi(tz) = \sum_0^\infty \frac{a_n t^n z^n}{n!}. \tag{4.87}$$

Then define the Borel sum of $\sum_0^\infty a_n z^n$ to be

$$S = \int_0^\infty e^{-t}\phi(tz)dt \tag{4.88}$$

provided the integral exists. This definition often extends the domain in which the sum "converges" beyond the normal radius of convergence. For example, for the series $\sum_0^\infty z^n$ we find $\phi(tz) = e^{tz}$ and $S = 1/(1-z)$, which is well defined everywhere but at $z = 1$, i.e. the Borel summation correctly extends the series to the whole complex plane.

4.6 The Special Functions of Mathematical Physics

The number and location of ordinary and singular points is commonly said to classify the functions appearing in mathematical physics, most of which are only second order. Since any second order equation can be put in the standard WKB form $y'' + p(x)y = 0$, the number of zeros of $p(x)$, known as turning points, is also relevant. The simplest examples were historically the first to be examined and solved. Since an analytic function cannot be bounded everywhere, there exist no analytic functions without at least having a singularity at some point. The most important functions of classical mathematical physics have an essential singularity only at infinity and are either analytic or possess a regular singular point at zero.

Equations with constant coefficients are the simplest of all, and give immediately the solutions $y = e^{r_k x}$ with r_k the roots of the associated polynomial.

The **Airy** function is analytic everywhere but infinity, where it has an essential singularity, and the equation has a single turning point, and is thus the simplest example of the classical functions. Nevertheless, the Bessel function was known before the Airy function.

The **Weber–Hermite** functions are analytic everywhere but infinity, and the equation has two turning points.

The **Bessel** functions have a regular singular point at $z = 0$, and an essential singularity at infinity, and the equation has two turning points.

The **Whittaker** functions have a regular singular point at $z = 0$ and an irregular singular point at infinity. The equation has two turning points.

The **Legendre** functions have two regular singular points at $z = \pm 1$.

The **Mathieu** equation in its standard form contains only functions that are analytic everywhere but infinity, and yet the Mathieu functions are the most complex of the classical functions, and have been incompletely investigated. Thus the cataloguing of the singularity points appears to be an insufficient guide to the nature of the functions. There is a transformation of variable that produces an equation with regular singular points at 0 and 1, and an essential singularity at infinity. The standard form $y'' + [a + 16qcos(2z)]y$, while containing only analytic functions, however possesses an infinity of WKB turning points.

The **Gamma** function has simple poles at all non-positive integers and an essential singularity at infinity. Having an infinity of singularities it is outside the class of solutions to differential equations known, and in fact satisfies no differential equation. However its ubiquitous appearance in the analysis of the solutions to simple differential equations makes its study essential. The reader can grasp the difficulty involved by examining the differential equation for $y = 1/sin(z)$, a simple function with an infinite number of poles on the real axis, $(1/y)'' + (1/y) = 0$.

The **Riemann Zeta** function has a simple pole at $z = 1$ and an irregular singular point at infinity. It satisfies no differential equation.

4.7 Problems

1. Confluent hypergeometric equation: Classify the point $x = 0$. Find two independent series solutions for 2μ non integer, and find the first two terms of each series.

$$\frac{d^2y}{dx^2} + \left(-\frac{1}{4} + \frac{\lambda}{x} + \frac{1/4 - \mu^2}{x^2}\right) y = 0.$$

Hint: Remove a factor of $e^{-x/2}$.

2. Classify the singular points of the following differential equations:

(a) $x^2y'' + x^2y' + (x^2 - \nu^2)y = 0$;
(b) $x^3y'' + (\frac{1}{4} + \kappa x)y = 0$.

3. Solve

$(x-1)xy' + y = 1$ with $y(0) = 1$.

4. (a) Consider the equation

$$\frac{d^3y}{dx^3} = x^n y,$$

for n an integer. For what values of n, if any, is the point at infinity: an irregular singular point; a regular singular point; an ordinary point?

(b) Now specialize to $n = 1$

$$\frac{d^3y}{dx^3} = xy.$$

Find the leading order behavior of each of the linearly independent solutions for $x \to \infty$.

5. Find the leading behavior of both solutions of

$$\frac{d^2y}{dx^2} - e^x y = 0,$$

for $x \to \pm\infty$. Also find the second terms in the divergent series.

6. Find the leading order behavior of each of the linearly independent solutions for $x \to \infty$.

$$\frac{d^4y}{dx^4} + \frac{y}{1+x^4} = x.$$

7. Find the leading order behavior of each of the linearly independent solutions for $x \to \infty$.

$$\frac{d^2y}{dx^2} + x^3\frac{dy}{dx} + xy = 2x^4e^{-x^2}.$$

8. Find the leading order behavior of each of the linearly independent solutions for $x \to \infty$.

$$\frac{d^2y}{dx^2} - \frac{y}{1+x} = x.$$

9. Find the first three terms for $x \to 0$ of a particular solution to

$$xy'' - 2y' + y = cosx.$$

10. Find the first three terms for $x \to 0$ of a particular solution to

$$y' + xy = x^{-3}.$$

11. Let $y(x)$ be such that $y(1) = 1$ and for $x \geq 1$ $y' = 1/(x^2 + y^2)$. Prove that $lim_{x\to\infty}y(x)$ exists and is less than $1 + \pi/4$.

12. Find the leading asymptotic behavior for $x \to \infty$ of the GENERAL solution to

$$y'' + x^2y = sinx.$$

13. Find the leading order behavior of each of the linearly independent solutions for $x \to \infty$.

$$\frac{d^2y}{dx^2} + \frac{3}{2x}\frac{dy}{dx} - \frac{3}{16x^2}y = 0.$$

14. Find the leading order behavior of each of the linearly independent solutions for $x \to \infty$.

$$\frac{d^3y}{dx^3} - \frac{3}{1+x^2}\frac{dy}{dx} - \frac{3}{1+x^3}y = 0.$$

15. Find the Taylor expansion about 0, given $y(0) = 2$, $y'(0) = 0$, for the equation

$$\frac{d^2y}{dx^2} - 3x\frac{dy}{dx} + 8y = 0.$$

16. Classify the points at 0 and at infinity

$$x^2\frac{d^2y}{dx^2} = e^{1/x}y.$$

17. Classify the point $x = 0$. Find two independent series solutions and find the first three terms of each series.

$$x(1-x)\frac{d^2y}{dx^2} + x\frac{dy}{dx} - y = 0.$$

18. Find the leading asymptotic behavior as $x \to \infty$ of

$x^2y'' + (1+2x)y' + y = 0.$

19. Find the asymptotic series for the solution to $xy'' + y' = y$ which decreases as $x \to \infty$. How many terms should you keep if $x = 10$?

20. Attempt Borel summation for the asymptotic series for the error function, Eq. 4.68, and show that the first term is given correctly, but that all other terms are infinite.

21. Examine the asymptotic expansion for the integral

$$G(x) = \int_0^\infty \frac{e^{-xt}}{1+t}dt.$$

How many terms should be kept and what is the accuracy for $x = 5, 10$?

Chapter 5

Phase-Integral Methods I

The techniques of analytic continuation in the complex plane of approximate solutions to differential equations were struggled with for over fifty years by a number of mathematicians and physicists, producing a long list of erroneous calculations and publications. The problem is that approximate solutions exist in disconnected domains of the real axis, and they must be connected across the intervening domains where they are not valid. Analytic continuation in the complex plane provides the means of connection, but although the exact solution to the differential equation may be analytic and thus valid everywhere, the approximate solutions are not, and have very different analytic properties from the exact solution, including the existence of cuts.

George Gabriel Stokes was the first to deal with the asymptotic approximations to the Airy equation, realizing in 1857 that discontinuities in the form of the asymptotic representation existed at what are now called Stokes lines, even though the functions themselves are continuous (Stokes [1857]). Lord Raleigh considered such approximate solutions in 1912 but did not succeed in making the connection across the gap. More systematic approaches to matching the solutions were obtained by Wentzel, Kramers, and Brillouin in 1926, when the interest was primarily in obtaining solutions to bound state problems in quantum mechanics. The theory was finally put in complete form by H. Jeffreys and J. Heading in 1962. It offers a very powerful means of obtaining reasonably good approximate solutions to differential equations, and is used in electromagnetic wave theory, quantum mechanics, and other disciplines.

5.1 Introduction

The power and simplicity of Phase-Integral Methods (Heading [1962]) for the approximate solution of differential equations make them a common tool in many branches of physics, where the equations are often too cumbersome to solve by standard exact methods. Many of the differential equations of interest can be put in the form

$$\frac{d^2\psi}{dz^2} + Q(z, w)\psi = 0. \tag{5.1}$$

In fact, any second order differential equation of the form $\psi'' + p_1(x)\psi' + p_0(x)\psi = 0$ can be put in this form using the substitution $z = \int dxW(x)$ with $W(x) = exp(-\int p_1 dx)$ the Wronskian. In this case the equation takes the form of Eq. 5.1 with $Q(z) = p_0 e^{2\int p_1 dx}$. In these cases, the existence of solutions, determination of transmission and reflection coefficients for cases with freely propagating asymptotic waves, and the approximate complex eigenfrequencies ω for cases of bound states or instabilities, can often be determined by phase-integral methods. It is not possible to cover all aspects of the method of phase integrals here. We restrict ourselves to deriving the most essential results for solving one-dimensional problems.

We take Eq. 5.1 as standard form for the differential equation to be examined. The complex frequency ω plays the role of an unknown eigenvalue in the case of an instability or bound state problem, or as a given imposed wave frequency in the case of a scattering problem, and $z = x + iy$ is a complex variable. The physical problem is initially defined on the real axis and the equation has been analytically continued into the complex plane. The physical problems include searching for the existence of an instability, in which case outgoing wave boundary conditions are imposed for a growing mode and ω becomes an eigenvalue to be determined, or finding the amplitudes and phases of reflected and transmitted waves given an incoming wave of frequency ω.

Briefly, the WKBJ approximate solutions of Eq. 5.1, so named after Wentzel [1926], Kramers [1926], Brillouin [1926], and Jeffries [1923], are constructed by assuming the form $\psi = e^S$, substituting into Eq. 5.1, and assuming the dominance of $(S')^2$ over S''. Lowest order gives $S' = \pm iQ^{1/2}$. To obtain one higher order write $S' = \pm iQ^{1/2} + g'$, giving

$$\pm\frac{iQ'}{2Q^{1/2}} + g'' \pm 2iQ^{1/2}g' + (g')^{1/2} = 0. \tag{5.2}$$

Assuming dominance of the first and third terms we find a second order asymptotic expression for ψ of the form

$$\psi_{\pm} = Q^{-1/4} e^{\pm i \int^{z} Q^{1/2} dz}. \tag{5.3}$$

This dominance as well as $(S')^2 \gg S''$ are both guaranteed provided

$$\left| \frac{dQ}{dz} Q^{-3/2} \right| \ll 1. \tag{5.4}$$

A general solution of Eq. 5.1 can then be approximated by

$$\psi = a_{+}\psi_{+} + a_{-}\psi_{-}. \tag{5.5}$$

The solutions $\psi_{\pm}$ are local, not global solutions of Eq. 5.1. Clearly, inequality (5.4) is not valid in the vicinity of a zero of $Q(z, \omega)$, commonly called a turning point. Aside from this, however, $\psi_{\pm}$ are not approximations of a continuous solution of Eq. 5.1 in the whole z plane; i.e. if ψ is to approximate a continuous solution of Eq. 5.1, then the coefficients $a_{\pm}$ are not fixed over the whole z plane. The Method of Phase Integrals consists in relating, for a given solution of Eq. 5.1, the WKBJ approximation in one region of the z plane to that in another.

These regions are separated by the so-called Stokes and anti-Stokes lines (Stokes, [1857], [1899]) associated with $Q(z, \omega)$, and thus the qualitative properties of the solution are determined once these lines are known. The Stokes (anti-Stokes) lines associated with $Q(z, \omega)$ are paths in the z plane, emanating from zeros or singularities of $Q(z, \omega)$, along which $\int Q^{1/2}(z, \omega) dz$ is imaginary (real). We review first the characteristic properties of these lines and then the way in which they determine the global nature of a WKBJ solution.

Define a local anti-Stokes line to be for any z_0 an infinitesimal path dz emanating from z_0 along which $Q^{1/2} dz$ is real. Along this path $|\psi_{\pm}|$ are essentially constant; i.e. the solutions are oscillatory. If $Q(z_0, \omega)$ is finite and well behaved, the local anti-Stokes line is given by setting dz equal to a real number times $\pm Q(z_0)^{-1/2}$ i.e. from z_0 there issue two oppositely directed lines. Points at which $Q(z_0)$ is zero or infinity must be analyzed with more care. When the zero is first order, consider an infinitesimal line emanating from z_0, with $dz = (z - z_0)$. Then write $Q(z) \simeq Q'(z_0) dz$, and require that dz be a local anti-Stokes line, i.e. that $Q(z)^{1/2} dz = Q'(z_0)^{1/2} dz^{3/2}$ be real. Since dz is then proportional to $Q'(z_0)^{-1/3} e^{i(2n\pi)/3}$ with n integer, we find that three anti-Stokes lines emanate from z_0. Similarly, one finds that from a double root there issue four anti-Stokes lines, from a simple pole a single line, etc. It is thus quite easy to read the locations of zeros, poles, etc., of a function from a plot of the z plane upon which are displayed the local anti-Stokes lines, which we will refer to as a Stokes diagram. An example is shown in Fig. 5.1 for a function Q which possesses simple zeros in the first and third quadrants, a second order zero in the fourth quadrant, a pole in the second quadrant, and no other zeros or singularities. In referring to Stokes diagrams, we will call both zeros and singularities of $Q(z)$ singular points since it is the function $Q^{1/2}$ which is relevant in this diagram.

A display of this nature allows a qualitative survey of the analytic structure of a function, without the numerical complication of an actual search for roots (White [1979], [2000]).

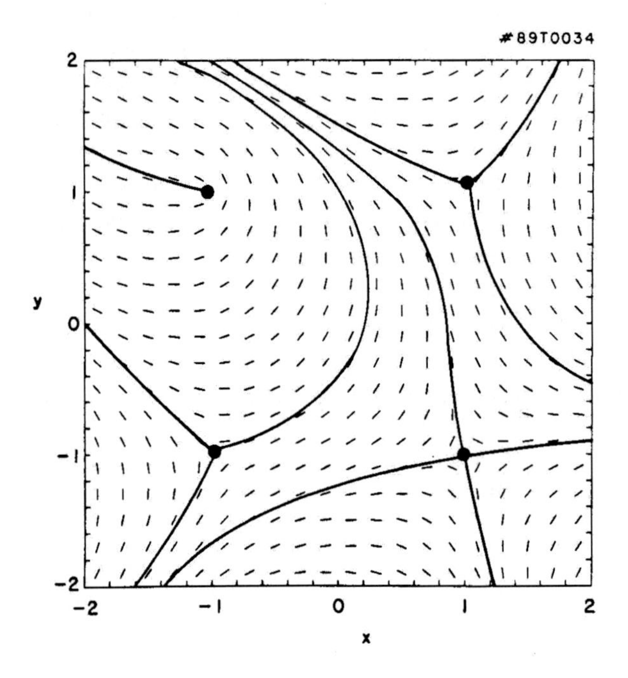

Fig. 5.1 Stokes diagram for $Q = (z - z_l)(z - z_2)(z - z_3)^2/(z - z_4)$ with $z_1 = l + i$, $z_2 = -1 - i$, $z_3 = 1 - i$, and $z_4 = -1 + i$.

Using the local anti-Stokes lines as guides, we can form global, continuous anti-Stokes lines for those particular lines which emerge from the singular points of the Stokes plot, and these lines have been added to Fig. 5.1. Along the global anti-Stokes lines the functions $\psi_{\pm}$ are, within the validity of the WKBJ approximation, of constant amplitude, i.e. oscillatory. We similarly define local and global Stokes lines to be lines emerging from the singular points for which the integral $\int Q^{1/2}dz$ is imaginary. Along the Stokes lines the WKBJ solutions are exponentially increasing or decreasing with fixed phase. Except at singular points the Stokes and anti-Stokes lines are orthogonal. The global anti-Stokes and Stokes lines which are attached to the singular points of the Stokes diagram, along with the Riemann cut lines, determine the global properties of the WKBJ solutions.

In the notation of Heading, including the slow $Q^{-1/4}$ dependence, a WKBJ solution is denoted by

$$(a, z)_s = Q^{-1/4}e^{i\int_a^z Q^{1/2}dz}, \tag{5.6}$$

where the subscript $s(d)$ indicates that the solution is subdominant (dominant); i.e. exponentially decreasing (increasing) for increasing $|z - a|$ in a particular region of the z plane, bounded by Stokes and anti-Stokes lines. The point a is taken to be a nearby singular point to which the dominancy or subdominancy refers. The two independent local WKBJ approximate solutions of Eq. 5.1 in this notation are given by (z, a) and (a, z). Clearly if (z, a) is subdominant, then (a, z) is dominant. It is readily verified that upon crossing an anti-Stokes line these two solutions reverse character. Thus we find that upon crossing an anti-Stokes line we must make the change $(a, z)_d \rightarrow (a, z)_s$ and $(z, a)_s \rightarrow (z, a)_d$. This is the first of the connection formulae, a collection of rules for continuing a solution through the z plane in the presence of cuts, Stokes lines, and anti-Stokes lines.

5.2 Solutions far from Singularities

In a domain far from any zeros or singularities of $Q(z)$ the solution to the differential equation in WKB approximation is simply given by the eikonal representation Eq. 5.6, where it is necessary to take either the real or imaginary part, to keep the solution real. If $Q^{1/2}$ is positive and real the solutions are oscillatory. It is then obvious that if the solution is required to satisfy boundary conditions such as $y(a) = y(b) = 0$ there is a condition imposed on $Q(z)$, namely

$$\int_a^b Q^{1/2}dz = n\pi \tag{5.7}$$

with n an integer. Similar conditions can be derived if the boundary conditions concern $y'(b)$. If $Q(z)$ is negative in the domain of interest boundary conditions can be imposed on the magnitude of $y(a)$ and the magnitude of $y(b)$, but of course the function cannot have more than one zero in the domain.

5.3 Connection Formulae: Isolated Zero

We next consider the continuation of a solution about an isolated turning point, located at $z = 0$. Later we show how one can pass from one turning point to another, allowing continuation through the entire z plane. The connection formulae depend on the nature of the turning point, and for simplicity we first consider a first order turning point associated with the Airy equation, $Q \sim -z$, with the associated Stokes diagram shown in Fig. 5.2.

First consider crossing a cut. Analytically continuing the solution, Eq. 5.6, counterclockwise around the turning point a we find that $\psi_{\pm} \to -i\psi_{\mp}$. Thus in crossing the cut in a clockwise sense, in order to insure continuity of our continued solution, we must make the changes

$$(0, z) \to i(z, 0), \qquad\qquad (z, 0) \to i(0, z). \tag{5.8}$$

Dominancy (or subdominancy) is not changed.

Now consider the process of crossing an anti-Stokes line, where dominant and subdominant solutions exchange character. Begin in the vicinity of a nearby Stokes line, and suppose the solution to Eq. 5.1 is approximated by a dominant expression $(z, 0)_d$ given by Eq. 5.6. A small subdominant part could also be present, so to speak, lost in the noise of the WKBJ approximation.

Trying to continue the solution past an anti-Stokes line creates a problem, because the previously small subdominant part, with an unknown coefficient, becomes dominant, making our solution totally inaccurate. To correct this one must, in the vicinity of the Stokes line, choose the coefficient of the subdominant solution so that the continuation to a nearby anti-Stokes line will give the correct solution. The necessary coefficient of the subdominant solution is called the Stokes constant. For a first order turning point the Stokes constant can be derived simply by requiring that the solution be single valued upon continuation about the turning point. We know this is true because for $Q = -z$ the point $z = 0$ is a regular point of the differential equation and the solution is representable by a Taylor series with infinite radius of convergence. This is not the case for all forms of Q; in some cases the solution itself may possess cuts originating at zeros

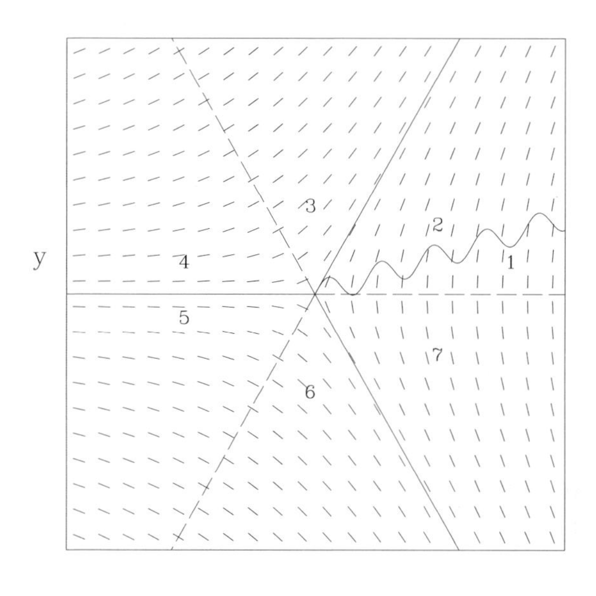

Fig. 5.2 Stokes diagram for a first order turning point.

or singularities of Q. Begin with a subdominant solution $(0, z)_s$ along the positive real axis of Fig. 5.2, a Stokes line. We deliberately begin with a subdominant solution so that the solution is small and cannot contain any dominant part due to the approximate nature of the WKBJ solution. Now continue this solution in both directions about the turning point.

Continue upward into domain 2. Crossing the cut changes the solution to $-i(z, 0)_s$. Passing upward into domain 3, or downward from domain 7 into domain 6, an anti-Stokes line is passed and the solution becomes dominant, reaching maximal dominancy along the Stokes lines separating domains 3,4 and domains 5,6. On crossing these Stokes lines we must add subdominant parts to make up for any lost by the WKBJ approximation. Thus on passing into domain 4 the solution becomes $-i[(z, 0)_d + T_1(0, z)_s]$ and in domain 5 it becomes $(0, z)_d - T_2(z, 0)_s$ where T_1 and T_2 are Stokes constants, and the signs are chosen to reflect the fact that the Stokes line is crossed clockwise in one case and counterclockwise in the other. Now continue from domain 5 into domain 4, giving $(0, z)_s - T_2(z, 0)_d$. Equating the two solutions in domain 4 determines the Stokes constants to be $T_1 = T_2 = i$.

In addition to the rules given by Heading, the notion of causality can in some cases provide additional information. To make this apparent, the time dependence of the solution must be introduced, which is most easily demonstrated for the case of quantum mechanics. The Schroedinger equation in one dimension is

$$E\psi = i\partial_t\psi = -\psi'' + V(x)\psi. \tag{5.9}$$

or $\psi'' + Q(x)\psi = 0$ with $Q = E - V(x)$. Define the flux

$$j = \psi^*\psi' - \psi'^*\psi. \tag{5.10}$$

Using Eq. 5.9 we find

$$\nabla \cdot j = \psi^*\psi'' - \psi''^*\psi = -\psi^*\psi(Q^* - Q). \tag{5.11}$$

But combining these equations we find that if Q is real

$$\frac{d}{dt}(\psi^*\psi) + \nabla \cdot j = 0, \tag{5.12}$$

which is the flux conservation equation, and can also be expressed as

$$\frac{d}{dt}\int_a^b \psi^*\psi dx = j(a) - j(b) \tag{5.13}$$

proving that $j(x)$ is the probability flux at point x, and that it is conserved provided Q is real on the real axis. Note that this is exact, it is not a result depending on WKB approximation.

In quantum mechanics this is the flux of probability density, and corresponds to the Poynting flux in electromagnetic theory. Now consider the solution $(0, z) = e^{i\frac{2}{3}z^{3/2}}/z^{1/4}$ for z negative and real in Fig. 5.2. We have $(0, z)' = ie^{i\frac{2}{3}z^{3/2}}z^{1/4}$ plus a higher order term in $1/z$, giving for the flux $S = 1$ for large negative z. Now continue to large positive x. In domain 5 this solution is exponentially increasing so we have $(0, z)_d$, and the same in domain 6. Finally in domain 7 we have $(0, z)_d - i(z, 0)_s$. Now continue across to domain 1, giving $(0, z)_d$. Now we recognize a problem, since the solutions in domains 1 and 7 disagree by the presence of the subdominant term. The correct expression on the real line can be obtained by considering the flux, since the flux for the correct solution must be 1, and this is an exact result. Write the solution as $(0, z)_d - T(z, 0)_s$. The flux is then $T - T^*$ and thus the correct value for T is $i/2$, not i. Thus the correct value on the real line, a Stokes line, is just the average of the values obtained in domains 1 and 2, above and below the line. We thus obtain a special rule regarding Stokes lines lying on the real axis. Use half the Stokes constant to step on to the Stokes line, and again half to step off. The value exactly on the line is given by the mean of the values above and below the line. See also section 12.2 for an additional explanation of this rule.

5.4 Derivation of Stokes Constants

It is possible to find the Stokes constant for a turning point of arbitrary order by analytically solving the exact differential equation in the vicinity of the turning point. Consider the vicinity of a turning point of order n, where the differential equation has the form

$$\frac{d^2\phi}{dw^2} + w^n\phi = 0. \tag{5.14}$$

Substitute $\phi = w^{1/2}u$, $z = 2w^{(n+2)/2}/(n+2)$, giving Bessel's equation of order $1/(n+2)$,

$$u'' + \frac{1}{z}u' + \left(1 - \frac{\nu^2}{z^2}\right)u = 0, \tag{5.15}$$

with $\nu = 1/(n+2)$. To find the Stokes constants, we analytically continue the solution to the Bessel equation. The canonical form, Eq. 5.1, can be obtained in the variable z by letting $u = \psi/z^{1/2}$, giving $\psi'' + Q(z)\psi = 0$, with $Q(z) = 1 + (1/4 - \nu^2)/z^2$. For large $|z|$ the solutions have the form

$$u \simeq \frac{1}{\sqrt{z}}e^{\pm iz}. \tag{5.16}$$

Shown in Fig. 5.3 is the asymptotic Stokes plot for the Bessel equation. There is actually structure near the origin, which will be discussed in section 13.2, but for the present purposes the asymptotic behavior is sufficient.

Begin in domain 1 with $(0,z)_s$. Continuing to domain 2 there is no change. In domain 3 we have $(0,z)_d$. Continuing to domain 4 we find $(0,z)_d + T(z,0)_s$. Note that the origin is a regular singular point of the differential equation, so the expression in domain 4 cannot be continued across the real axis and equated to the expression in domain 1, because the solution itself possesses a cut.

A general solution of the Bessel equation can be written

$$u = Az^{\nu}P_1(z^2) + Bz^{-\nu}P_2(z^2) \tag{5.17}$$

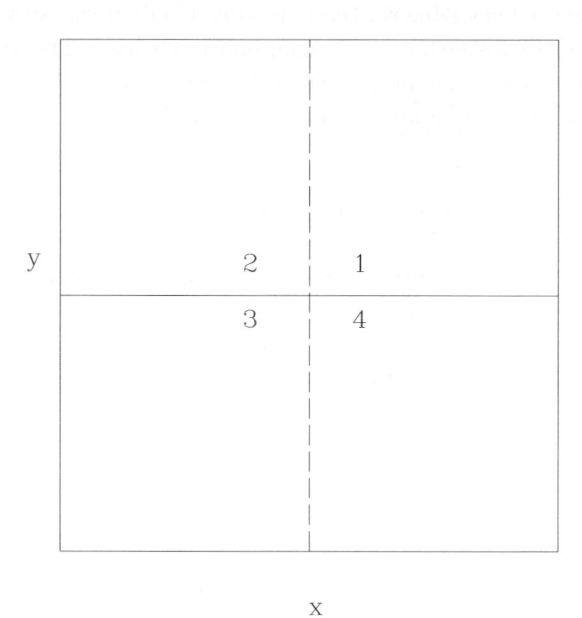

Fig. 5.3 Large scale Stokes diagram for the Bessel function.

where P_1 and P_2 are convergent series in z^2, as can be shown by a local expansion about $z = 0$, where ν is assumed to be finite and non integer, i.e. $n \neq -2$. We choose A and B so that the asymptotic form of u matches the expressions above. Along the real axis in domain 1 we have for large $x = r$

$$Ar^{\nu} P_1(r^2) + Br^{-\nu} P_2(r^2) = \frac{1}{\sqrt{r}} e^{ir}. \tag{5.18}$$

Continuing counterclockwise by taking $z = re^{i\pi}$ and matching to the subdominant form in domain 2 we find for large r

$$Ae^{i\nu\pi} r^{\nu} P_1(r^2) + Be^{-i\nu\pi} r^{-\nu} P_2(r^2) = \frac{e^{-i\pi/2}}{\sqrt{r}} e^{-ir}. \tag{5.19}$$

Similarly letting $z = re^{i2\pi}$ and matching to the form in domain 4 we find

$$Ae^{i2\nu\pi} r^{\nu} P_1(r^2) + Be^{-i2\nu\pi} r^{-\nu} P_2(r^2) = \frac{e^{-i\pi}}{\sqrt{r}} e^{ir} + T \frac{e^{-i\pi}}{\sqrt{r}} e^{-ir}. \tag{5.20}$$

These three equations can be written as a matrix equation for the vector $(Ar^\nu P_1(r^2), Br^\nu P_2(r^2), r^{-1/2}e^{ir})$. There is a solution only if

$$det \begin{pmatrix} 1 & 1 & -1 \\ e^{i\nu\pi} & e^{-i\nu\pi} & e^{-i2r} \\ e^{i2\nu\pi} & e^{-i2\nu\pi} & 1 + Te^{-i2r} \end{pmatrix} = 0, \tag{5.21}$$

which gives $T = 2icos\nu\pi$ for the Stokes constant, the r dependence dropping out of the equation. Returning to Eq. 5.14 we find the Stokes constant for a turning point of order n to be given by

$$T = 2icos\left(\frac{\pi}{n+2}\right). \tag{5.22}$$

The form of the WKBJ solutions in the w plane is $\phi \sim e^{\pm i \int w^{n/2} dw}$, and in the z plane $u \sim e^{\pm i \int dz}$. Since $dz \sim w^{n/2}dw$ the Stokes and anti-Stokes lines in the two planes correspond as they must. The Stokes constant for crossing Stokes lines in the complex w plane of Eq. 5.14 is thus given by Eq. 5.22.

5.5 Rules for Continuation

These results can be simply summarized. They prescribe a set of rules, first given by Heading, for obtaining a globally defined WKBJ solution which corresponds to the approximation of a single solution of the differential equation.

Begin with a particular solution in one region of the z plane, choosing that combination of subdominant and dominant solutions which gives the desired boundary conditions in this region. The global solution is obtained by continuing this solution through the whole z plane, effecting the following changes:

(1) **If a_d and a_s are respectively the coefficients of the dominant and subdominant terms of a solution, then upon crossing a Stokes line in a counterclockwise sense a_s must be replaced by $a_s + Ta_d$ where T is called the Stokes constant. When the Stokes line originates at an isolated zero of order n, $T = 2icos(\pi/(n+2))$.**

(2) **Upon crossing a cut in a counterclockwise sense, the cut originating from a first order zero of Q at the point a, we have**

$$\begin{aligned}(a, z) &\to -i(z, a)\\ (z, a) &\to -i(a, z).\end{aligned} \tag{5.23}$$

The property of dominancy or subdominancy is preserved in this process.

(3) **Upon crossing an anti-Stokes line, subdominant solutions become dominant and *vice versa*.**

(4) **Reconnect from singularity a to singularity b using $(z, a) = (z, b)[b, a]$ with $[b, a] = e^{i\int_b^a Q^{1/2}dz}$. If a, b are joined by a Stokes line, reconnect while on the line, using 1/2 the usual Stokes constant to step on the line, and again 1/2 to step off.**

Using these rules we can pass from region to region across the cuts, Stokes and anti-Stokes lines emanating from a turning point. Beginning with any combination of dominant and subdominant solutions in one region, this process leads to a globally defined single valued approximate solution of Eq. 5.1. Although it would appear that the first rule gives rise to a discontinuous solution, this is not the case. At the Stokes line, in the presence of a dominant solution, the discontinuity produced is small compared to the error due to the WKBJ approximation itself. As one continues further away from the Stokes line, however, the subdominant term will begin to be important, and the modified coefficient is the correct one.

A Stokes structure consisting of more than an isolated singularity is more complicated, in that the Stokes constants are modified by the proximity of the other singularities. However, the modification is normally exponentially small, and one can in most cases use the values of the Stokes constants obtained for isolated singularities also for complex structures. In the case of a bound state the Bohr–Sommerfeld condition, which guarantees matching of the solutions for positive and negative x, can be derived without finding the Stokes constant. For scattering problems the use of the values for isolated singularities produces only exponentially small errors, but this results in a violation of flux conservation. Thus for the scattering problem it is possible to find the correct reflection and transition amplitudes using flux conservation. In Chapter 12 we will calculate the modified Stokes constants associated with diagrams consisting of two singular points.

5.6 Causality

To properly understand the choice of boundary conditions in solving differential equations arising from physical problems it is necessary to invoke causality, namely to require an outward group velocity in directions from which there is assumed to be no incoming wave. Assume for simplicity that Q tends to a constant k^2 for large $|x|$. The WKBJ solutions take the form $\Psi_\pm = e^{\pm ikx}$. In a physics problem the time dependence is also relevant, and the wave has a frequency $\omega(k)$ with the k dependence given by properties of the physical system. Form a wave packet

$$\Psi_+(x,t) = \int dk f(k) e^{i(kx-\omega t)}, \tag{5.24}$$

where $f(k)$ is localized about k_0. Expand $\omega(k) \simeq \omega_0 + d\omega/dk(k-k_0)$, giving

$$\Psi_+(x,t) = e^{i(k_0 x - \omega_0 t)} \int dk f(k) e^{i(k-k_0)(x-td\omega/dk)}. \tag{5.25}$$

The integral is very small unless $x \simeq (d\omega/dk)t$, so Ψ_+ forms a wave packet, localized at this value of x, giving the usual identification of $d\omega/dk$ as group velocity, and thus Ψ_+ is outgoing for $x > 0$ if $d\omega/dk > 0$. Now consider the spatial dependence of Ψ_+ with a complex frequency. If the mode frequency is $\omega = \omega_r + i\gamma$ expanding $e^{ik(\omega)x}$ gives

$$\Psi_+ \simeq e^{(ik_0 x - \gamma x dk/d\omega)}. \tag{5.26}$$

We then obtain the rules for the asymptotic behavior of the solution. If $\gamma > 0$ and $d\omega/dk > 0$ the solution is decreasing (subdominant) for large x. Physically the spatial behavior can be simply understood in terms of information propagation. If the mode is growing the news of its growth propagates outward at the group velocity, so the mode is largest for small $|x|$, i.e. it is damped in the direction of propagation. In the following we will use the introduction of a small dissipation to clarify the choice of the solution.[1]

[1]Note that the convention $e^{-i\omega t}$ is opposite from that used by Heading, which switches upper and lower half planes in some arguments.

5.7 Bound States and Instabilities

The bound state problem, or the search for instability, is generically given by a function $Q(z)$ which is real on the real axis, with two first order zeros at points a and b, and with Q positive between a and b and negative otherwise. The Stokes diagram is shown in Fig. 5.4. We will find that the boundary conditions immediately give the Bohr–Sommerfeld condition, which determines the energy of the bound state, or equivalently, the growth rate of the instability. In this chapter we will use the approximate values for Stokes constants given by isolated singularities, but in fact, as we will see in Chapter 12, the Bohr–Sommerfeld condition is independent of this approximation.

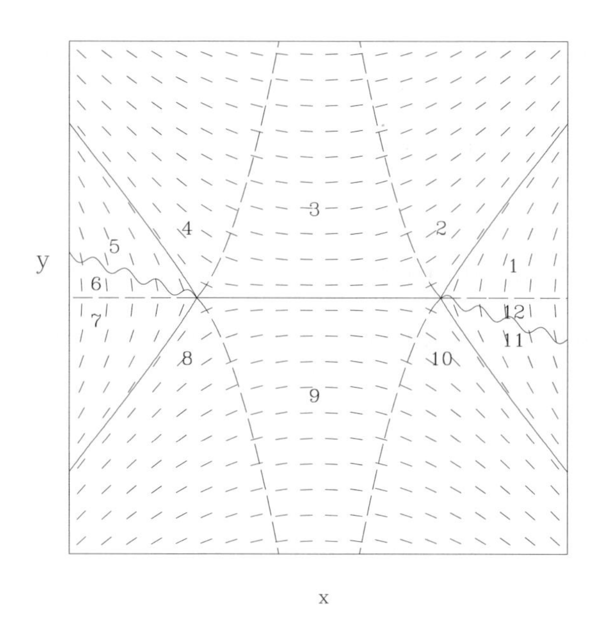

Fig. 5.4 Stokes plot for the bound state problem.

Begin at large positive x with a subdominant solution. Define $[a, b] = e^{iW}$. Along the real axis between a and b, Q is real and positive. Choose the branch of the square root making also $\sqrt{Q}$ real and positive, and thus W is real and positive. $Q(z)$ is a given function, but we must determine at each location the phase of $Q^{1/2}$ and $Q^{1/4}$. With the choice of cuts as shown, and choosing $Q^{1/2}$ positive and real for $a < x < b$ we have for $x > b$ and for $x < a$ that $Q^{1/2} \sim e^{-i\pi/2}$. Thus for $z > b$ and real the solution (b, z) is dominant. Begin with a subdominant solution at large positive x and continue, using the Stokes constant associated with a first order zero.

(1) $(z, b)_s$
(2) $(z, b)_d$
(3) $(z, b)_d + i(b, z)_s = e^{iW}(z, a)_d + e^{-iW} i(a, z)_s$
(4) $e^{iW}(z, a)_d + i[e^{iW} + e^{-iW}](a, z)_s$
(5) $e^{iW}(z, a)_s + 2icos(W)(a, z)_d$
(6) $-ie^{iW}(a, z)_s + 2cos(W)(z, a)_d$

Requiring the dominant part of the solution to be zero gives $W = (n + 1/2)\pi$, the Bohr–Sommerfeld quantization condition,

$$\int_a^b Q^{1/2} dx = (n + 1/2)\pi. \tag{5.27}$$

It is easy to see that this result is independent of the choice of orientation of the cuts.

5.8 Scattering, Overdense Barrier

The overdense barrier is given by a function $Q(z)$ which is real on the real axis, with two first order zeros at real points a and b, and Q is negative between a and b and positive otherwise. The Stokes diagram is shown in Fig. 5.5, and propagating oscillatory solutions exist for large positive and negative x.

We consider an incident wave from the left, the problem being to determine the reflected and transmitted waves. In classical physics problems the absolute square of these coefficients gives the reflected and transmitted power, and in quantum mechanical problems the probability of reflection and transmission. Thus causality requires outgoing boundary conditions at the far right. Define W through $[a, b] = e^{-W}$. Take Q to be real and positive for $b < x$, then with the choice of the cuts as shown in Fig. 5.5 along the real axis between a and b, Q has phase $Q \sim e^{i\pi}$ and W is real and positive, and for $x < a$ Q is again real and positive. Requiring outgoing wave conditions for large positive x gives boundary conditions of a subdominant solution in domain (1). Continuation through the upper half plane gives:

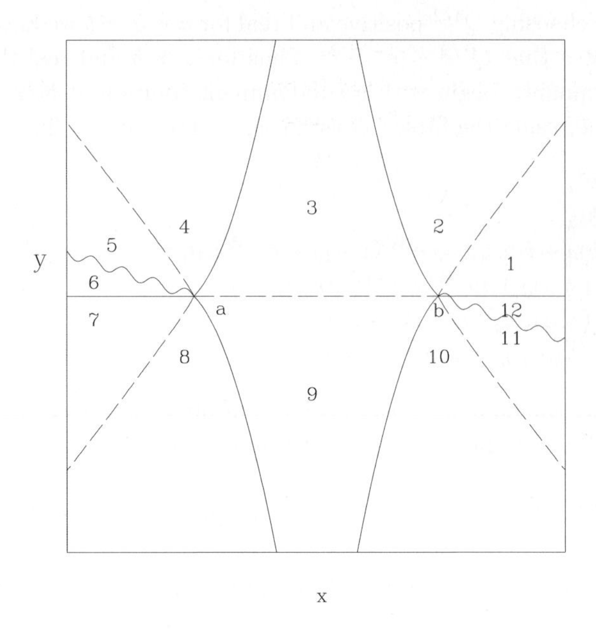

Fig. 5.5 Stokes plot for the overdense barrier.

(1) $(b, z)_s$
(2) $(b, z)_s$
(3) $(b, z)_d = e^W (a, z)_s$
(4) $e^W (a, z)_d$
(5) $e^W (a, z)_d + S e^W (z, a)_s$
(6) $-ie^W (z, a)_d - iS e^W (a, z)_s$

We have left the Stokes constant undetermined. For some problems the approximation of isolated singularities is too strong, one would like to know the result even if the singularities are closely spaced. This problem will be examined in more detail in Chapter 12. For this particular problem the magnitude of the Stokes constant, but not its phase, can be determined by flux conservation. Consider the flux on the real axis, where Q is real. In domain 1 we have $flux = Im(\psi^* \psi') = 1$. In domain 6 we find $flux = SS^* e^{2W} - e^{2W}$. Equating these expressions we can without loss of generality write

$$S = i\sqrt{1 + e^{-2W}} e^{i\delta}, \tag{5.28}$$

with δ undetermined. This Stokes constant differs from the isolated singularity only by a term exponentially small in W, but without this correction the flux is not conserved. We make no assumption about the magnitude of W. The solution in domain (6) becomes

$$(6)\ \sqrt{1+e^{-2W}}e^{i\delta}e^{W}(a,z)_s - ie^{W}(z,a)_d.$$

In domain (6) we must identify the incoming and reflected waves. Follow the phase of $\sqrt{Q}$ through the Stokes plot, taking $\sqrt{Q}$ to be positive real at the far right, so $\int^z \sqrt{Q}dz$ increases with z and $(b,z) \sim e^{i\int^z \sqrt{Q}dz - i\omega t}$ is a right moving wave. Then circling b CCW gives $\sqrt{Q} \sim e^{i\pi/2}$ along (a,b), and circling a CW gives $\sqrt{Q}$ positive real for large negative z. Thus $\int^z \sqrt{Q}dz$ decreases as $|z|$ increases and (b,z) is a right moving and thus incoming wave. For example, taking $Q = (z-a)(z-b)$ one finds on the left $\sqrt{Q} \sim -z$ and $\int^z \sqrt{Q}dz \sim -z^2/2$.

Using the edges of the propagation domains, a and b as the reference points for phase changes, we find reflection and transmission coefficients of

$$R = -\frac{ie^{-i\delta}}{\sqrt{1+e^{-2W}}}, \qquad T = \frac{e^{-W-i\delta}}{\sqrt{1+e^{-2W}}}. \tag{5.29}$$

We thus find the probabilities for reflection and transmission to be

$$|R|^2 = \frac{1}{1+e^{-2W}}, \qquad |T|^2 = \frac{e^{-2W}}{1+e^{-2W}}, \tag{5.30}$$

and we see that $|R|^2 + |T|^2 = 1$ for any W. Examine the limit $a \to b$. The Stokes diagram collapses to a single second order zero, and the Stokes plot is shown in Fig. 5.6. We retain the cuts to make the results coincide exactly with those resulting from the limit $a \to b$ in Fig. 5.5. In this case the Stokes constant is $\sqrt{2}i$ giving for the reflection and transmission coefficients $R = -i/\sqrt{2}$, $T = -1/\sqrt{2}$, and $|R|^2 = |T|^2 = 1/2$.

5.9 Scattering, Underdense Barrier

The underdense barrier problem is given by a function $Q(z)$ which is real on the real axis, with two first order zeros at points a and b which are pure imaginary, and Q is positive everywhere on the real axis. The Stokes diagram is shown in Fig. 5.7. Again define W through $[a,b] = e^{-W}$. Along the imaginary axis between a and b Q is real and positive and W is real and positive. Consider an incident wave from the left, giving again outgoing boundary conditions at the far right. For this problem Heading discusses

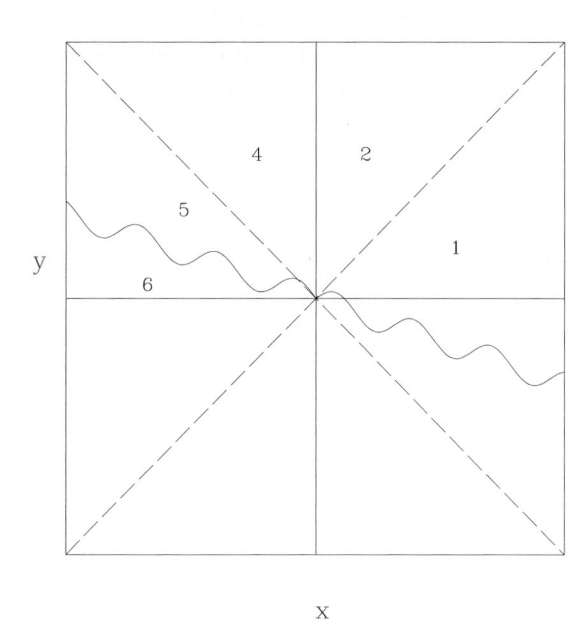

Fig. 5.6 $Q = z^2$, Stokes plot.

two methods which he discards as approximate. The first method is continuation in the lower half plane alone. This gives a transmission coefficient of 1 and zero reflection. It is used by Berry [1990] in an analysis of the birth point of reflected waves.

Continuation in the upper half plane gives a transmission coefficient of 1 and a reflection coefficient of $R = -ie^{-W}$. This continuation involves the singular point $z = b$ only, and the correct Stokes constant is that associated with an isolated first order zero, $S = i$. This is because, as far as the continuation is concerned, the second singularity at $z = a$ does not exist, and the Stokes diagram consists only of the structure in the upper half plane.

However, neither of these continuations give solutions which conserve flux and neither can be considered correct. A third method is given by Heading, who describes it as more accurate, but does not give a justification for its use, which can be obtained from the consideration of causality. Not only is it necessary to consider outgoing wave boundary conditions, a small dissipation must be considered so as to cause waves to damp in the direction of propagation. This is done by multiplying $Q(z)$ by $e^{i\nu}$ with ν small and positive. Since Stokes lines are given by $dz \sim 1/\sqrt{Q(z)}$ this

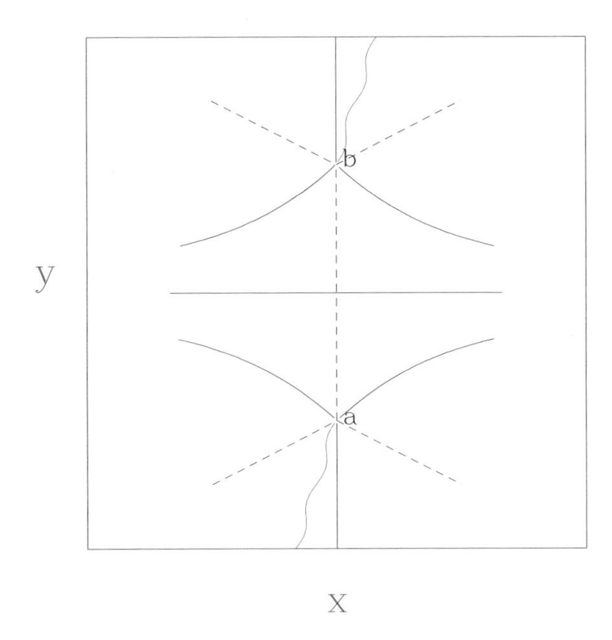

Fig. 5.7 Stokes plot for the underdense barrier.

results in a rotation of the Stokes plot. Shown in Fig. 5.8 is the resulting plot (the rotation has been exaggerated for clarity). Now the only possible continuation from large real positive z to large real negative z is clear. It is necessary to begin in domain 1 and continue to domain 7, a process which necessarily involves both singular points.

Continuation then gives:

(1) $(b,z)_s$
(2) $(b,z)_s$
(3) $-i(z,b)_s$
(4) $-i(z,b)_d$
(5) $-i(z,b)_d - iS(b,z)_s$
(6) $-i(z,b)_s - iS(b,z)_d$
(6) $-i(z,a)_d[a,b] - iS[b,a](a,z)_s$
(7) $-i(z,a)_s[a,b] - iS[b,a](a,z)_d$

Take the reference point for the phase to be $z = 0$. The incoming solution is $(0,z)$, and the reflected wave is $(z,0)$. Then in (1) we have $(b,z) = [b,0](0,z)$, and in (7) we have $-i(z,0)[0,b] - iS[b,0](0,z)$ giving for

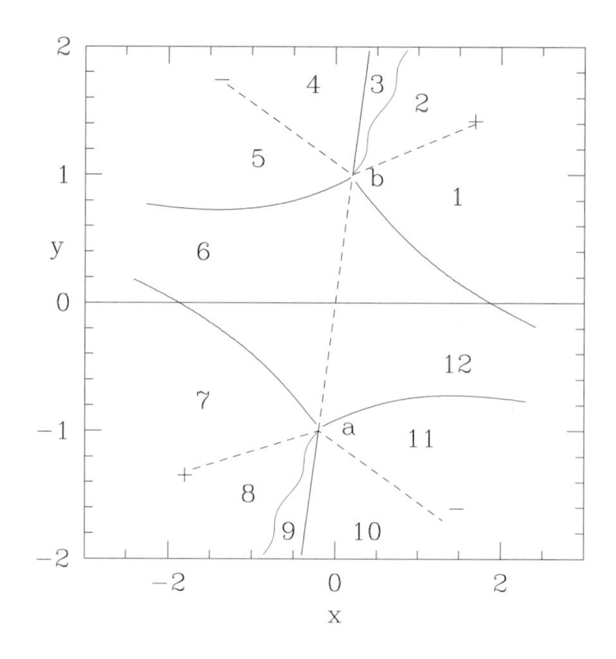

Fig. 5.8 Underdense barrier, rotated Stokes plot.

reflection and transmission

$$R = \frac{e^{-W}}{S}, \qquad T = \frac{i}{S}, \tag{5.31}$$

and equating flux for $x \to -\infty$, $x \to \infty$ gives again $S = i\sqrt{1+e^{-2W}}e^{i\delta}$, but again δ is undetermined. This gives

$$|R|^2 = \frac{e^{-2W}}{1+e^{-2W}}, \qquad |T|^2 = \frac{1}{1+e^{-2W}}. \tag{5.32}$$

A comparison with numerical integration for the case $Q = z^2 + b^2$ is shown in Fig. 5.9, obtained numerically at $|x| = 5$ using 100 terms in the Taylor series, with the constants in the series fixed by requiring outgoing wave conditions at $x = 5$. Solid points show the data from the Taylor expansion and open points the data from the WKBJ continuation using Eqs. 5.32.

The phase δ can again be determined in two limits. First examine the limit $a \to b$. The Stokes diagram collapses to a single second order zero, and the Stokes plot is shown in Fig. 5.10, the same as in the case of the overdense barrier in Fig. 5.6 except for cut placement. We retain the cuts

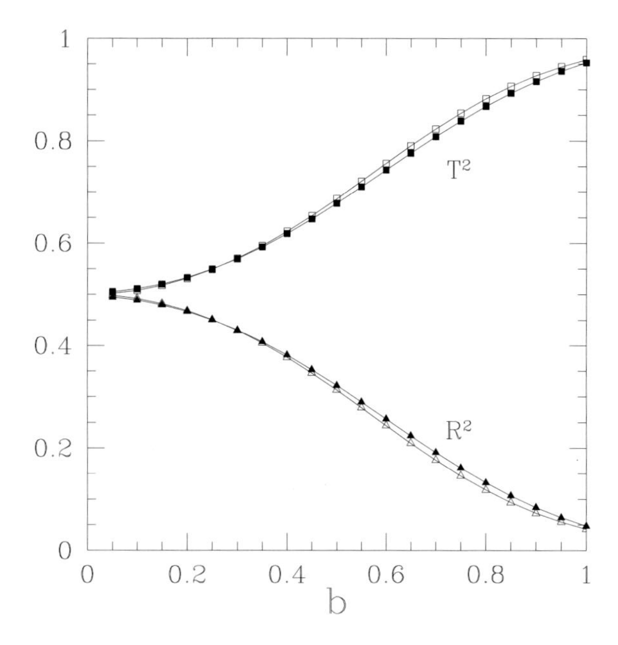

Fig. 5.9 $Q = z^2 + b^2$, transmission and reflection vs b.

to make the results coincide exactly with those resulting from the limit $a \to b$ in Fig. 5.8. Again the Stokes constants have a continuous limit. In this case the Stokes constant is $\sqrt{2}i$ and the continuation is given by the above with $W = 0$, giving for the reflection and transmission coefficients $R = -i/\sqrt{2}$, $T = -1/\sqrt{2}$, and $|R|^2 = |T|^2 = 1/2$.

5.10 The Budden Problem

Consider a normally incident plane polarized electromagnetic wave entering a magnetized plasma with the field given by $\vec{B}_0 = B_0\hat{z}$ where the plasma frequency $\omega_p(x)$ increases with x. If $\vec{k}$ and $\vec{E}$ of the wave are perpendicular to $\vec{B}_0$, $k_y = k\hat{y}$, $\vec{E} = E\hat{y}$ the y component of $\vec{E}$ satisfies

$$\frac{d^2E_y}{dx^2} + k_0^2\epsilon(x)E_y = 0. \tag{5.33}$$

The dielectric function has a cutoff, $\epsilon = 0$, and resonance $\epsilon = \infty$. Model this dielectric through $k^2\epsilon(x) = 1 + a/x$. This is called the Budden problem. (See Budden [1979], White and Chen [1974].)

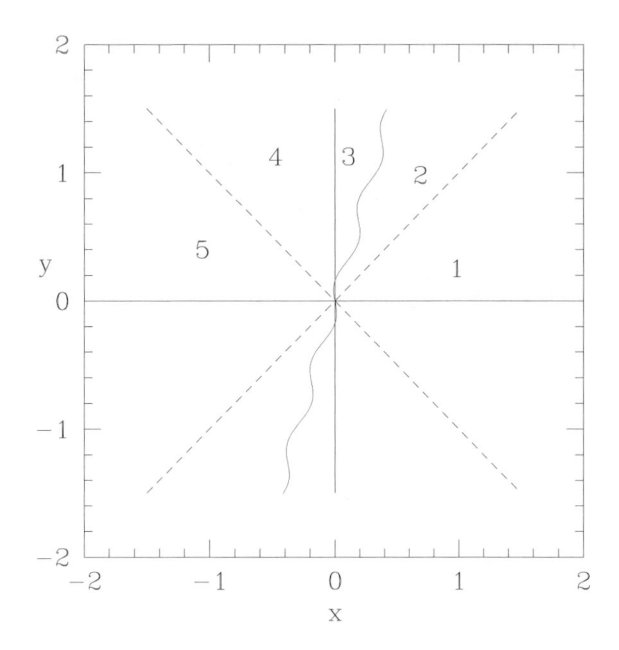

Fig. 5.10 $Q = z^2$, Stokes plot.

Figure 5.11 shows the resulting Stokes diagram, with cuts shown as wavy lines, anti-Stokes lines as solid lines, and Stokes lines with dashed lines. Also shown with short dashes are the local anti-Stokes lines. The cut at the pole, $z = 0$, must be taken upwards as shown. This condition is equivalent to the Landau prescription for the pole to describe Landau damping [1946]. To see this consider the flux $S = Im\psi^*\psi'$. Near a pole $\psi'' = a/x$ and thus integrating past the pole we find $\psi'_+ - \psi'_-$ equal to πia integrating above the pole and $-\pi ia$ integrating below the pole. Thus there is a net loss of flux of πa if the cut is taken upwards, i.e. the pole is a sink of flux, whereas if the integration is taken above, the pole acts as a source of flux. Causality requires the former.

Now let us find the reflection and transmission coefficients for a wave incident from the left using phase-integral methods. We will use the Stokes constants for isolated singularities. This restriction means that the results will be valid only for large separation of the zero and pole. The general problem, for arbitrary separation, will be done in Chapter 12. It is not possible to use the technique of flux conservation to treat general separation, because in this problem, unlike the previously treated scattering problems,

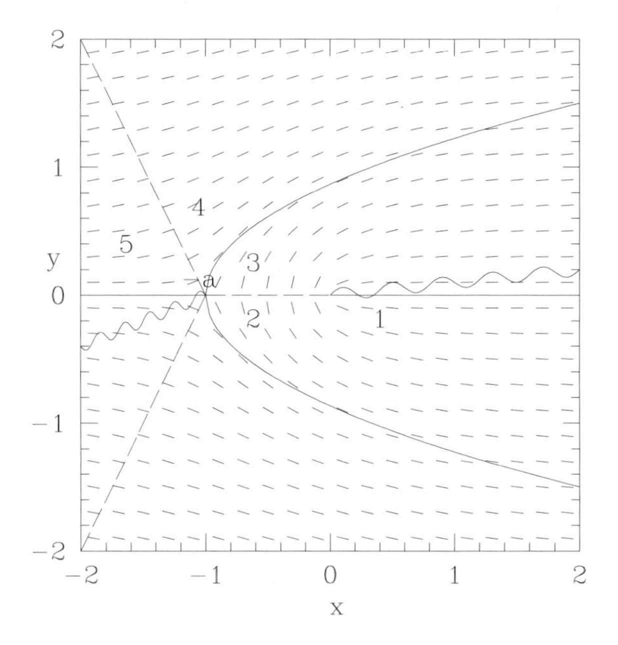

Fig. 5.11 Stokes diagram for Budden problem.

flux is absorbed at the pole. Begin at the far right with an outgoing transmitted wave of the form e^{iz} and take Q positive and real. With the choice of cuts as shown we then find $Q \sim e^{i\pi}$ on the real axis for $-a < x < 0$ and $Q \sim e^{2i\pi}$ on the real axis for $x < -a$. In domain 1 below the real axis this solution is dominant. Write it as $(0, z)_d$. Now step onto the Stokes line, using minus half the Stokes constant for the pole, since we are moving clockwise, giving $(0, z)_d + i(z, 0)_s$. Reconnect while on the line, giving $[0, -a](-a, z)_s + i(z, -a)_d[-a, 0]$. Then step off the line to domain 3, giving $i(z, -a)_d[-a, 0] + (-a, z)_s([0, -a] - [-a, 0]/2)$. Then move to domain 4 giving $i(z, -a)_s[-a, 0] + (-a, z)_d([0, -a] - [-a, 0]/2)$. Finally step to domain 5 giving $(-a, z)_d([0, -a] - [-a, 0]/2) + i(z, -a)_s([0, -a] + [-a, 0]/2)$.

Now identify the incoming and reflected waves. At the far left an incoming wave is of the form e^{iz} which is subdominant in domain 5. To see that this is indeed $(z, -a)$ note $(z, -a) = exp(i \int_z^{-a} \sqrt{Q} dz)$. But from the cut structure we have that on the real axis at the far left $Q \sim e^{2\pi i}$ and thus $(z, -a) \sim e^{iz}$. We then have

$$R = \frac{[0, -a] - [-a, 0]/2}{i([0, -a] + [-a, 0]/2)}, \qquad T = \frac{1}{i([0, -a] + [-a, 0]/2)}. \tag{5.34}$$

To evaluate this expression let $u = \sqrt{-1 - a/x}$ giving

$$\int_0^{-a} \sqrt{Q} dx = -i \int_{-\infty}^{\infty} \frac{au^2 du}{(u^2+1)^2}, \tag{5.35}$$

which is readily evaluated by closing the contour in the upper half plane, giving $[0, -a] = e^{\pi a/2}$, and

$$R = \frac{e^{\pi a/2} - e^{-\pi a/2}/2}{i(e^{\pi a/2} + e^{-\pi a/2}/2)}, \qquad T = \frac{1}{i(e^{\pi a/2} + e^{-\pi a/2}/2)}. \tag{5.36}$$

Note that $|R|^2 + |T|^2 = (e^{\pi a} + e^{-\pi a}/4)/(1 + e^{\pi a} + e^{-\pi a}/4)$, which is less than 1, there is absorption at the pole. An impressive feature of this asymptotic analysis is the fact that the magnitude of this absorption is completely determined, it is independent of the nature of the physics describing the wave generation or kinetic dissipation occurring in the vicinity of the pole. However, the limit $a \to 0$ is clearly wrong; in this limit $Q = 1$ and there should be no reflection or absorption. The error is due to the approximation of using Stokes constants for isolated singularities. In Chapter 12 this problem will be solved exactly.

5.11 The Error Function

The error function

$$erf(z) = \frac{2}{\sqrt{\pi}} \int_0^z e^{-t^2} dt \tag{5.37}$$

satisfies the differential equation

$$f'' + 2zf' = 0, \tag{5.38}$$

and thus has one solution given by $f(z) = C$, in addition to the integral given above. Asymptotically for large z writing $f = e^S$ dominant balance gives $S = const$ and $S = -z^2 - ln(z)$, giving the two solutions. However, this differential equation cannot be put in the form $f'' + Q(z)f = 0$, it is a degenerate case, so the standard WKBJ theory cannot be applied and we do not know the Stokes constant. Nevertheless the Stokes diagram is simply obtained by examining the asymptotic behavior of the solutions. There are Stokes lines along the real and imaginary axes, and anti-Stokes lines along the 45 degree lines, $z \sim \pm e^{\pm i\pi/4}$. Now we are in a position to understand the asymptotic expansion, Eq. 4.68. Along the positive real axis the error function has the form

$$erf(z) \simeq 1 - \frac{2}{\sqrt{\pi}z} e^{-z^2}, \tag{5.39}$$

with the exponential term the subdominant solution and 1 the dominant solution. Continue counterclockwise around the complex plane. Crossing the anti-Stokes line at $z = e^{i\pi/4}$ the exponential term becomes dominant and 1 becomes subdominant. Crossing the Stokes line at the imaginary axis the coefficient of the subdominant term changes due to the Stokes phenomenon. In fact, according to Eq. 4.68 the Stokes constant must be -2, since the coefficient changes from 1 to -1 upon crossing the line, and is zero on the line.

5.12 Eigenvalue Problems

Normally in electrical circuit theory or in physics problems the function Q is a complex function of the frequency of the mode ω. For an arbitrary guess for ω the Stokes plot will be quite complicated. To find an unstable mode one must identify turning points and check whether there exists a value of ω giving a solution which satisfies the required boundary conditions. This is normally best accomplished numerically. Unless ω is chosen to be the eigenvalue, pairs of turning points will not exhibit the canonical attractive well Stokes structure shown in Fig. 5.4. A typical example of what is seen is shown in Fig. 5.12. In addition there may be other nearby singularities and zeros of $Q(z, \omega)$, making it difficult to recognize possible unstable mode structures. This is especially true if the physical problem possesses several parameters which must be explored.

As an example we give a problem which arose in the theory of the collisionless drift wave. The differential equation is given on the real axis by

$$Q = -\frac{\omega L}{2}\left(K^2 - 1 - \frac{1}{\omega}\right) + \frac{x^2}{4} + \frac{L}{2}(1-\omega)AZ(A) \tag{5.40}$$

where L, K, R are real parameters, $A = (\omega RL)^{1/2}/|x|$ and Z is the plasma dispersion function. A more complete description of the derivation and

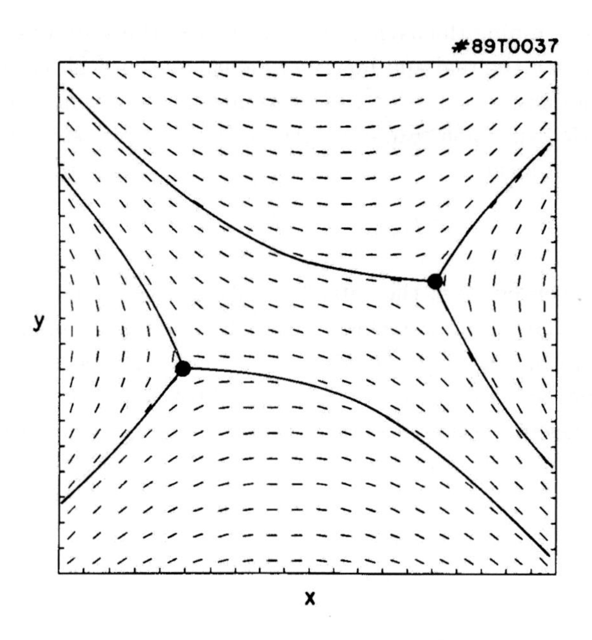

Fig. 5.12 Stokes diagram during search for eigenvalue.

solutions of this equation has been reported elsewhere (Chen *et al.* Phys. Rev. Lett **41**, 649 [1978]). The expression $|x|$ arises in a derivation valid on the real axis, and the analytic continuation into the complex plane is provided by $|x| = (z^2)^{1/2}$ with an appropriate choice of cuts.

The physical real axis is determined by $|x| \geq 0$, and thus the physical plane is divided into two parts by branch cuts which can be taken along the imaginary axis. Thus in the physical plane for $Rez > 0$, $|x|$ is replaced by z, and for $Rez < 0$, $|x|$ is replaced by $-z$. The continuation of these functions through the cuts defines a second plane which we will refer to as the nonphysical plane. It is also divided into two distinct parts. For values of K much less than a critical value K_c which depends on R and L (for $R = 1/1837$ and $L = 50$, $K_c = 0.36$) there are two turning points located approximately along the line $x = -y$ in the physical plane. As K increases, these turning points begin to migrate toward the imaginary axis.

Figure 5.13 shows the location of the turning points (v and $-v$) for a mode with $L = 50$, $K = 0.26$, and $R = 1/1837$. Note the presence of another pair of turning points (g and $-g$), located in the nonphysical plane. Assume a decaying mode, in which case the solution for $x \to \infty$ must be

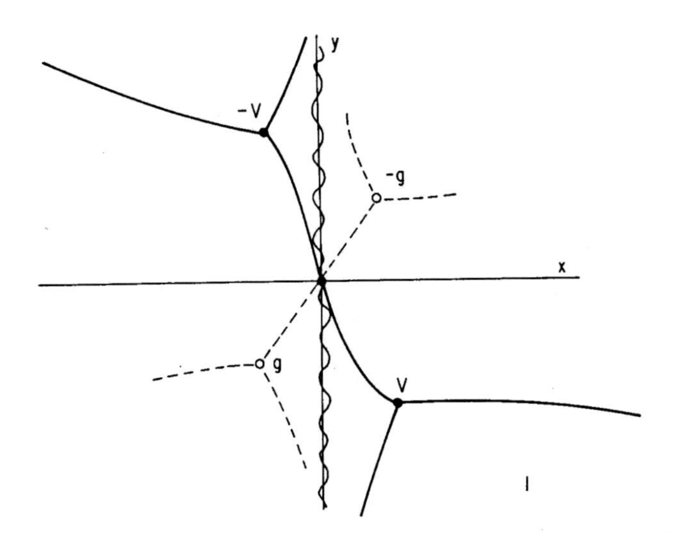

Fig. 5.13 Stokes structure for $K < K_c$.

dominant, $\psi(z) = (z, v)_d$. Continuing in toward $x = 0$ we cross a Stokes line associated with v and thus the solution becomes $\psi(z) = (z, v)_d + i(v, z)_s$. The differential equation and the turning points under consideration are symmetric about $z = 0$, thus we can choose a solution with a particular parity with respect to z.

For an odd solution we require $\psi(0) = 0$, or $(0, v) + i(v, 0) = 0$. This has the solution $2\int_0^v Q^{1/2}dz = [n + 1/2]\pi$, n odd. For the even solutions we require that $d\psi/dz$ vanish at $z = 0$, or $i(0, v) + (v, 0) = 0$. This has the solution $2\int_0^v Q^{1/2}dz = [n + 1/2]\pi$, n even. Thus we obtain the standard connection formula between $v, -v$ which for the parameters given above gives $\omega = 0.908 - 0.008i$. The mode is decaying in agreement with our initial assumption of a purely dominant solution for large x. It is readily verified that the anti-Stokes line emanating from v toward positive x does not cross the real axis.

As the parameter K increases, provided $L > 3R^{-1/4}$, the turning points $v, -v$ as well as the turning points $g, -g$ approach the imaginary axis and coalesce for $K = K_c$, after which the Stokes diagram takes the form of Fig. 5.14. The role of turning point for the determination of the solution has been passed on from v to g. Once again the analysis presented above carries through, only now the connection formula is determined by performing the integral from g to $-g$. We then discover that the growth rate $Im(\omega)$ is zero.

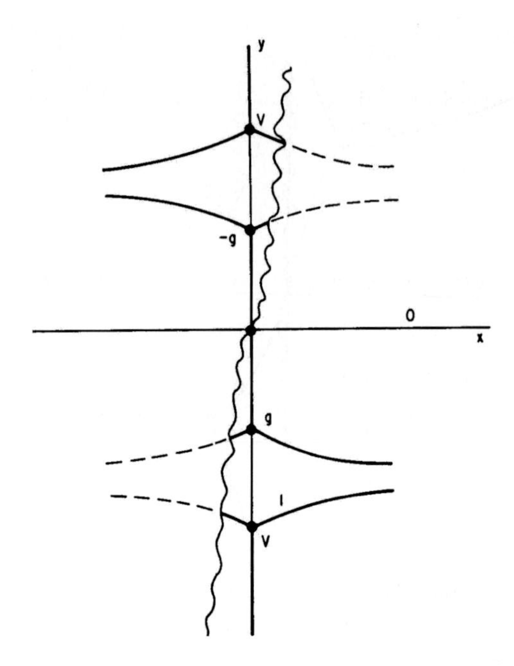

Fig. 5.14 Stokes structure for $K > K_c$.

If $L < 3R^{-1/4}$ this coalescense does not occur for any value of K, and the mode continues to be determined by the turning point v.

For K less than the critical value, i.e. with Stokes structure as shown in Fig. 5.13, imposing the connection formula between g and $-g$ gives rise to a growing mode $Im(w) > 0$. However, this ghost mode remains in the nonphysical plane, interpretable as an outwardly propagating solution only in this plane. The critical value of K is seen to be the coalescense of the turning points of a nonphysical growing mode and a physical damped mode. The nonphysical turning point then dominates to produce a marginally stable mode for all larger values of K.

5.13 Problems

1. Sketch the anti-Stokes lines associated with $d^2\psi/dz^2 + Q\psi = 0$ for $|z| < 4$:

$$Q = \frac{z - i\pi}{z + i\pi} sinz.$$

2. Show that for a transition point z^n the rule for crossing a branch cut is

$$(a, z) \to i(-1)^{(n+1)/2}(z, a) \qquad n \; odd$$
$$(a, z) \to i(-1)^{n/2}(a, z) \qquad n \; even.$$

3. Consider the eigenvalue problem

$$\frac{d^2y}{dx^2} + \lambda^2 q(x)y = 0,$$

with $q(x) = (3\pi/7)^2(x+1)^4$ and $y(0) = y(1) = 0$. Find the first three eigenvalues using WKB theory, and compare to the numerical values given by Bender and Orszag, $\lambda^2 = .924915, 3.89727, 8.88444$.

4. Sketch the anti-Stokes and Stokes lines and find the Stokes constant T for a second order zero, $Q = z^2$, by directly continuing the solution around the turning point and requiring the solution to be single valued. Check your result using Eq. 5.22.

5. Consider the eigenvalue problem $y'' + Ecos(x)y = 0$ with $y(\pi) = 0$, and $y'(0) = 0$, $E > 0$. Sketch the Stokes plot for y.

a. Use WKB to find the three smallest eigenvalues E, approximating Stokes constants as those from isolated singularities.

Hint 1: The boundary condition $y(\pi) = 0$ is satisfied ON a Stokes line.

Hint 2: Continue the WKB solution to $x = 0$ and require $y'(0) = 0$, giving a transcendental equation for $w = \int_0^{\pi/2} \sqrt{Ecos(x)}dx$. Solve this iteratively. Note

$$\int_0^{\pi/2} \sqrt{cos(x)}dx = \int_{\pi/2}^{\pi} \sqrt{-cos(x)}dx \simeq 1.19814025.$$

b. Integrate the equation numerically using the WKB values as first approximations to E. Adjust E to give $y(\pi) = 0$. What is the accuracy of the WKB eigenvalues?

6. Consider the differential equation

$$\frac{d^2y}{dx^2} + \left(\omega^2 - \frac{x^2}{4}\right) y = 0.$$

a. Find the asymptotic behavior of the two solutions for $x \to \pm\infty$.

b. Write the solutions as $y \sim e^S W(x)$ and find a series representation for $W(x)$ for the solution tending to zero at ∞. Find the radius of convergence of the series.

c. Draw the Stokes diagram. Use WKB analysis to find ω such that the solution tends to zero at $\pm\infty$. These are the eigenvalues given by the WKB analysis.

d. What happens to the series for $W(x)$ for these values of ω?

e. What is the accuracy of the eigenvalues and eigenfunctions given by the WKB method?

7. The y component of the electric field in a dielectric medium satisfies the equation

$$E'' + (1 + x^2)E = 0$$

and it must vanish at a metallic conducting surface.

If one metal wall is placed at $x = 0$ use the WKB approximation to find the position $x = b$ of the nearest possible wall that will allow a standing wave, i.e. $E(0) = E(b) = 0$.

8. A simple model for nuclear decay estimation by perturbation analysis. Consider the equation

$$\frac{d^2y}{dx^2} + Q(x)y = 0$$

with $Q = A(1 - x^2)$ for $x < 1$, and $Q = (1 - x)(1 - x/b)$ for $x > 1$, with $b >> 1$ giving the nuclear potential tunneling distance.

a. First take $b = \infty$, draw the Stokes diagram, and find the eigenvalues A so that $y \to 0$ for $x \to \pm\infty$.

b. Now suppose b finite but large. Draw the Stokes diagram, and continue the solution of part *a* to find the solution on the real axis for $x > b$.

c. Calculate the flux of the obtained solution on the real axis for $x > b$.

d. This flux is the probability of decay of the original solution in one time interval, so its inverse is the nuclear lifetime. How does it depend on b?

Chapter 6

Perturbation Theory

Richard Feynman was born in Manhattan in 1918, and grew up in Far Rockaway. He had an early interest in science, repairing radios, devouring mathematics, and taking on a healthy skepticism regarding knowledge consisting of empty naming with no factual content. He was socially awkward with little interest in music, literature, or art. After finishing high school he entered MIT with extraordinary test scores in mathematics and physics but dismal in everything else. At MIT he quickly showed enormous intuitive ability for solving difficult problems. His thesis was on the thermal expansion of quartz, and resulted in his first publication, on forces in molecules. From MIT he went to Princeton, where he worked with John Wheeler, and developed a formulation of electromagnetic interaction of electrons which was invariant under time reversal. Before finishing his thesis he became involved in the war-related effort to separate Uranium-235 from Uranium-238, a project which was finally cancelled at Princeton in favor of the calutron in use at Berkeley. At this point he finished his thesis and then joined the bomb project at Los Alamos, where he attained a special reputation not only for his ability in mathematics and physics but for demonstrating the insufficiency of the security by leaving cryptic messages inside locked safes. After the war he taught at Cornell, where he developed the most successful and most mysterious perturbation theory ever made, quantum electrodynamics, the terms of which were given by Feynman diagrams. He, along with Julius Schwinger and Shinichiro Tomonaga, was awarded the Nobel prize for this work in 1965. In 1952 he moved to Caltech, where he worked for the rest of his life.

He was reluctant to serve on the Presidential Rogers Commission to investigate the Challenger space shuttle disaster, but he complied, and discovered a history of corner-cutting and bad science. Management misunderstood the science, and he was tipped off by engineers at Morton Thiokol that they ignored it, most importantly when warned about a possible problem with an o-ring. He famously demonstrated the brittleness of the o-ring at low temperatures by dropping it into a glass of ice water during the commission meeting.

6.1 Introduction

By assuming that the solution of an equation involving a small parameter ϵ has a series solution in ϵ valid for all ϵ less than some limiting value, it is possible to insert the expansion and solve the resulting infinite set of equations order by order, reducing one difficult problem to an infinite number of simpler ones, but with the additional advantage that for a given accuracy only a small number of terms may be required. If the resulting series converges, a solution is obtained which is valid for a range of values of ϵ, not merely for a single value. When such a Taylor's series exists, the solution is analytic in ϵ. The analytic property of the solution as a function of ϵ determines the nature of the solution. For example, in Eq. 1.3,

$$1 - x - \epsilon x^2 - 2\epsilon^3 x^3 = 0, \tag{6.1}$$

set $x = \sum_0^\infty x_n \epsilon^n$, and substitute, giving upon solving order by order in ϵ

$$x = 1 - \epsilon + 2\epsilon^2 + \dots. \tag{6.2}$$

It takes very little effort to discover that this process is much more cumbersome than the Kruskal–Newton iteration. In addition, we have obtained only a single solution. To find the other solutions, we can see from the solutions obtained using the method of dominant balance that one must choose series of the form $x = \sum_0^\infty x_n \epsilon^{n-1}$, and $x = \sum_0^\infty x_n \epsilon^{n-2}$ to succeed. But without using the method of dominant balance, these series might have been difficult to guess. From the form of these series we see that two of the solutions are not analytic functions of ϵ. In this case we have encountered a singular perturbation series, the perturbation solution not existing in the limit of $\epsilon = 0$. A singular perturbation problem may also possess a series which proves to be divergent. The classification of the solutions according to their behavior in ϵ is similar to the classification of local expansions of solutions to differential equations, see section 4.2, and as seen there, even divergent series can be very useful. As the other examples in Chapter 1 show, it may be necessary to choose series in $\sqrt{\epsilon}$, possibly also beginning with a nonzero power, to find perturbation series which give the desired solution. Even after finding such a series, determining its convergence properties is also a difficult matter, whereas in the case of the Kruskal–Newton method it is fairly easy to discover whether the fixed point iteration converges. Nevertheless, perturbation theory provides a different means of attacking a problem, and can prove very useful. It is occasionally advantageous even to introduce a small parameter in a problem which does not initially possess one, in order to introduce perturbation methods.

6.2 Eigenvalues of a Hermitian Matrix

Perturbation theory for the eigenvalues of a Hermitian matrix is well known from quantum mechanics. For any vector space with an inner product for any two vectors (v, w) such that $(v, cw) = c(v, w)$ and $(cv, w) = c^*(v, w)$, for any complex constant c and c^* is the complex conjugate of c, an operator H is Hermitian if $(Hv, w) = (v, Hw)$. Consider a matrix H which can be written as a simpler matrix H_0 with eigenvalues E_n and corresponding eigenvectors v_n plus a small Hermitian perturbation H'. We then have

$$H_0 v_n = E_n v_n. \tag{6.3}$$

If the eigenvalues are nondegenerate we find immediately that the corresponding eigenvectors are orthogonal, since $(Hv_n, v_m) = E_n(v_n, v_m)$ but also $(Hv_n, v_m) = (v_n, Hv_m) = E_m(v_n, v_m)$, and thus unless $E_n = E_m$ we find $(v_n, v_m) = 0$. Furthermore by dividing each vector by $\sqrt{(v_n, v_n)}$ they can be chosen orthonormal. Now write a perturbation expansion for a particular eigenvalue W and eigenvector ψ satisfying $Hv = Wv$ using $H = H_0 + \epsilon H'$,

$$\begin{aligned} W &= W_0 + \epsilon W_1 + \epsilon^2 W_2 + \dots \\ \psi &= \psi_0 + \epsilon \psi_1 + \epsilon^2 \psi_2 + \dots . \end{aligned} \tag{6.4}$$

Substitute these into $H\psi = W\psi$ and equate terms of each order in ϵ, giving

$$\begin{aligned} H_0\psi_0 &= W_0\psi_0 \\ H_0\psi_1 + H'\psi_0 &= W_0\psi_1 + W_1\psi_0 \\ H_0\psi_2 + H'\psi_1 &= W_0\psi_2 + W_1\psi_1 + W_2\psi_0 \\ &\dots . \end{aligned} \tag{6.5}$$

Thus ψ_0 is any one of the unperturbed eigenvectors. Choose $\psi_0 = v_m$, giving $W_0 = E_m$, and thus calculate the change in this eigenvalue due to H'. Now expand ψ_1 in terms of the unperturbed eigenvectors,

$$\psi_1 = \sum a_n^1 v_n. \tag{6.6}$$

Substituting we find

$$\sum a_n^1 H_0 v_n + H' v_m = E_m \sum a_n^1 v_n + W_1 v_m. \tag{6.7}$$

Take the inner product with v_k and use orthonormality, giving

$$a_n^1 (E_m - E_k) + W_1 \delta_{km} = (v_k, H' v_m) \equiv H'_{km}. \tag{6.8}$$

Thus

$$W_1 = H'_{mm}, \qquad a_k^1 = \frac{H'_{km}}{E_m - E_k} \qquad k \neq m. \tag{6.9}$$

The next order is found in the same manner by writing $\psi_2 = \sum a_n^2 v_n$. Substituting and again using orthonormality we find

$$W_2 = \sum_{k \neq m} \frac{|H'_{km}|^2}{E_m - E_k}, \tag{6.10}$$

$$a_k^2 = \sum_{n \neq k} \frac{H'_{kn} H'_{nm}}{(E_m - E_k)(E_m - E_n)} - \frac{H'_{km} H'_{mm}}{(E_m - E_k)^2} + \frac{a_m^1 H'_{km}}{E_m - E_k} \qquad k \neq m.$$

Note that Eqs. 6.9 and 6.11 do not determine a_m^1, a_m^2 but these are given by normalizing ψ. Setting $(\psi, \psi) = 1$ gives to second order in ϵ

$$Re(a_m^1) = 0, \qquad 2Re(a_m^2) + \sum |a_n^1|^2 = 0. \tag{6.11}$$

There is no loss in generality is choosing the a_m to be real, giving $a_m^1 = 0$ and $a_m^2 = -\sum |a_n^1|^2$. The perturbed eigenvalues are independent of this choice of phase. We then have to second order the modification of eigenvalue

m and eigenvector m

$$
\begin{aligned}
W_m &= E_m + H'_{mm} + \sum_{k \neq m} \frac{|H'_{km}|^2}{E_m - E_k}, \\
\psi_m &= v_m + \sum_{k \neq m} \frac{H'_{km} v_k}{E_m - E_k} \\
&\quad + \sum_{k \neq m} \left[\sum_{n \neq m} \frac{H'_{kn} H'_{nm}}{(E_m - E_k)(E_m - E_n)} - \frac{H'_{km} H'_{mm}}{(E_m - E_k)^2} \right] v_k \\
&\quad - \sum_{k \neq m} \frac{|H'_{km}|^2}{2(E_m - E_k)^2} v_m,
\end{aligned}
\tag{6.12}
$$

and if the matrix elements H'_{mn} are small one can hope that this expansion converges.

If some of the original eigenvalues are equal, i.e. the matrix H_0 is degenerate, this perturbation theory fails. In this case it is necessary to solve the subspace involving the degeneracy exactly, without the perturbation expansion, or to include all the degenerate eigenvectors in the expression for ψ_0, leading to more complicated expressions but again to a solution (Schiff [1955]).

From the perturbation expansion above one can see that there is a hint of a problem when the eigenvalues approach one another. If the initial eigenvalues are equal the problem can be averted, but what if eigenvalues approach one another due to the perturbation? It is worth examining a problem which can be done exactly. Consider a diagonal 2x2 matrix M with initial unperturbed eigenvalues a, b, and add a small perturbation of the form $M_{11} = \epsilon x$, $M_{22} = \epsilon y$, $M_{12} = \epsilon z$, $M_{21} = \epsilon z^*$, with x, y real and z complex, giving

$$
M = \begin{pmatrix} a + \epsilon x & \epsilon z \\ \epsilon z^* & b + \epsilon y \end{pmatrix}. \tag{6.13}
$$

The eigenvalues are given by

$$
\lambda = \frac{a + b + \epsilon x + \epsilon y \pm r}{2}, \tag{6.14}
$$

with $r^2 = (a + \epsilon x - b - \epsilon y)^2 + 4\epsilon^2 |z|^2$. All eigenvalues of a Hermitian matrix are real, so it is no surprise that r^2 is positive, but it is perhaps surprising that unless z is zero, r cannot vanish. This means that the two eigenvalues can never be equal, and is referred to in quantum mechanics as the repulsion of the levels. An example is shown in Fig. 6.1, with $a = 1$, $b = 1.1$, $x = .2$, $y = -.1$, $z = .01$.

Considered as a function of ϵ, the eigenvalue equation $\lambda(\epsilon)$ has a branch cut at

$$\epsilon = \frac{a-b}{y-x \pm 2iz}, \tag{6.15}$$

and the series converges only to a distance given by the nearest singularity. If the unperturbed eigenvalues are degenerate the radius of convergence is zero.

That the eigenvalues tend to distance themselves from one another leads to the remarkable result (Mehta [1967]) that if one considers the ensemble of all $N \times N$ Hermitian matrices with the real and imaginary parts of each element chosen to be independent Gaussian random variables, then the probability of finding two eigenvalues with spacing xD, with D the average spacing level, is given by

$$p(x) = 1 - \left(\frac{sin\pi x}{\pi x}\right)^2. \tag{6.16}$$

This function is shown in Fig. 6.2. Note that the probability of eigenvalues being separated by much less than the mean separation is very small, and that there are small bumps at integer values of x, as though the eigenvalues attempt to distribute themselves at integer multiples of the mean spacing. Random Hermitian matrices are used with some success to approximate the energy levels of large nucleii such as uranium, where the large number of protons and neutrons makes it impossible to explicitly model the nucleus.

6.3 Broken Symmetry Due to Tunneling

Exponential tunneling often produces a very small effect, the result of which is given very well by lowest order perturbation theory. In the case of a potential which is perfectly symmetric under the transformation $x \to -x$ this can produce different energy levels for the eigenstates with even ($\psi(-x) = \psi(x)$) and odd ($\psi(-x) = -\psi(x)$) parity. This is one example of broken symmetry, i.e. the equation describing the system is invariant with respect to some transformation; nevertheless the solution is not.

Consider the one-dimensional quantum mechanics problem given by the differential equation

$$\frac{d^2\psi}{dx^2} + B^2(x^2 - x^4 + E - 1/4)\psi = 0, \tag{6.17}$$

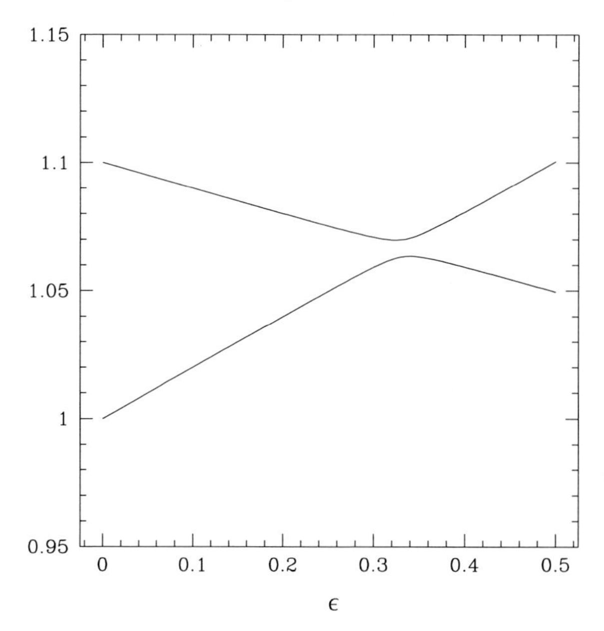

Fig. 6.1 Eigenvalues of a Hermitian matrix vs perturbation.

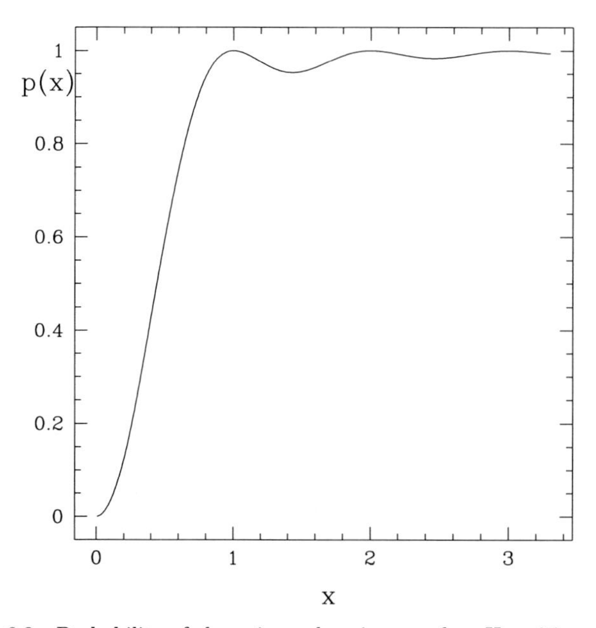

Fig. 6.2 Probability of close eigenvalues in a random Hermitian matrix.

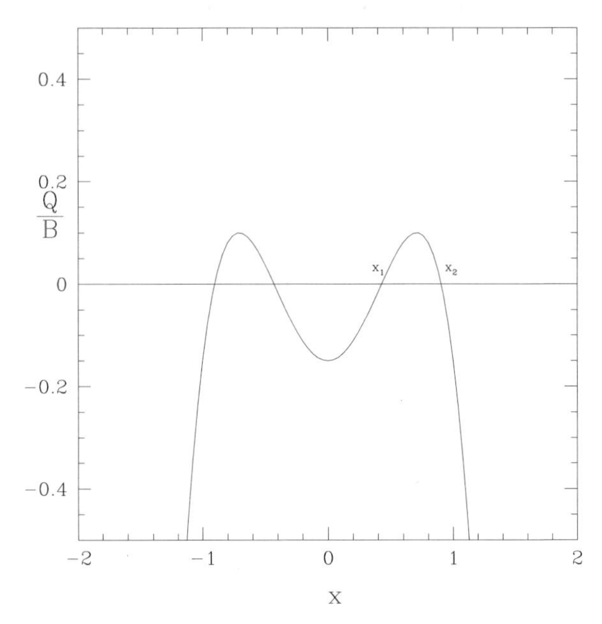

Fig. 6.3 Q function for the double potential.

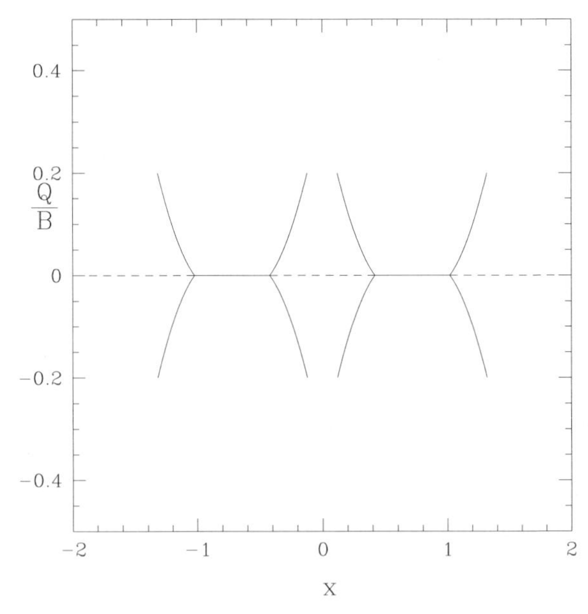

Fig. 6.4 Stokes plot for the double potential.

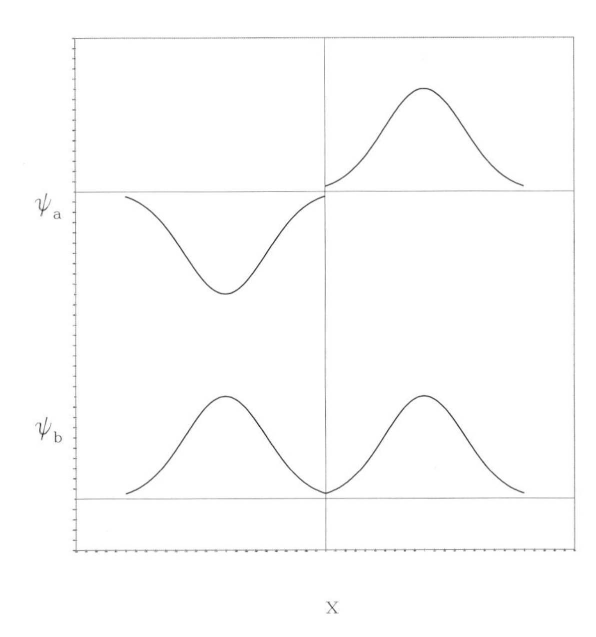

Fig. 6.5 Even and odd parity solutions.

with $B^2 \gg 1$ and E the energy eigenvalue. Using the usual WKB notation $d^2\psi/dx^2 + Q\psi = 0$, a plot of $Q(x)$ on the real axis is shown in Fig. 6.3 and the Stokes plot is shown in Fig. 6.4. To begin we neglect the effect of tunneling between the two bound state wells, i.e. treat them as though they are infinitely far apart. Approximating the local potentials by expanding to second order around the local maxima of Q, we then find two eigenstates for each n given by integrating between the turning points, with energy $E_n = 4(n+1/2)/B$, one state localized between the two right-hand turning points and one localized between the two left-hand turning points. By taking the sum and the difference of these two solutions as shown in Fig. 6.5 we can construct solutions which are symmetric (ψ_s) and antisymmetric (ψ_a) functions of x. Here we take the lowest energy states $n = 0$ for simplicity. Now calculate the splitting of this degeneracy due to tunneling effects. Perform the usual WKB analysis, beginning with a subdominant solution at $x = +\infty$. Perform the usual continuation to the neighborhood of $x = 0$, giving

$$\psi(z) = -i[x_1, x_2](z, x_1)_s + ([x_1, x_2] + [x_2, x_1])(x_1, z)_d. \qquad (6.18)$$

For an even solution require $\psi'(0) = 0$ and for an odd solution $\psi(0) = 0$. Let $[x_1, 0] = e^A$ where $A = B\int_0^{x_1}(x^4 - x^2 - E + 1/4)^{1/2}dx > 0$ since $Q < 0$ in the domain $0 < x < x_1$. Similarly

$$[x_1, x_2] = e^{i\int_{x_1}^{x_2} Q^{1/2}dx}. \tag{6.19}$$

But $\int_{x_1}^{x_2} Q^{1/2}(E)dx = \pi/2 + \epsilon$, where ϵ is due to the change in the energy caused by the tunneling, with $\epsilon \ll 1$. Thus write

$$\int_{x_1}^{x_2} Q^{1/2}(E)dx - \int_{x_1}^{x_2} Q^{1/2}(E_0)dx = \epsilon. \tag{6.20}$$

Expanding $Q(E)$ around E_0 and keeping only the lowest order term we find

$$BdE\int_0^{x_1} \frac{dx}{(x_1^2 - x^2)^{1/2}} = \epsilon, \tag{6.21}$$

giving $dE = 2\epsilon/(\pi B)$.

Now find how ϵ is related to the parameter of the potential by calculating the eigenfunction at $x = 0$. For an odd eigenfunction

$$0 = -i[x_1, x_2]e^{-A} + ([x_1, x_2] + [x_2, x_1])e^A. \tag{6.22}$$

Use first order perturbation theory to write $[x_1, x_2] = e^{i(\pi/2+\epsilon)} \simeq i(1 + i\epsilon)$, giving $[x_1, x_2] + [x_2, x_1] = -2\epsilon$. Thus we find that $\epsilon = e^{-2A}/2$ and thus for an odd eigenfunction

$$dE = \frac{e^{-2A}}{\pi B}. \tag{6.23}$$

For an even eigenfunction $\psi'(0) = 0$, giving

$$0 = -i[x_1, x_2](-i\sqrt{Q})e^{-A} + ([x_1, x_2] + [x_2, x_1])(i\sqrt{Q})e^A. \tag{6.24}$$

Thus we find that $\epsilon = -e^{-2A}/2$ and so for an even eigenfunction

$$dE = -\frac{e^{-2A}}{\pi B}. \tag{6.25}$$

A determination of the energy of the system necessarily results in an eigenstate of definite parity, with the even parity state of lower energy than the odd parity state, but with the energy difference exponentially small due to the tunneling factor. The cause of this energy split is easily seen in Fig. 6.5. To connect the two solutions near $x = 0$ requires a small decrease in the second derivative in the case of odd parity, and a small increase in the case of even parity.

6.4 Problems

1. Find the eigenvalues of the matrix

$$M = \begin{Bmatrix} 1 & \epsilon & \epsilon \\ \epsilon & 2 & \epsilon \\ \epsilon & \epsilon & 3 \end{Bmatrix}$$

using a Kruskal–Newton graph, and to second order using perturbation theory.

2. Compare the methods of Kruskal–Newton graph iteration and perturbation theory expansion for the roots of the following equations for $\epsilon = 0.1$. Use two iterations, and second order in ϵ.

(a) $x^2 + \epsilon x + 6\epsilon = 0$
(b) $x^3 - \epsilon x - 2 = 0$
(c) $x^4 + \epsilon^2 x^2 - \epsilon = 0$
(d) $\epsilon x^3 + x^2 - 2x + 2 = 0$
(e) $\epsilon^2 x^8 - \epsilon x^6 + x - 3 = 0$

3. Find the first three terms of $y(x)$ expressed as a power series in ϵ with $y = x - \epsilon sin(y)$. Write also the third iteration of the Kruskal–Newton expansion.

4. Consider

$$y'' + y = \frac{sin(2x)}{4 + y^2 + cos^2 x}$$

in the range $0 < x < 1$ with boundary conditions $y(0) = 0$, $y'(0) = 1$. Since the right-hand side is always smaller than $1/4$, treat it perturbatively, by adding a factor of ϵ. Solve the equation to first order in ϵ and compare the result to a numerical integration.

5. Find perturbative solutions to

$$x sin(x) = 1$$

for large x.

6. Find the first three terms to a perturbative solution to

$$y = x - \epsilon sin(3y).$$

7. Find the first three terms to a perturbative solution to

$$\frac{d^2}{dx^2}y = \epsilon y sin(x)$$

with $y(0) = 1$, $y'(0) = 0$.

Chapter 7

Asymptotic Evaluation of Integrals

Pierre-Simon Laplace was born in Normandy, France in 1749. He was expected to follow a career in the church, but during his two years at the University of Caen, Laplace discovered his mathematical talents. He then left for Paris, and was helped by d'Alembert to find a position there. He became professor of mathematics at the École Militaire in Paris in 1768. In 1784 he was appointed examiner at the Royal Artillery Corps, and the next year examined and passed the 16-year-old Napoleon Bonaparte. He made significant contributions to celestial mechanics, probability theory, and the solution of differential equations. His five volume *Celestial Mechanics* changed the geometrical study of classical mechanics to one based on calculus. Both he and Newton developed proofs that the law of gravitational attraction applied between uniform spherical bodies, not only mass points. He also invented the concept of a gravitational potential to replace the idea of instantaneous action at a distance. He found a means of evaluating the asymptotic value of an integral. Laplace served on a committee which created the metric system and is responsible for the terms meter, centimeter and millimeter. He restated the nebular hypothesis for the origin of the solar system and was one of the first scientists to postulate the existence of black holes and gravitational collapse. He pioneered the Laplace transform, which apppears in many branches of mathematics and physics. He was refered to as the French Newton. In *A Philosophical Essay on Probabilities* he stated causal determinism, i.e. that a great intellect with information regarding the positions and velocities of all objects at any one time could predict the entire future of the universe.

7.1 Introduction

Having found an integral solution to a differential equation, using methods to be discussed in Chapter 9, it is often necessary to evaluate asymptotic limits, either to determine arbitrary constants in the integral representation by matching to desired boundary conditions, or in order to compare different limits of the same solution, for example, the behavior at $z = 0$ and $z = \infty$. If the variable z appears in the integration limits, the problem is easily solved. For example, to find the limit for small z of the integral $I = \int_z^\infty f(t)dt$ we write $I = \int_0^\infty f(t)dt - \int_0^z f(t)dt$. Then the integrand in the second integral can be approximated for small argument, giving for example a series in t allowing direct integration leading to a series in z. To evaluate $I(z)$ for large z, the same method can be employed to the first integral form if $f(t)$ can be expanded in powers of $1/t$. If instead $I(t)$ decreases faster than any power, say $f(t) = e^{-t^3}$, then integration by parts can be used. Write

$$I = -\int_z^\infty \frac{1}{3t^2}\frac{d}{dt}e^{-t^3}dt, \tag{7.1}$$

and integrate by parts, giving

$$I = \frac{1}{3z^2}e^{-z^3} + \int_z^\infty \frac{2}{3t^3}e^{-t^3}dt. \tag{7.2}$$

This process can be continued to give a series in inverse powers of z.

It may occur that the integral is dominated by the contribution from a small range of the integration path as $z \to \infty$. This may happen at the ends of the integration path, or at some point in the interior of the path. An example of this is given by integrals of the form

$$y(z) = \int_C f(t)e^{\phi(t,z)}dt, \tag{7.3}$$

with C a given contour. Integrals of this type are generalizations of Laplace integrals, $\int_0^\infty e^{-xt}f(t)dt$. Only analytic properties of $\phi(t,z)$ in the variable t are relevant, and we will sometimes suppress the z dependence. If $\phi(t)$ is analytic in t any local maxima of the integrand has the form of a saddle point. To see this, write the Taylor expansion about a point t_0 where $\phi'(t_0) = 0$ and thus $\phi(t) \simeq \phi_0 + \phi_0''(t-t_0)^2/2$. Thus for $t - t_0 \sim \pm s/\sqrt{-\phi_0''}$ with s real, the magnitude of the function $e^{\phi(t)}$ is exponentially decreasing away from the point t_0 as e^{-s^2}, but in the orthogonal directions $t - t_0 \sim \pm s/\sqrt{\phi_0''}$ it is exponentially increasing as e^{s^2}, forming a saddle point at

$t = t_0$. The paths leading away from the saddle along which the integrand is exponentially decreasing are called lines of steepest descent. These results are easily generalized to higher order zeros.

Away from a saddle point a maximum of the integrand can occur only at an end point of the integration. If at an end point $t = a$ we have $\phi' \neq 0$, then the function is exponentially decreasing in the direction $\phi'(a)(t - a) < 0$. Depending on the integration path, this can make the integrand either a minimum or a maximum at the end point.

If the magnitude of ϕ', ϕ'' is large, this exponential behavior dominates over the variation of the function $f(t)$ near the point of interest (t_0 in the case of a saddle, or $t = a$ in the case of an end point). In particular we are interested in the case that ϕ', ϕ'' tend to infinity in the limit of large z, the simplest case being that in which the function ϕ is proportional to z, which is the Laplace form. In this case dominant contributions to this integral in the limit of large z can be due to contributions from end points or from one or more saddle points through which the integration contour can be made to pass.

7.2 End Point

Consider an integral of the form

$$y(z) = \int_a^b f(t)e^{\phi(t,z)}dt, \tag{7.4}$$

with a, b complex end points of the contour of integration. Evaluate $\phi'(a)$, $\phi'(b)$, or if these are zero, the first nonzero terms in the Taylor expansions. Now look for deformations of the integration contour to arrive at the end points such that the end point contribution is either small or large. If $\phi'(a)dt < 0$ for dt leading away from a along the contour, i.e. the value of ϕ at the end point is large, then there can be a significant contribution from this end point.

To find an asymptotic expression for the contribution from end point a to all orders we choose a path in the t plane defined by $\phi(t) = \phi(a) - u$ with u positive real, and change integration variables to u, with $u = 0$ corresponding to $t = a$. Writing $t = x + iy$ this gives two real equations determining $x(u)$, $y(u)$ which describe the path of steepest descent in the complex t plane. It is easy to see that this path either approaches infinity or approaches a singularity of ϕ as $u \to \infty$.

It is not always easy to carry out this change of variables. If this proves difficult, first obtain the leading order behavior, by approximating ϕ using a Taylor expansion. Choose the path with $dt \sim -1/\phi'(a)$ and write $\phi(t) \simeq \phi(a) + \phi'(a)(t-a)$, valid for some small domain of magnitude ϵ in t. The integrand then becomes exponentially small outside this ϵ domain for large $\phi'(a)$. A change of variable gives for the contribution to the integral near the end point $y(z) \simeq -e^{\phi(a,z)} \int_0^c e^{-u}(f(t)/\phi'(t,z))du$ with $u = -\phi'(a,z)(t-a)$ and $c = \epsilon\phi'(a,z) \gg 1$ giving in the case that $f(t)$ and $\phi'(t,z)$ are nonzero and finite at $t = a$

$$y_a(z) \simeq -f(a)\frac{e^{\phi(a,z)}}{\phi'(a,z)}. \tag{7.5}$$

If instead $f(t)$ is either zero or possesses a singularity at $t = a$ then the leading order expansion of $f(t)$ near $t = a$ must be used in the integrand. Similarly if $\phi'(t,z)$ is either zero or possesses a singularity at $t = a$ the leading order expansion of ϕ' must be used. In any case the integral gives a Gamma function of a z dependent argument. If z is taken so large that $c = \epsilon\phi'(a,z) \gg 1$ and at the same time $\epsilon \ll 1$, the Taylor expansion is valid over the entire (asymptotically infinitesimal in t, asymptotically infinite in u) range of integration and this expression gives the dominant end point contribution. Higher order corrections can be obtained by using the full integration path $\phi(t) = \phi(a) - u$, expanding the function $f(t)$ in the neighborhood of the point a, and expressing all functions of t in terms of u, the resulting integrals giving Gamma functions.

If the function $f(t)$ has only an asymptotic expansion about the point a, one can still use this expansion to integrate term by term, the result being an asymptotic expansion for $y(z)$. For example, if

$$f(t) \simeq \sum_0^\infty a_n t^{\beta n+\alpha} \tag{7.6}$$

for small t, and $\phi(t,z)$ has the simple form $-zt$, then the evaluation takes the form

$$\int_0^b dt f(t)e^{-zt} \simeq \sum_0^\infty \frac{a_n\Gamma(\alpha+\beta n+1)}{z^{\alpha+\beta n+1}}, \tag{7.7}$$

for large z. This is known as Watson's lemma.

To complete the determination of the asymptotic value in the general case, evaluate $y_b(z)$, connect the integration path leading from the end point to the rest of the integration contour, and add the contributions from saddle and end points. We give here a few examples, and others will be encountered in the course of the following chapters.

Consider the integral

$$g(z) = \int_0^1 dt e^{it^3 z}, \tag{7.8}$$

and examine the limit $z \to \infty$. There are no saddle points away from the end points, only the third order zero at $t = 0$. Consider the paths of steepest descent emanating from an end point $t = c$ of the function $e^{\phi(t)}$,

$$\int_c^t dt\phi' = \phi(t) - \phi(c) = -u, \tag{7.9}$$

with u a positive real. From the point $t = 0$ we find the equation $it^3 z = -u$, giving the straight line $t = (u/z)^{1/3} e^{i\pi/6}$. (There are two other paths, but they do not point in directions of interest for the completion of the integral.) From the point $t = 1$ we find the curve $it^3 - i = -u/z$, which for large t is asymptotic to the line $t = (u/z)^{1/3} e^{i\pi/6}$, so there is a continuous path from $t = 0$ to $t = 1$ connecting the two end points as shown in Fig. 7.1.

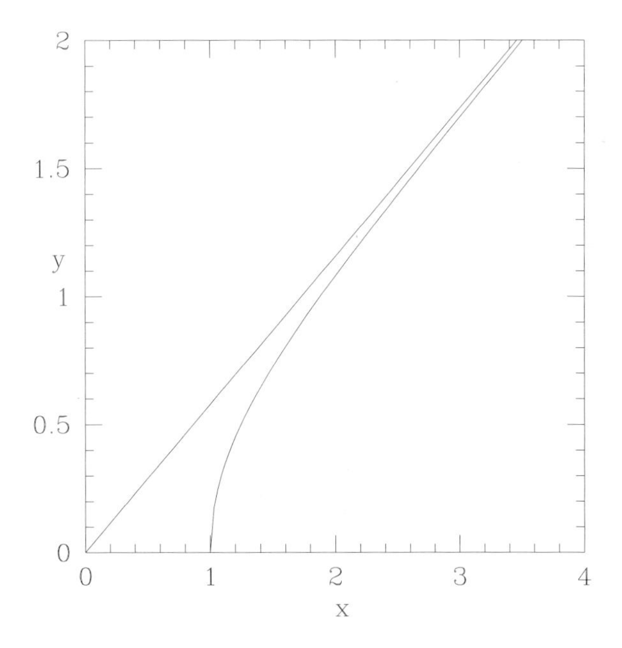

Fig. 7.1 Integration path for Eq. 7.8.

Furthermore for large positive z both exponentials have widths approaching zero so the integration ranges can be arbitrarily small in t. Thus the contribution from the range of the integral far from 0 and 1 is not significant; what is essential is that the path exists connecting these points over which the integrand is vanishingly small compared to the end value contributions. For the point $t = 0$, letting $u = -it^3 z$, positive real, with the integration path determining the sign of the cube root $dt \sim e^{i\pi/6}$ we have

$$g_0 \simeq z^{-1/3} e^{i\pi/6} \int_0^\infty du e^{-u} \frac{1}{3u^{2/3}}, \tag{7.10}$$

giving $g_0 \simeq z^{-1/3} e^{i\pi/6} \Gamma(1/3)/3$. At the second end point we set $u/z = i - it^3$ or $t = (1 + iu/z)^{1/3}$ giving

$$g_1 \simeq \frac{i}{3} e^{iz} \int_0^\infty du e^{-u} (1 + iu/z)^{-2/3}. \tag{7.11}$$

Expanding $(1 + iu/z)^{-2/3}$ using the binomial expansion we find

$$g_1 \simeq \frac{1}{\Gamma(-1/3)} e^{iz} \sum_0^\infty \Gamma(n + 2/3)(-iz)^{-n-1}, \tag{7.12}$$

and finally for $z \to \infty$ in S_Δ

$$g(z) \simeq z^{-1/3} e^{i\pi/6} - \frac{e^{iz}}{\Gamma(-1/3)} \sum_0^\infty \Gamma(n + 2/3)(iz)^{-n-1}, \tag{7.13}$$

with S_Δ given by $|arg(z)| < \pi/2 - \Delta$ with $\Delta > 0$. Clearly the first term dominates over the contribution of the sum for large z, which, however, is a divergent asymptotic expansion and must be truncated appropriately if it is to be kept.

If it proves impossible to carry out the change of variables to follow the path of steepest descent to all orders, find only the leading asymptotic behavior $y \sim e^S$ using a Taylor expansion. Then find a differential equation for $W(z)$ using $y \sim e^{S(z)} W(z)$, and finally an asymptotic series for $W(z)$.

7.3 Saddle Point

Consider first a simple integral of the form

$$y(z) = \int_C e^{\phi(t,z)} dt, \tag{7.14}$$

with C a given contour. Suppose there exists a saddle point at t_0 where $\phi'(t_0) = 0$. If the integration path can be made to pass through the saddle along the direction of exponential decrease away from t_0, the dominant contribution to the integral along this path comes from the vicinity of the saddle point. Continue to follow the path emanating from t_0 defined by $\phi'(t)dt$ = *negative real* in both directions. Unless another saddle point is encountered the integrand must continue to decrease in magnitude. In the simplest case, this path can be made to coincide with the integration contour. In more complex cases, locate all saddle points of the function $e^{\phi(t,z)}$ and draw the associated lines of steepest descent. Then attempt to deform the contour of integration so as to pass along these lines through whatever combination of saddle points is required in order to reach the end points of the contour.

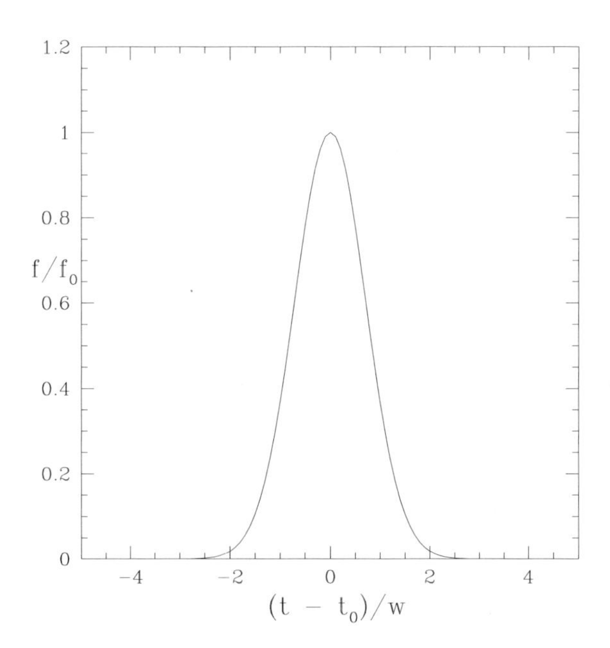

Fig. 7.2 Integrand for a saddle point of width w.

Now evaluate the integral along this path. In the vicinity of a saddle point, along the path $t - t_0 \sim \pm 1/\sqrt{-\phi_0''}$, except for the overall phase e^{ϕ_0}, the integrand has the form of a Gaussian, with half width $w = \sqrt{2}/|\sqrt{\phi_0''}|$ as shown in Fig. 7.2. Integrate over the range $-\epsilon < t < \epsilon$ with $w \ll \epsilon \ll 1$.

Use the symmetry of the integrand about t_0 and change integration variable to $u = -\phi_0''(t-t_0)^2/2$, with u positive real. The contribution to the integral in the vicinity of the saddle reduces to

$$y_s \sim 2e^{\phi_0(z)} \int_0^c du \frac{e^{-u}}{\sqrt{-2u\phi_0''(z)}}, \tag{7.15}$$

with the sign of the square root chosen to agree with the sense of the integration contour through the saddle (i.e. the differential $du = (-2u\phi_0'')^{1/2}dt$ must be positive real, and the phase of dt is fixed by the contour) and $c = \epsilon^2\phi_0''(z) \gg 1$. Contributions to the integral for $u > c$ are exponentially small in w. If z is taken so large that $c \gg 1$ and at the same time $\epsilon \ll 1$, the expansions are valid over the entire (asymptotically infinitesimal in t, asymptotically infinite in u) range of integration and this expression gives the dominant saddle point contribution, with corrections being exponentially small, giving $\Gamma(1/2)$ and a leading order contribution from the saddle of

$$y_s(z) = \frac{\sqrt{2\pi}}{\sqrt{-\phi_0''(z)}} e^{\phi_0(z)}. \tag{7.16}$$

This method of obtaining the asymptotic limit of the integral using the path of steepest descent is due to Riemann. It is clearly an adaptation of the method of Laplace. Note that the orientation of each saddle point is a function of the variable z, so that the whole analysis depends on the limiting value of z sought.

In general one finds the slightly more complicated case

$$y(z) = \int_C e^{\phi(t,z)} f(t) dt. \tag{7.17}$$

The treatment is essentially the same as above, only now one must also expand the function $f(t)$ about the saddle point. The same substitution then leads to a series of Gamma functions with different arguments. If the function $f(t)$ has only an asymptotic expansion about the point t_0, one can still use this expansion to integrate term by term, the result being an asymptotic expansion for $y(z)$.

Consider $Z(x)$ for $x \to +\infty$, with

$$Z(x) = \int_0^\infty e^{-(t+x/t)} dt, \tag{7.18}$$

so $\phi = -(t + x/t)$. We find that saddle points exist at $t = \pm\sqrt{x}$, with $\phi'' = -2x/t^3$. The saddle at $t = -\sqrt{x}$ is vertical and of no use. The saddle at $t = \sqrt{x}$ has a horizontal orientation and the integration path passes directly through it. At this point $\phi'' = -2/x^{1/2}$, producing a local Gaussian with half width $w = x^{1/4}$. At the saddle point $\phi_0 = -2\sqrt{x}$. For large x the ratio of the Gaussian width to its location shrinks, so that even though the width is increasing, the scale of the function, defined by the saddle location, increases more rapidly, and the usual analysis holds. Using Eq. 7.16 we find for the leading order behavior

$$Z(x) \simeq \sqrt{\pi}x^{1/4}e^{-2\sqrt{x}}. \tag{7.19}$$

If a moving saddle point is not desirable, a simple transformation of variables $s = t/\sqrt{x}$ moves the saddle to $s = 1$, and changes the integral into the form

$$Z(x) = \int_0^\infty \sqrt{x}e^{-\sqrt{x}(s+1/s)}ds. \tag{7.20}$$

The saddle is now at $s = 1$, with $\phi' = -\sqrt{x}(1 - 1/s^2)$ and $\phi'' = -2\sqrt{x}/s^3$. At the saddle point $\phi_0 = -2\sqrt{x}$. Using Eq. 7.16 we find again the same result.

This function has no other saddle points and it is easy to see that it decreases in magnitude all the way from the saddle to the end points, and thus there can be no contribution from them.

Consider $G(x)$ for $x \to +\infty$, with

$$G(x) = \int_{-\infty}^\infty e^{x(it-t^2/2)}t^2dt. \tag{7.21}$$

Writing $\phi = x(it - t^2/2)$ we find that a saddle point exists at $t = i$, with $\phi'' = -x$ and $\phi_0 = -x/2$. For x positive the direction of decrease is $dt = real$, so the contour can be deformed to pass through the point $t = i$. Using Eq. 7.16 we find

$$G(x) \simeq -\sqrt{2\pi}x^{-1/2}e^{-x/2}. \tag{7.22}$$

Figure 7.3 shows are contours of equal magnitude of the exponential in the integrand, decreasing horizontally away from $t = i$ and increasing vertically. Shown also is the deformed integration contour and the saddle point.

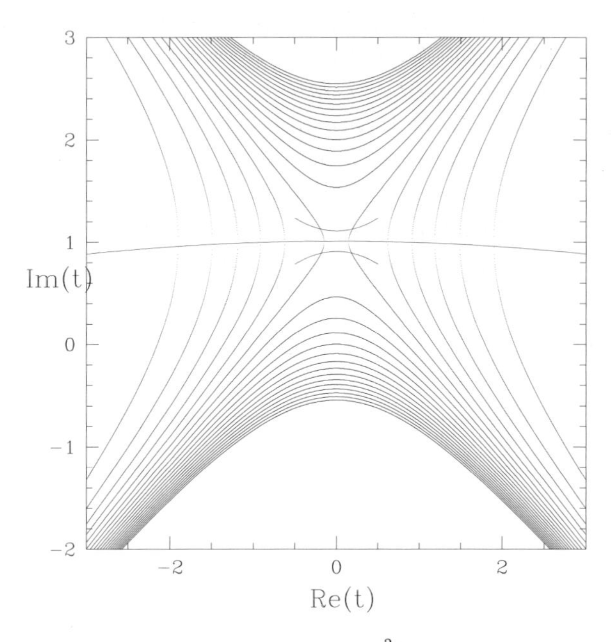

Fig. 7.3 Contours of equal magnitude of $e^{ixt-xt^2/2}$, showing the integration contour and the saddle.

7.4 Finding an Integration Path

For a given integral it is necessary to discover an integration path leading from one end point to the other along which the integrand is small everywhere except at end points and saddle points, allowing its asymptotic evaluation. Consider the integral

$$I(z) = \int_0^1 dte^{-4zt^2} cos(5zt - zt^3) \tag{7.23}$$

where the limit for large z is desired. First make use of the fact that the cosine is an even function of its argument to write

$$I(z) = \int_{-1}^1 dte^{\phi} \tag{7.24}$$

with $\phi = z(-4t^2 + 5it - it^3)$. Find that $\phi' = 0$ at $t = i,\ 5i/3$, with $\phi'' = z(-8 - 6it)$, so $\phi''(i) = -2z$ and $\phi''(5i/3) = 2z$. The orientation of the saddles depends on the phase of z, and we will first consider z to be real and positive. Then the saddle at $t = i$ is horizontal, and that at

$t = 5i/3$ is vertical. Next look at the behavior of the integrand at the end points. At $t = -1$ we have $\phi' = z(8+2i)$ and we require $\phi' dt$ to be negative real giving the line of steepest descent from $t = -1$ to be in the direction $-4+i$. Similarly the path of steepest descent from $t = 1$ gives the direction $4 + i$. Now look at where in the t plane the integrand is asymptotically small, i.e. where $\phi \to -\infty$. For large t we have $\phi \sim -izt^3$ and thus the integrand is small for large t with phase $-\pi/6 + 2n\pi/3$, $n = 0, 1, 2$. Now follow the path of steepest descent starting at $t = -1$. It is important to

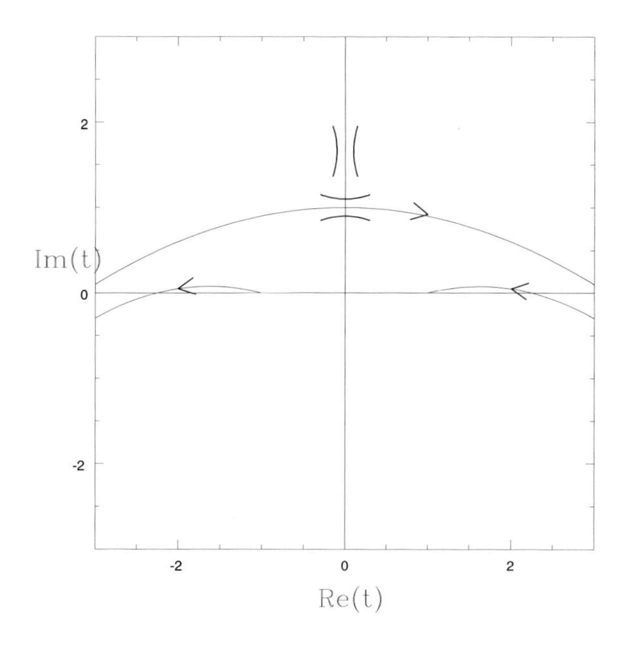

Fig. 7.4 Integration path for Eq. 7.24.

realize that the saddle locations give a global idea of where the function is large. It is easy to see that the path of steepest descent starting at $t = -1$ must asymptotically approach infinity along the phase $7\pi/6$ ($= -5\pi/6$). Similarly the path of steepest descent starting at the saddle at $t = i$ and going to the left must end up along the same asymptote. The path starting at the end point $t = 1$ and that beginning at the saddle at $t = i$ both end up along the asymptote with phase $-\pi/6$. Thus the path shown in Fig. 7.4 connects the two end points entirely along paths of steepest descent from the left end point, through the saddle, to the right end point. The saddle at $t = 5i/3$ is useless. Crossing it leads to the asymptote with phase $\pi/2$ and there is no way to return.

Now evaluate the integral. The left end point gives for large x

$$I_{-1}(z) \simeq \frac{e^{-4z(1+i)}}{4z(4+i)}. \tag{7.25}$$

The right end point gives

$$I_1(z) \simeq \frac{e^{-4z(1-i)}}{4z(4-i)} \tag{7.26}$$

and the saddle point

$$I_s(z) \simeq \frac{\sqrt{\pi}e^{-2z}}{2\sqrt{z}} \tag{7.27}$$

so

$$I(z) \simeq \frac{e^{-4z(1+i)}}{4z(4+i)} + \frac{e^{-4z(1-i)}}{4z(4-i)} + \frac{\sqrt{\pi}e^{-2z}}{2\sqrt{z}} \tag{7.28}$$

and the first two terms are complex conjugates of each other, so the result is real as it must be.

This result is analytic in z, and we can examine what happens for large z of different phase, and see how the result exhibits Stokes phenomena. For positive real z the last term dominates over the first two. Let $z = r(cos\theta + isin\theta)$ giving magnitudes of

$$I_1 \simeq \frac{e^{-4r(cos\theta - sin\theta)}}{4\sqrt{17}r}, I_2 \simeq \frac{e^{-4r(cos\theta + sin\theta)}}{4\sqrt{17}r}, I_s \simeq \frac{\sqrt{\pi}e^{-2rcos\theta}}{2\sqrt{r}}. \tag{7.29}$$

The function is subdominant for $-\pi/4 < \theta < \pi/4$ where I_s is larger than either I_1 or I_{-1}. In the interval $\pi/4 < \theta < \pi$ the behavior is dominant given by I_1, and in the interval $-\pi < \theta < -\pi/4$ the behavior is dominant given by I_2.

Of course as the phase of z changes, the orientation of the steepest descent paths and the saddle shift, and it is interesting to see how the path is deformed, but the general topology of Fig. 7.4 is preserved.

7.5 Problems

1. Find the first two terms in the asymptotic expansion of the following integrals as $x \to +\infty$.

(a)

$$\int_x^\infty \frac{dt}{1-e^t}.$$

(b)

$$\int_x^\infty e^{-t^4} dt.$$

(c)

$$\int_0^1 ln(1+t)e^{ixsin^2 t} dt.$$

(d)

$$\int_0^{\pi/4} dtcos(xt^2)sin^2 t.$$

2. Find the leading behavior of

$$\int_{1/2}^1 \frac{e^{i\alpha t^2}}{1+t} dt,$$

for α large and positive.

3. Find the leading behavior of

$$\int_0^{\pi^2/2} ds \int_0^{\pi^2/2} dte^{\alpha cos(\sqrt{s+t})},$$

for $\alpha \to \infty$.

4. Evaluate $I(z)$ for $z \to \infty$.

$$I(z) = \int_0^1 e^{-zt^2} cos(zt + zt^3) dt.$$

5. Find the leading behavior of

$$\int_0^{\pi/4} \sqrt{tan(3t)}e^{-at^2}dt, \qquad a \to +\infty.$$

6. Find the leading behavior of

$$\int_0^{\pi/2} \sqrt{t}e^{-asin^4 t}dt, \qquad a \to +\infty.$$

7. Find the leading behavior of

$$\int_0^1 [t(1-t)]^{3/2}(t+a)^{-x}dt,$$

for $x \to \infty$ with $a > 0$.

8. Find the leading behavior of

$$\int_1^\infty dt \frac{cos^3(xt)}{t}, \qquad x \to +0.$$

9. Find the first two terms in the asymptotic expansion of the following integral as $x \to +\infty$.

$$\int_0^{\pi/4} dtcos(2xt^2)tan^2 t.$$

10. Find the first leading order in the asymptotic expansion of the following integral as $x \to +\infty$.

$$\int_0^1 dte^{-4t^2} \frac{cos(4xt - xt^3)}{(1-t^2)^{1/2}}.$$

11. Show that for $x \to +\infty$

$$\int_0^{\pi/2} tsin(xcos\, t)dt \simeq \frac{\pi/2 - cos\; x}{x}.$$

12. Show that for $x \to +\infty$

$$\int_1^\infty (1-e^{1-t})e^{ixt(1-ln\, t)}dt \simeq -(i/x)e^{ix}.$$

13. Find the leading order asymptotic value for $x \to +\infty$.

$$\int_0^{\infty} \frac{e^{x\cos(t)}}{\sqrt{t}}\, dt.$$

Chapter 8

The Euler Gamma Function

Leonhard Euler was born in 1707 in Basel, Switzerland, where he attended University and became friends with the Bernoulli family. His father was a Calvinist pastor and wanted him to study theology, but the Bernoullis persuaded him that his son had a special talent for mathematics and he was allowed to change his studies. Daniel Bernoulli found a position for him at the St. Petersburg Academy in Russia. In 1741 Frederick the Great of Prussia invited him to the Berlin Academy. His first book, on the calculus of variations, was written while he was in Berlin. In 1766 Catherine the Great invited him back to Russia. He continued to do mathematics in his old age despite becoming totally blind, and a fantastic memory allowed him to keep and manipulate mathematical and physics formulas in his head, and to dictate articles at such a rate that it took the St. Petersburg Academy fifty years after his death to publish them all. He developed an algorithmic iterative procedure for calculating the phases of the moon far into the future with high accuracy, used by the British Admiralty for navigation. This was a perturbative calculation of the effect of the sun on the earth–moon system. One hundred years after the affirmation by Fermat that $x^n + y^n = z^n$ has no integer solutions for $n > 2$, Euler was able to make the first step toward a proof by showing this correct for $n = 3$. In order to extend his proof to the case $n = 4$ he had to introduce imaginary numbers. He standardized the notation for complex numbers by introducing $i = \sqrt{-1}$, and his most famous formula is $e^{i\theta} = cos(\theta) + isin(\theta)$ along with the equally famous $e^{i\pi} = -1$. He also gave the proof that e is irrational, and the choice of the symbol for this number is in his honor. The notation $n!$ for the denominators in the series for e is due to him. He proved many important theorems in number theory. The Euler ϕ function $\phi(N)$ is defined to be the number of positive integers less than N relatively prime to N. This function is of great use in cryptography for creating and breaking codes, and is used to make the public key codes, discussed in Chapter 17.

Notice that although the Gamma function appears repeatedly in expressions for the asymptotic analysis of differential equations, we have never written a differential equation for Gamma. This is because it is a transcendental function, satisfying no differential equation with rational coefficients, just as the transcendental numbers e, π, γ, etc. do not satisfy any algebraic equation. Because of its occurrence in various forms in the asymptotic expressions for solutions to differential equations there are a number of identities concerning Gamma functions which are indispensable for such studies.

8.1 Introduction

The Euler Gamma function is defined through

$$\Gamma(z) = \int_0^\infty dt e^{-t} t^{z-1}, \tag{8.1}$$

for $Rez > 0$, from which it follows that

$$z\Gamma(z) = \Gamma(z+1). \tag{8.2}$$

This relation can be used to analytically continue $\Gamma(z)$ to values of z with $Rez < 0$.

We also find that $\Gamma(1) = 1$, and at the positive integers $\Gamma(n) = (n-1)!$. Now calculate $\Gamma(1/2)$. Write

$$\Gamma^2(\tfrac{1}{2}) = \int_0^\infty \int_0^\infty e^{-(t+s)} \frac{1}{\sqrt{ts}} dt ds. \tag{8.3}$$

Change variables to polar coordinates, using $\sqrt{t} = rcos\theta$, $\sqrt{s} = rsin\theta$, giving

$$\Gamma^2(\tfrac{1}{2}) = 4 \int_0^\infty \int_0^{\pi/2} e^{-r^2} r dr d\theta, \tag{8.4}$$

and finally $\Gamma(1/2) = \sqrt{\pi}$.

Further, using Eq. 8.2 in the form $\Gamma(z) = \Gamma(z+1)/z$ and taking the limit $z \to 0$ we find that $\Gamma(z)$ has a simple pole at $z = 0$. In the same manner we find the behavior of $\Gamma(z)$ at any negative integer, and in fact

$$\Gamma(z-n) \simeq \frac{(-1)^n}{n!z}, \tag{8.5}$$

for $z \to 0$.

Figure 8.1 shows a plot of the Gamma function for real x.

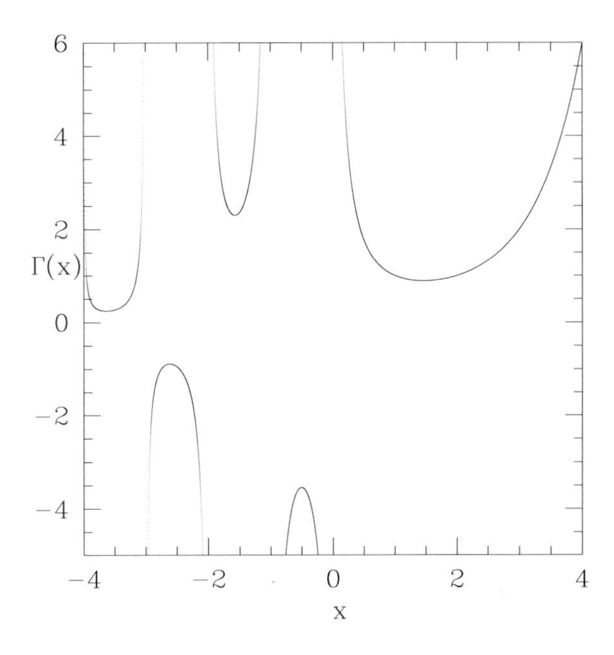

Fig. 8.1 Plot of $\Gamma(x)$.

8.2 The Stirling Approximation

Now we find an asymptotic approximation for the Gamma function for large positive x. Starting with Eq. 8.1 for $z = x + 1$, real,

$$\Gamma(x+1) = x\Gamma(x) = \int_0^\infty e^{-t} t^x dt, \tag{8.6}$$

change variables using $t = sx$, giving

$$\Gamma(x) = x^x \int_0^\infty e^\phi ds, \tag{8.7}$$

with $\phi = x(lns - s)$. Now look for saddle points of the integrand in the s plane. We have $\phi' = x(1/s - 1)$, and the only zero occurs at $s = 1$ with $\phi'' = -x$, and $\phi^{(n)} = (n-1)!(-1)^{n-1}x$. Then write

$$\int e^{\phi} ds \simeq e^{\phi_0} \int ds e^{-x(s-1)^2/2} e^{x[(s-1)^3/3-(s-1)^4/4+\ldots]}. \tag{8.8}$$

Orders of $s-1$ higher than $(s-1)^2$ cannot be retained in the exponent, since we know that for s real ϕ has a single maximum at $s = 1$. Write $t = s - 1$, and $a = x[t^3/3 - t^4/4 + t^5/5 - \ldots]$ giving

$$\int e^{\phi} ds \simeq e^{\phi_0} \int dt e^{-xt^2/2} \sum \frac{a^n}{n!}, \tag{8.9}$$

this expression to be integrated term by term. To convert the integral to Gamma functions we substitute $u = xt^2/2$ giving

$$\int_{-\infty}^{\infty} e^{\phi} ds \simeq 2e^{\phi_0} \int_0^{\infty} \frac{du}{\sqrt{2ux}} e^{-u} \sum \frac{a^n}{n!}, \tag{8.10}$$

with, however, the understanding that in converting the integral from $-\infty, \infty$ to $0, \infty$ we must drop all odd powers of t in the sum. Now look order by order in x, using $t = (2u/x)^{1/2}$. Terms of order $1/x$ are given by xt^4 and x^2t^6, and terms of order $1/x^2$ are given by xt^6, x^2t^8, x^3t^{10}, and x^4t^{12}. Keeping terms up to a^4 and dropping all terms odd in t we find

$$e^a \simeq 1 + \frac{1}{x}\left[-u^2 + \frac{4}{9}u^3\right] + \frac{1}{x^2}\left[-\frac{4}{3}u^3 + \frac{47}{30}u^4 - \frac{4}{9}u^5 + \frac{8}{243}u^6\right]. \tag{8.11}$$

Combining we find

$$\Gamma(x) \simeq x^x e^{-x}\sqrt{\frac{2}{x}}\left[\Gamma(\tfrac{1}{2}) + \frac{A}{x} + \frac{B}{x^2}\right] \tag{8.12}$$

with

$$A = -\Gamma(\tfrac{5}{2}) + \frac{4}{9}\Gamma(\tfrac{7}{2}), \tag{8.13}$$

$$B = -\frac{4}{3}\Gamma(\tfrac{7}{2}) + \frac{47}{30}\Gamma(\tfrac{9}{2}) - \frac{4}{9}\Gamma(\tfrac{11}{2}) + \frac{8}{243}\Gamma(\tfrac{13}{2}). \tag{8.14}$$

Using Eq. 8.2 repeatedly we find

$$\Gamma(x) \simeq x^x e^{-x}\sqrt{\frac{2\pi}{x}}\left[1 + \frac{1}{12x} + \frac{1}{288x^2} + \ldots\right], \tag{8.15}$$

and in spite of how rapidly these terms appear to be decreasing, this is a divergent asymptotic series, as can be seen from the appearance of terms of the form $\Gamma(n + 1/2)$ in the coefficients of the inverse powers of x, which eventually become very large. The complete series will be derived in section 17.6.

8.3 The Euler–Mascheroni Constant

The Euler–Mascheroni constant is defined to be $\gamma = -\frac{d\Gamma}{dz}(1)$. Differentiate Eq. 8.2 to find

$$\Gamma'(z) = -\Gamma(z)/z + \Gamma'(z+1)/z. \tag{8.16}$$

Now iterate this equation:

$$\Gamma'(z) = -\frac{\Gamma(z)}{z} - \frac{\Gamma(z)}{z+1} + \frac{\Gamma'(z+2)}{z(z+1)}, \tag{8.17}$$

giving after m iterations

$$\begin{aligned}\Gamma'(z) = -\Gamma(z)\left[\frac{1}{z} + \frac{1}{z+1} + \ldots \frac{1}{z+m-1}\right] \\ + \frac{\Gamma'(z+m)}{z(z+1)(z+2)\ldots(z+m-1)}.\end{aligned} \tag{8.18}$$

Take the limit of large m for fixed z, using the asymptotic form Eq. 8.15 for the last term, giving $\Gamma'(m) \simeq \Gamma(m)ln(m)$ and thus

$$\Gamma'(1) = \lim_{m\to\infty}\left[-1 - \frac{1}{2} - \frac{1}{3}\cdots - \frac{1}{m} + ln(m)\right], \tag{8.19}$$

giving a value $\gamma = -\Gamma'(1) = 0.5772157\ldots$.

To prove convergence of this expression let $u_n = \int_0^1 dt \frac{t}{n(n+t)} = \frac{1}{n} - ln(\frac{n+1}{n})$. Then $u_n < \int_0^1 dt/n^2 = 1/n^2$ and $lim(1 + 1/2 + 1/3 + \ldots 1/m - ln(m)) = lim(\sum_1^m 1/n + ln((m+1)/m) = \sum_1^\infty u_n$ which converges.

Whether γ is a transcendental number, or is in fact the root of a polynomial equation with integer coefficients is still unknown. See Davis ([1959]) for a history of the Gamma function.

8.4 Sine Product Identity

Now prove the identity

$$\frac{sin(\pi z)}{\pi z} = \prod_{1}^{\infty} \left(1 - \frac{z}{n}\right)\left(1 + \frac{z}{n}\right). \tag{8.20}$$

To prove this, let $f(z) = sin(\pi z)$, analytic with simple zeros at the integers $z = n$. Then $f'(z)/f(z)$ is analytic except for simple poles at the integers, and near an integer $z = n$ we have $f'/f \simeq 1/(z-n)$.

Evaluate the integral

$$I = \frac{1}{2\pi i} \int_{C_m} \frac{dz}{z-w} \frac{f'(z)}{f(z)}, \tag{8.21}$$

with the contour C_m given in Fig. 8.2. This method is due to Mittag-Lefler.

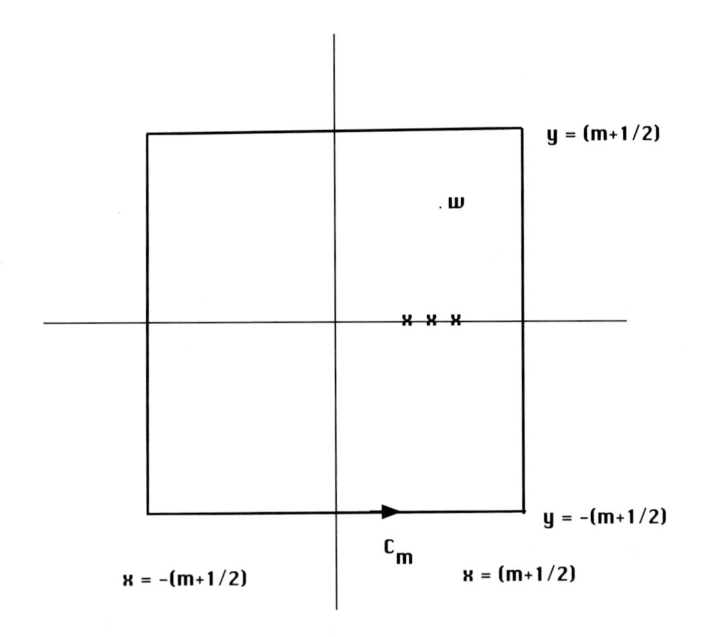

Fig. 8.2 Integration path for evaluation of I.

Note that $f'(z)/[f(z)(z-w)]$ is analytic everywhere except at $z = w$ and the integers $z = n$. Deform the contour into a sum of small circles around w and each integer n, giving

$$\frac{1}{2\pi i}\int_{C_m}\frac{dz}{z-w}\frac{f'(z)}{f(z)}=\frac{f'(w)}{f(w)}+\sum_{n=-m}^{m}\frac{1}{n-w}. \tag{8.22}$$

The left-hand side can be evaluated directly. Write $1/(z-w)=1/z+w/[z(z-w)]$. Then the left-hand side becomes

$$lhs=\frac{1}{2\pi i}\int_{C_m}\frac{dz}{z}\frac{f'(z)}{f(z)}+\frac{w}{2\pi i}\int_{C_m}\frac{dz}{z(z-w)}\frac{f'(z)}{f(z)}. \tag{8.23}$$

But near $z=0$

$$\frac{f'}{f}=\frac{\pi cos(\pi z)}{sin(\pi z)}\simeq\frac{\pi(1-\pi^2z^2/2)}{\pi z(1-\pi^2z^2/6)}\simeq\frac{1}{z}\left(1+O(z^2)\right), \tag{8.24}$$

and thus the contribution of the first term from the point $z=0$ is of the form $\oint dz/z^2(1+O(z^2))=0$, giving

$$\frac{1}{2\pi i}\int_{C_m}\frac{dz}{z}\frac{f'(z)}{f(z)}=\sum_{-m}^{m}\frac{1}{n}=0, \tag{8.25}$$

where the sum does not include $n=0$.

Now consider the second term in Eq. 8.23, bounded in magnitude by

$$\left|\int_{C_m}\frac{dz}{z(z-w)}\frac{f'(z)}{f(z)}\right|<\int_{C_m}\left|\frac{dz}{z(z-w)}\right|\left|\frac{f'(z)}{f(z)}\right|. \tag{8.26}$$

Evaluate f'/f along the four sides of the contour, with $z=x+iy$. Along the top and the bottom $e^{\pm\pi y}$ dominates and $|f'/f|\simeq\pi$. Along the right-hand edge $e^{i\pi x}=i(-1)^m$ giving $|f'/f|<\pi$, and similarly along the left-hand edge. The integration path is $8m$ long, and $|z|>m$ giving

$$\left|\int_{C_m}\frac{dz}{z(z-w)}\frac{f'(z)}{f(z)}\right|<\frac{8\pi}{|m-w|}\to 0, \qquad m\to\infty. \tag{8.27}$$

Thus for large m we have

$$0=\frac{f'(w)}{f(w)}+\sum_{-m}^{m}\frac{1}{n-w}, \tag{8.28}$$

or

$$\frac{f'(w)}{f(w)}=\frac{1}{w}+\sum_{1}^{m}\left[\frac{1}{w-n}+\frac{1}{w+n}\right]. \tag{8.29}$$

Integrate this equation, giving

$$ln\left(\frac{f(w)}{w}\right) = \sum_{1}^{m} ln((w-n)(w+n)) + K_m, \tag{8.30}$$

and the integration constant K_m can be evaluated by taking the limit $w \to 0$. Substituting the value for $f(w)$, combining K_m with the sum of log terms, and taking the limit $m \to \infty$ we find

$$\frac{sin(\pi w)}{\pi w} = \prod_{1}^{\infty}\left(1-\frac{w}{n}\right)\left(1+\frac{w}{n}\right). \tag{8.31}$$

8.5 Continuation of $\Gamma(z)$

A convergent integral representation provides a means of analytically continuing the Gamma function to all z. Consider the definition

$$\Gamma(z) = \frac{1}{2isin(\pi z)}\int_C dte^{-t}(-t)^{z-1}, \tag{8.32}$$

with the contour C defined as in Fig. 8.3.

If $Rez > 0$ we can move the contour to the real axis, giving

$$\Gamma(z) = \frac{1}{2isin(\pi z)}\int_0^{\infty} d\rho\left[e^{-i\pi(z-1)} - e^{i\pi(z-1)}\right]e^{-\rho}\rho^{z-1}, \tag{8.33}$$

which reduces to the usual definition of $\Gamma(z)$ for $Rez > 0$. But Eq. 8.32 is defined for all z, and thus provides an analytic continuation of Gamma.

8.6 Asymptotic $\Gamma(z)$

Now use the continuation in Eq. 8.32 to find $\Gamma(z)$ for large $|z|$. Write

$$z\Gamma(z) = \frac{i}{2sin(\pi z)}\int_C dte^{\phi}, \tag{8.34}$$

with $\phi = -t + zln(-t)$. Then ϕ has a saddle point at $t_0 = z$ with $\phi'' = -1/z$. For $Rez > 0$ deform the path of integration through this saddle point as shown in Fig. 8.4.

The dominant contribution from each side of the cut is

$$z\Gamma(z) \sim \frac{i2\pi}{2sin(\pi z)}\frac{e^{\phi_0}}{\sqrt{\pi}(-2\phi_0'')^{1/2}}, \tag{8.35}$$

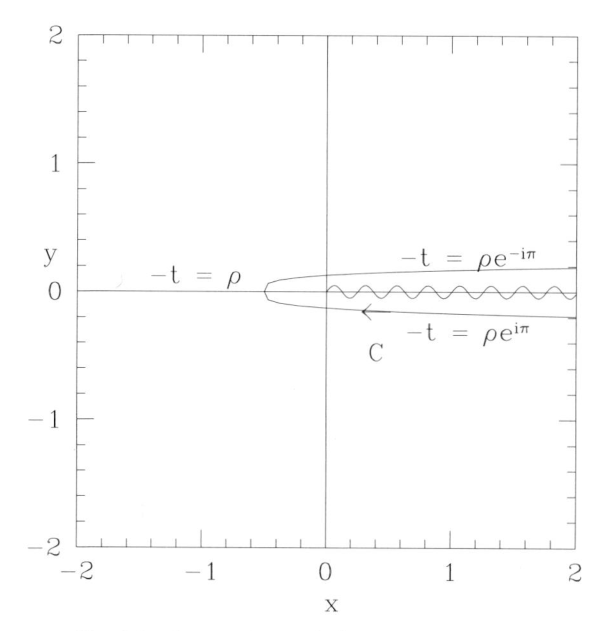

Fig. 8.3 Integration path for continuation of Γ.

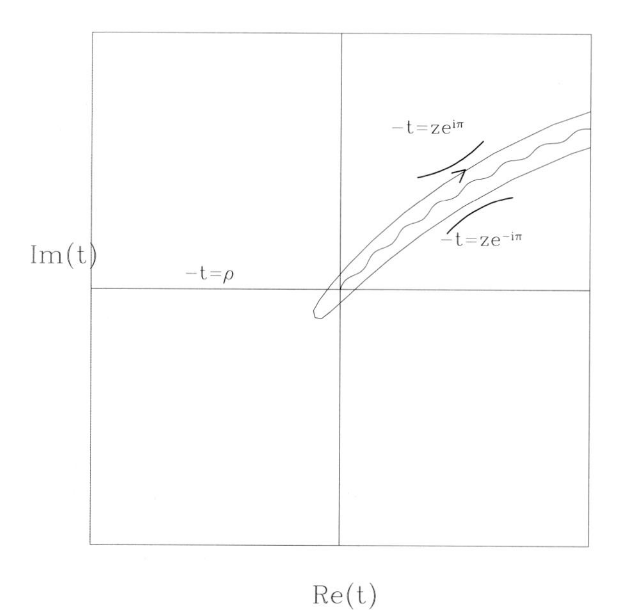

Fig. 8.4 Integration path for evaluation of Γ, $Rez \to +\infty$.

and adding contributions from above and below the cut we find $-2ie^{-z}z^z sin(\pi z)$ giving $\Gamma(z) \simeq \sqrt{2\pi/z}e^{-z}z^z$, which is the first term of the Stirling approximation. But note that for $Imz >> 1$ the upper contribution dominates, the lower being subdominant, and thus the part of the integration path below the cut can be deformed to directly go to $+\infty$ without passing through the saddle.

For $Rez < 0$ the position of the saddle point is shown in Fig. 8.5, giving

$$z\Gamma(z) \sim \frac{i2\pi}{2sin(\pi z)} \frac{e^{\phi_0}}{\sqrt{\pi}(-2\phi_0'')^{1/2}}, \tag{8.36}$$

which reduces to

$$z\Gamma(z) \sim \frac{\sqrt{\pi}e^{-z}(-z)^z}{sin(\pi z)(-2z)^{1/2}}. \tag{8.37}$$

Note that this asymptotic form displays the poles at negative integers. The asymptotic representations for $Rez > 0$ and $Rez < 0$ match at large positive Imz where $sin(\pi z) \simeq ie^{-i\pi z}/2$.

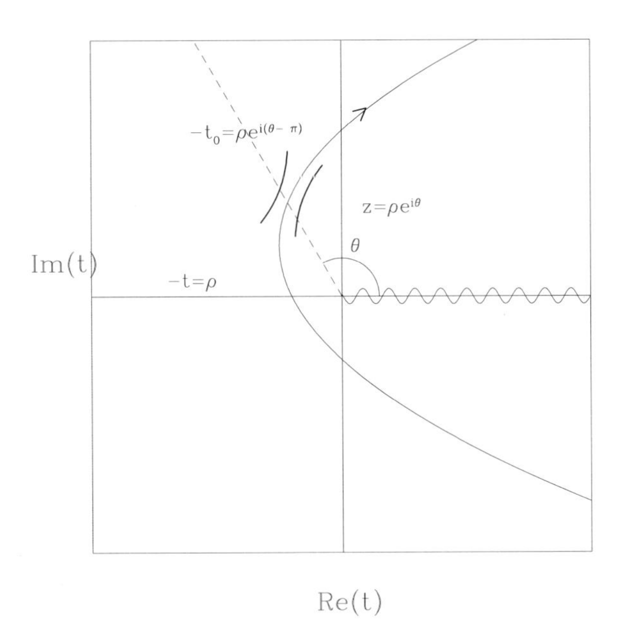

Fig. 8.5 Integration path for evaluation of Γ, $Rez \to -\infty$.

8.7 Euler Product for Γ

Let $f(z) = 1/\Gamma(z)$, again as in section 8.4 we have from Eq. 8.5 near the integers $f'/f \simeq 1/(z-n)$, but only for $n = 0, -1, -2, \ldots$. Again integrate over the contour C_m shown in Fig. 8.2, giving

$$\frac{1}{2\pi i}\int_{C_m} \frac{dz}{z-w}\frac{f'(z)}{f(z)} = \frac{f'(w)}{f(w)} + \sum_{0}^{m} \frac{1}{-n-w}. \tag{8.38}$$

Now evaluate the left-hand side. Write $1/(z-w) = 1/z + w/z(z-w)$. Then the left-hand side becomes

$$\frac{1}{2\pi i}\int_{C_m} \frac{dz}{z}\frac{f'(z)}{f(z)} + \frac{w}{2\pi i}\int_{C_m} \frac{dz}{z(z-w)}\frac{f'(z)}{f(z)}. \tag{8.39}$$

Now we need f'/f near $z = 0$. But

$$\frac{f'(z)}{f(z)} = -\frac{\Gamma'(z)}{\Gamma(z)} = -\frac{\Gamma'(z+1)}{\Gamma(z+1)} + \frac{1}{z}. \tag{8.40}$$

Thus near $z = 0$ we have

$$\frac{f'(z)}{f(z)} = \gamma + \frac{1}{z}, \tag{8.41}$$

and thus

$$\frac{1}{2\pi i}\int_{C_m} \frac{dz}{z}\frac{f'(z)}{f(z)} = -\sum_{1}^{m} \frac{1}{n} + \gamma. \tag{8.42}$$

Now use the asymptotic values to evaluate the contributions on the contour. For $Re z < 0$ use Eq. 8.37 to find $\Gamma'(z) \simeq \Gamma(z) ln(-z) - \pi cos(\pi z)\Gamma(z)/sin(\pi z)$, and thus on the path between the integers $\Gamma'(z)/\Gamma(z) \simeq ln(-z)$. For $Re z > 0$ we have $\Gamma'(z)/\Gamma(z) \simeq ln(z)$ and thus for $r = |z|$

$$\left|\int_{C_m} \frac{dz}{z(z-w)}\frac{f'(z)}{f(z)}\right| < \frac{ln(r)}{r} \to 0, \qquad r \to \infty. \tag{8.43}$$

Combining we find

$$-\sum_{n=1}^{m} \frac{1}{n} + \gamma = \frac{f'(w)}{f(w)} - \sum_{n=1}^{m} \frac{1}{n+w} - \frac{1}{w}. \tag{8.44}$$

Now substitute the expression for γ ($m \to \infty$ understood) to find

$$-ln(m) = \frac{f'(w)}{f(w)} - \sum_{n=1}^{m} \frac{1}{n+w} - \frac{1}{w}. \tag{8.45}$$

Integrate, and use $f(1) = 1$ to determine the integration constant, giving

$$f(w) = wm^{1-w} \prod_{n=1}^{m} \left(\frac{n+w}{n+1} \right). \tag{8.46}$$

But $m = \prod_{n=1}^{m-1}(n+1)/(n)$ so

$$\frac{1}{\Gamma(w)} = f(w) = w \left(\frac{m}{m+1} \right) \prod_{n=1}^{\infty} \frac{(1+\frac{w}{n})}{(1+\frac{1}{n})^{w}}, \tag{8.47}$$

and taking the limit $m \to \infty$ gives finally the expression due to Euler

$$\Gamma(w) = \frac{1}{w} \prod_{1}^{\infty} \frac{(1+\frac{1}{n})^{w}}{(1+\frac{w}{n})}. \tag{8.48}$$

Now from the Euler expression we find

$$\Gamma(z)\Gamma(1-z) = \frac{1}{z(1-z)} \prod_{1}^{\infty} \frac{(1+\frac{1}{n})^{z}\left(1+\frac{1}{n}\right)^{1-z}}{[1+z/n][1+(1-z)/n]}, \tag{8.49}$$

which reduces to

$$\Gamma(z)\Gamma(1-z) = \frac{1}{z} \prod_{1}^{\infty} \frac{n^2}{(n+z)(n-z)} = \frac{1}{z} \prod_{1}^{\infty} \frac{1}{(1+z/n)(1-z/n)}, \tag{8.50}$$

and using Eq. 8.31 we find the identity

$$\Gamma(z)\Gamma(1-z) = \frac{\pi}{sin\pi z}. \tag{8.51}$$

8.8 Integral Representation for $1/\Gamma(z)$

Consider

$$I(z) = \int_{C} t^{-z} e^{t} dt, \tag{8.52}$$

for the contour C shown in Fig. 8.6. Moving the contour to the real axis for $z < 1$ we find

$$I(z) = (e^{i\pi z} - e^{-i\pi z}) \int_0^\infty \rho^{-z} e^{-\rho} d\rho, \tag{8.53}$$

or $I = 2isin\pi z\Gamma(1-z)$. But using Eq. 8.51 we find

$$\frac{1}{\Gamma(z)} = \frac{1}{2\pi i} \int_C t^{-z} e^t dt, \tag{8.54}$$

which is analytic for all z, and thus $\Gamma(z)$ has no zeros. This integral representation can be used to find the asymptotic value of $\Gamma(z)$ in any direction in the z plane, in the same manner as in section 8.6.

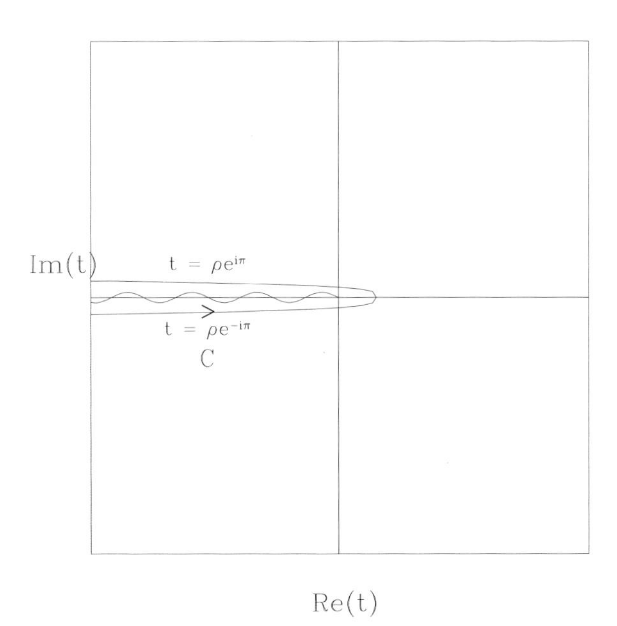

Fig. 8.6 Integration path for $1/\Gamma$.

8.9 $\Gamma(nx)$

To prove the identity

$$\Gamma(2x) = \frac{2^{2x-1}}{\sqrt{\pi}}\Gamma(x)\Gamma(x+1/2), \tag{8.55}$$

let

$$A = \frac{2^{2x}\Gamma(x)\Gamma(x+1/2)}{\Gamma(2x)}. \tag{8.56}$$

Now use the Euler product Eq. 8.48, writing m factors for each of the Gamma functions in the numerator, and $2m+1$ factors for $\Gamma(2x)$, giving

$$A = \frac{2^{2x}2x(2m+1)^{-2x}\prod_1^{2m+1}(1+2x/n)}{xm^{-x}\prod_1^m(1+x/n)(x+1/2)m^{-(x+1/2)}\prod_1^m(1+(x+1/2)/n)}. \tag{8.57}$$

Rearranging terms and cancelling we find

$$A = 2\sqrt{m}\left(\frac{m}{m+1/2}\right)^{2x}\frac{\Gamma(m+1)\Gamma(1/2)}{\Gamma(m+3/2)}. \tag{8.58}$$

Now take the limit $m \to \infty$ using $\Gamma(x) \simeq x^x\sqrt{2/\pi x}e^{-x}$, and $(1+x/m)^m \to e^x$ we find $A = 2\sqrt{\pi}$, proving the identity as stated.

In a similar manner one can prove the more general identity

$$\Gamma(nx) = (2\pi)^{(1-n)/2}n^{nx-1/2}\prod_{k=0}^{n-1}\Gamma(x+k/n). \tag{8.59}$$

8.10 The Euler Beta Function

Introduce the Euler beta function[1]

$$B(\alpha,\beta) = \int_0^1 u^{\alpha-1}(1-u)^{\beta-1}du. \tag{8.60}$$

[1]String theory began when G. Veneziano ([1968]) recognized that data for particle scattering could be fit with an amplitude which is a sum of Euler beta functions. It was then recognized that this representation naturally occurred in a depiction of particles as strings rather than points.

Now examine

$$\Gamma(\alpha)\Gamma(\beta) = \int_0^\infty e^{-t}t^{\alpha-1}dt \int_0^\infty e^{-s}s^{\beta-1}ds. \tag{8.61}$$

Change to polar coordinates using $\sqrt{s} = rsin\theta$, $\sqrt{t} = rcos\theta$, and let $\rho = r^2$, giving

$$\Gamma(\alpha)\Gamma(\beta) = 2\int_0^\infty e^{-\rho}\rho^{\alpha+\beta-1}d\rho \int_0^{\pi/2} cos^{2\alpha-1}\theta sin^{2\beta-1}\theta d\theta. \tag{8.62}$$

Letting $u = sin^2\theta$ we find

$$\Gamma(\alpha)\Gamma(\beta) = \Gamma(\alpha+\beta)B(\alpha,\beta), \tag{8.63}$$

giving a simple expression for $B(\alpha,\beta)$ in terms of Gamma functions.

8.11 Problems

1. The digamma function is defined by $\Psi(z) = \Gamma'(z)/\Gamma(z)$. Show that $\Psi(z+1) = \Psi(z) + 1/z$.

2. The Euler expression for $\Gamma(z)$ is

$$\Gamma(z) = \frac{1}{z}\prod_{1}^{\infty}\frac{(1+\frac{1}{n})^z}{1+\frac{z}{n}}.$$

Verify using this representation that $\Gamma(z+1) = z\Gamma(z)$.

3. The digamma function $\Psi(z) = \Gamma'(z)/\Gamma(z)$ has the integral representation

$$\Psi(z) = ln(z) - \frac{1}{2z} - \int_0^{\infty}[(e^t-1)^{-1} - t^{-1} + 1/2]e^{-tz}dt.$$

Use this expression to generate the first three terms of the asymptotic expansion of $\Psi(z) - lnz + 1/(2z)$ as $z \to \infty$.

4. Show that

$$\Gamma(\tfrac{1}{2}-i\tfrac{\lambda}{2})\Gamma(\tfrac{1}{2}+i\tfrac{\lambda}{2}) = \frac{\pi}{cosh(\pi\lambda/2)}.$$

5. Show that

$$\frac{\Gamma'(n)}{\Gamma(n)} = -\gamma + \sum_{1}^{n-1}\frac{1}{k}.$$

6. For n integer, write $\Gamma(2n) = (2n-1)(2n-2)\ldots 2\cdot 1$. Remove a 2 from each factor, rearrange and show that

$$\Gamma(2n) = \frac{2^{2n-1}}{\sqrt{\pi}}\Gamma(n)\Gamma(n+1/2).$$

Why is this not a proof of Eq. 8.55 ?

7. Show that

$$\frac{\Gamma'(1)}{\Gamma(1)} - \frac{\Gamma'(1/2)}{\Gamma(1/2)} = 2ln(2). \qquad (Jesus, 1903)$$

8. If

$$f(z) = \int_z^{z+1} ln(\Gamma(t))dt,$$

show that

$$df(z)/dz = ln(z). \qquad \text{(Raabe, } \textit{Journal fur Math xxv}\text{)}$$

9. Show that

$$\prod_{s=1}^{\infty} \frac{s(s+a+b)}{(s+a)(s+b)} = \frac{\Gamma(a+1)\Gamma(b+1)}{\Gamma(a+b+1)},$$

provided that neither a nor b is a negative integer.

10. Show the asymptotic value for large x to be

$$\int_a^{\infty} \frac{x^{s-1}}{\Gamma(s)} ds \simeq e^x,$$

with $a > 0$.

Chapter 9

Integral Solutions

Joseph Fourier was born in Auxerre, France in 1768. He trained for the priesthood, but instead became a teacher at the Benedictine college where he had studied. He became involved in the revolution and narrowly avoided the guillotine. He then studied at the École Normale, where he was taught by Lagrange. After beginning to teach at the École Centrale des Travaux Publiques, which later became the École Polytechnique, he was again arrested but released. His teaching career at the École Polytechnique was interrupted when he accompanied Napoleon to Egypt in 1798. He collected an enormous amount of scientific and cultural material on Egyptian history which laid the foundations for modern Egyptology. Napoleon then requested him to take up the position of Prefect in Grenoble, an administrative position which, however, left him time for mathematical research. His research concentrated on heat conduction in solids, and he was the first to define thermal conductivity and write the diffusion equation

$$\frac{dT}{dt} = \frac{K}{CD}\left[\frac{d^2T}{dx^2} + \frac{d^2T}{dy^2} + \frac{d^2T}{dz^2}\right] \tag{9.1}$$

in a treatise in 1822, with T the temperature, t time, K thermal conductivity, C specific heat, and D density of the solid, and for the solution he invented the method of expanding the periodic solution in trigonometric series, the coefficients of the series being determined by integration. He later extended this method to integration over a continuum of frequencies. He was also the first to obtain the solution of the diffusion equation with initial conditions of a point source, see Eq. 2.37.

9.1 Constructing Integral Solutions

Integral solutions of differential equations are not always possible, but when possible they provide global solutions from which all asymptotic limits can be obtained, and hence connect solutions from one domain to another, as well as giving a means of evaluating the solution for general complex z. The asymptotic limits are obtained by finding the associated limiting values of the integral, normally due to either saddle points or to contributions from end points. The Stokes phenomenon comes about from the motion or change in orientation of the saddle points as z is moved through the complex plane. Sometimes the saddle points shift position so as to modify the path of integration, as occurs with the Airy function, and sometimes a saddle point moves past the origin of a cut, necessitating an addition to the integration path around the cut, as occurs for the parabolic cylinder function.

Writing the solution $y(z)$ as an integral of a given kernel $K(z,t)$ times an unknown function of t, $y(z) = \int_C K(z,t)f(t)dt$, the function $f(t)$ is obtained either through the derivation of a differential equation by integration by parts or, depending on the nature of the kernel, through a difference equation. For some choices of the kernel it is also possible to derive the integral representation beginning with a convergent series representation of $y(z)$, without making use of a differential equation.

9.1.1 *Integration by parts*

Attempt a solution of the form

$$y(z) = \int_C K(z,t)f(t)dt, \tag{9.2}$$

with a given kernel $K(z,t)$. Substituting this expression for $y(z)$ into the original differential equation we integrate by parts in such a way as to obtain a differential equation for $f(t)$, plus a condition for the vanishing of the terms resulting from the integration by parts. The method is successful if the resulting differential equation for $f(t)$ is simpler than the original equation for $y(z)$ and if a contour C can be found so that the extra terms created through integration by parts vanish. In the case of an inhomogeneous differential equation these extra terms can be chosen, in fact, not to vanish but to reproduce the inhomogeneous terms. If the inhomogeneous terms are periodic functions, a Fourier–Laplace kernel can lead to an $f(t)$ which is a sum of delta functions, giving a closed form analytic solution. Some examples of kernels are Fourier–Laplace with $K(z,t) = e^{zt}$, Euler with $K(z,t) = (z-t)^{\mu}$, Sommerfeld with $K(z,t) = e^{zsinh(t)}$, and Mellin with $K(z,t) = z^t$. In addition to there being many different kernels to choose from, even with a given kernel the integration contour is not unique, and the different contour choices can provide solutions with different properties under analytic continuation, or with different boundary conditions.

Consider a differential equation of the form

$$L_z y = p(z)\frac{d^2y}{dz^2} + q(z)\frac{dy}{dz} + r(z)y = 0. \tag{9.3}$$

Write the solution in the form $y(z) = \int K(z,t)f(t)dt$ and set $L_z y = 0$, or

$$L_z y = \int L_z K(z,t)f(t)dt = 0, \tag{9.4}$$

subject of course to the condition that the integral converges. Now it is essential that the kernel be chosen so that the operator L_z on the kernel $K(z,t)$ can be written as another differential operator M_t acting on K, so that

$$L_z y = \int M_t(K)f(t)dt. \tag{9.5}$$

The necessity of this equivalence between an operator in z and an operator in t severely restricts the choice of the kernel $K(z,t)$. Now through integration by parts we move the operator M_t from acting on the kernel to make it act on the function $f(t)$, giving

$$\begin{aligned}\int M_t(K)f(t)dt &= \int \left[K\overline{M}_t(f) + \frac{d}{dt}P(f,K)\right] dt \\ &= \int K\overline{M}_t(f)dt + P(f,K)|_a^b,\end{aligned} \tag{9.6}$$

with a, b the end points of the contour, and, if

$$M_t(K) = \alpha(t)\frac{d^2K}{dt^2} + \beta(t)\frac{dK}{dt} + \gamma(t)K, \tag{9.7}$$

then

$$\overline{M}_t(f) = \frac{d^2}{dt^2}(\alpha f) - \frac{d}{dt}(\beta f) + \gamma(t)f, \tag{9.8}$$

and the term resulting from the integration by parts is

$$P(f,K) = \alpha f\frac{dK}{dt} - K\frac{d(\alpha f)}{dt} + \beta f K. \tag{9.9}$$

If now the contour is chosen so that $P(f,K)|_a^b = 0$ then this term vanishes and we are left with

$$L_z y = \int K\overline{M}_t(f)dt, \tag{9.10}$$

and if $f(t)$ is a solution to $\overline{M}_t(f) = 0$ then y is a solution to the original differential equation. With some luck in the choice of the kernel, this differential equation is simpler than the original, and can be solved.

It is possible to use the form of the equation for $K(z,t)$ to determine a useful kernel for the differential equation in question. An example of this is given in section 10.1.4, where the Laplace integral representation of the Legendre function is determined in this manner.

9.1.2 *Asymptotic value of a series*

Consider the convergent series

$$y(x) = \sum_0^\infty \frac{x^n}{(n!)^2}, \tag{9.11}$$

where the asymptotic value for large x is desired. Differentiating we find

$$y'(x) = \sum_1^\infty \frac{x^{n-1}}{n!(n-1)!}, \tag{9.12}$$

and then $(xy'(x))' = y(x)$, and thus we have the differential equation

$$xy''(x) + y'(x) - y(x) = 0. \tag{9.13}$$

Now find the asymptotic behavior using dominant balance. Writing $y = e^S$ we find $x[(S')^2 + S''] + S' - 1 = 0$. Dominant balance is given by $x(S')^2$ and 1, and going to next order we find $S' = x^{-1/2} - x/4$. To complete the asymptotic analysis write $y = e^S W(x)$, giving

$$xW'' + 2\sqrt{x}W' + \frac{W}{16x} = 0. \tag{9.14}$$

Substituting $W = \sum a_n x^{-n/2}$ and using $W(0) = 1$ we find the divergent series given by $a_0 = 1$ and

$$a_n = \frac{1}{4^n n!}\prod_1^n (k^2 - 3/4) \qquad n > 0 \tag{9.15}$$

with $y(x) \sim C\frac{e^{2\sqrt{x}}}{x^{1/4}}W(x)$, but the constant of normalization is unknown.

To find the normalization we must find a way to connect the asymptotic behavior at $x = 0$ to that at $x = \infty$, and an integral representation provides this. Attempt a Fourier–Laplace solution,

$$y(x) = \int_C e^{xt} f(t)dt. \tag{9.16}$$

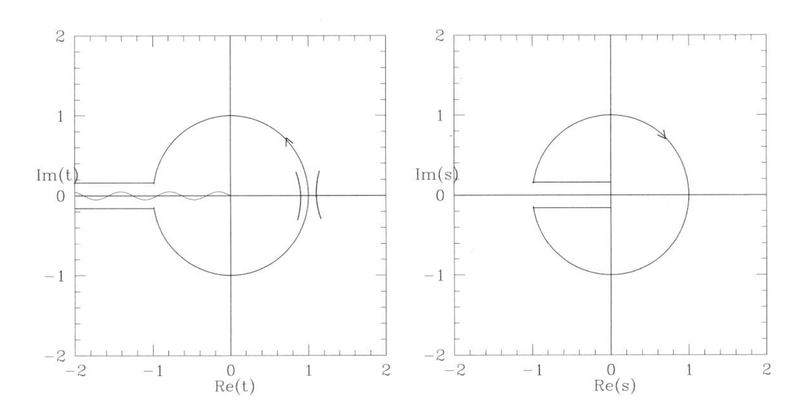

Fig. 9.1 Integration contour in the t and s planes.

Substituting into the differential equation, using $xe^{xt} = de^{xt}/dt$ and integrating by parts we find

$$t^2 f' + (t+1)f = 0, \qquad t^2 f e^{xt}|_a^b = 0 \tag{9.17}$$

where a, b are the end points of the contour C, to be determined so that the term produced by integrating by parts vanishes. The first equation gives $f(t) = Ae^{1/t}/t$, and since we are interested in large positive x the integration end points must be at $t = -\infty$. The function $f(t)$ is not single valued, place the cut extending from $t = 0$ to $t = -\infty$. Thus we have

$$y(x) = A \int_C \frac{e^{xt+1/t}}{t} dt, \tag{9.18}$$

with the contour shown in Fig. 9.1. Saddle points are located at $t = \pm 1/\sqrt{x}$, and the saddle at positive t is shown in the figure for $x = 1$. First find the normalization by setting $y(0) = 1$. Making the substitution $t = 1/s$ the contour in the s plane is shown in Fig. 9.1, the contributions from the straight line segments cancel, and we find $A = 1/(2\pi i)$. Writing $\phi = xt + 1/t$ we find at the saddle point $\phi_0 = 2\sqrt{x}$ and $\phi_0'' = 2x^{3/2}$.

Evaluating the saddle point contribution with Eq. 7.16 we then find that for large x

$$y(x) \simeq \frac{1}{2\sqrt{\pi}} \frac{e^{2\sqrt{x}}}{x^{1/4}}. \tag{9.19}$$

A numerical verification of this behavior is seen in Fig. 9.2. Note that the leading order asymptotic expression is reasonably valid for very small x. In fact the discrepancy in the plot for large x is due to the difficulty in evaluating many terms in the series.

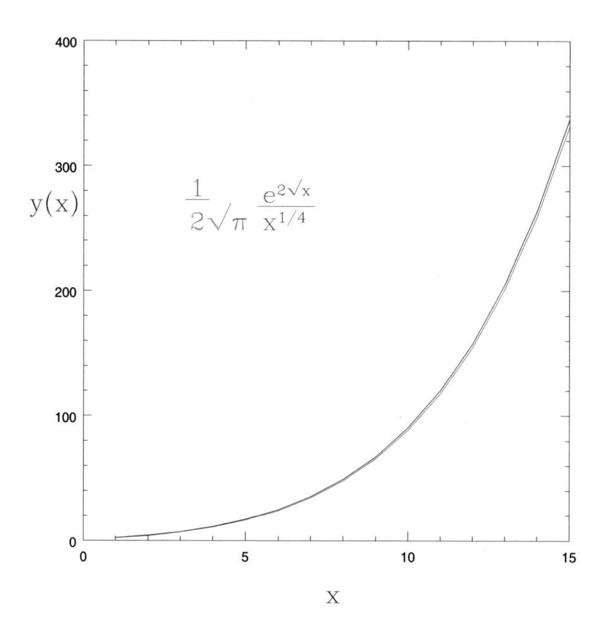

Fig. 9.2 Asymptotic expression and a numerical summation of Eq. 9.11.

Higher order corrections are of course given by $W(x)$, with the first few terms given by

$$W(x) = 1 + \frac{x^{-1/2}}{16} + \frac{13x^{-1}}{512} + \frac{429x^{-3/2}}{24576} + \frac{26169x^{-2}}{1572864} + \ldots. \tag{9.20}$$

Even for $x = 1$ one keeps three terms in the series and the error in the approximation is less than 2%.

9.1.3 *Finding a discrete difference equation*

The Mellin representation is instead obtained by solving a difference equation. Assume the representation

$$y = \frac{1}{2\pi i}\int_C M_s z^{-s} ds, \tag{9.21}$$

and substitute into the differential equation Eq. 9.3, obtaining

$$s(s+1)p(z)z^{-s-2}M(s) - sq(z)z^{-s-1}M(s) + r(z)z^{-s}M(s) = 0. \tag{9.22}$$

Now to obtain a Mellin representation the functions $p(z), q(z), r(z)$ must be such that there appear in this equation only two powers of z, i.e. that the equation can, for example, be reduced to the form

$$(s-a)(s-b)z^{-s}M(s) + (s-c)Kz^{-s+j}M(s) = 0. \tag{9.23}$$

If this form can be obtained we then find a difference equation for the $M(s)$

$$(s-a)(s-b)M(s) + (s+j-c)KM(s+j) = 0, \tag{9.24}$$

with solutions

$$M(s) = Nj^{s/j}\frac{\Gamma\left(\frac{s-a}{j}\right)\Gamma\left(\frac{s-b}{j}\right)}{\Gamma\left(\frac{s-c+j}{j}\right)(-K)^{s/j}}, \tag{9.25}$$

and possible variations such as

$$M(s) = Nj^{s/j}\frac{\Gamma\left(\frac{s-a}{j}\right)\Gamma\left(\frac{c-s}{j}\right)}{\Gamma\left(\frac{j-s+b}{j}\right)(-K)^{s/j}}, \tag{9.26}$$

where the variations can be obtained by substituting $\Gamma(x)$ with $1/\Gamma(1-x)$. Here N is a normalization independent of s. All these possibilities satisfy the original difference equation, as can be verified by direct substitution. Note, however, that they have very different properties. In particular, noting that $\Gamma(z)$ has poles at negative integers, but no zeros, the first expression has poles at $s = b - nj$, but the second expression has none. Thus these solutions to the initial difference equation are really different.

If the original differential equation does not allow reduction to a two term difference equation for $M(s)$ the extraction of asymptotic behavior of the solution can often produce this, i.e. writing $y = e^{S(z)}f(z)$ with $S(z)$ giving the asymptotic behavior, and then finding a Mellin representation for $f(z)$.

9.1.4 *Construction from a series*

The Mellin representation can instead be constructed given only the convergent series representation of the function (Dingle [1973]). Consider the convergent series

$$y = \sum_{0}^{\infty} a_n z^n, \tag{9.27}$$

and represent this function with a contour integral of the form

$$y = \frac{1}{2\pi i} \int_C M_s z^{-s} ds, \tag{9.28}$$

with the contour enclosing the entire left half plane, and possessing a vertical line to the right of $s = 0$. Depending on the behavior of M_s for large s the contour may consist entirely of the vertical line to the right of $s = 0$ for $|z| < 1$. For this integral to reproduce the series we need $M(s)$ to have first order poles at $s = 0, -1, -2, \ldots$ with residue a_{-s}, and for the integral to converge. A solution is given by

$$M(s) = a_{-s}(-1)^s \Gamma(1-s)\Gamma(s), \tag{9.29}$$

provided the integral converges. For example, in this case since $\Gamma(1-s)\Gamma(s) = \pi/sin(\pi s)$, writing $s = t + iu$, this product gives for large u a factor $exp(-\pi|u|)$. Writing $z = re^{i\phi}$ we have $\Gamma(1-s)\Gamma(s)(-1)^s z^{-s} \sim exp(\pm\phi|u| - 2\pi|u|)$ with the $\pm\phi$ coming from the two limits of the integral, and the ambiguity of the factor $(-1)^{iu}$ has been resolved by continuing from the positive real axis, i.e. take $-1 = e^{i\pi}$ for $u > 0$ and take $-1 = e^{-i\pi}$ for $u < 0$. Combining this with the behavior of a_{-s} determines convergence and gives a restriction on the phase of z in which this expansion is valid. The restriction of the phase is related to the Stokes phenomenon, i.e. as a particular phase line is passed a previously subdominant asymptotic form becomes dominant, changing the form of the asymptotic expansion.

In some cases it helps to change the asymptotic behavior of the series. For example, given the series

$$y = \sum_{0}^{\infty} \frac{\Gamma(n+a)}{n!\Gamma(n+b)} z^n, \tag{9.30}$$

we note that asymptotically the series becomes that of the exponential, $\sum z^n/n!$. Thus if it is possible to extract this exponential behavior by writing

$$y = e^z \sum_0^\infty b_n z^n, \tag{9.31}$$

the new series will tend to a constant at infinity, and can be used to generate the Mellin representation. This extraction is easily performed if we have given a differential equation for $y(z)$. Simply write $y(z) = e^z g(z)$ and find the new differential equation for $g(z)$ and a series representation. This method will be demonstrated in Chapter 15.

9.2 Causal Solutions

In physical problems involving space and time one often encounters singularities in the formulation of integral solutions. The treatment of the singularities is determined by the imposition of causality, most simply formulated as the condition that information cannot propagate backwards in time, and in relativistic problems, cannot propagate more rapidly than the speed of light. To illustrate how this principle determines integration contours, consider the propagation of a scalar wave function produced by a given source $S(x)$ which acts only at $t = 0$. The wave equation with this source is

$$(\partial_t^2 - c^2\partial_x^2)\phi(x,t) = S(x)\delta(t) \tag{9.32}$$

with c the speed of light. First find the Green's function for this equation, i.e. solve

$$(\partial_t^2 - c^2\partial_x^2)G(x, x_0, t, t_0) = \delta(x - x_0)\delta(t - t_0). \tag{9.33}$$

Use a Fourier–Laplace representation,

$$\phi(x,t) = \int e^{i(kx-\omega t)} f(k,\omega) d\omega dk, \tag{9.34}$$

$$G(x, x_0, t, t_0) = \int e^{i(kx-\omega t)} g(k, \omega, x_0, t_0) d\omega dk, \tag{9.35}$$

with all integrals extending from $-\infty$ to $+\infty$, and use the integral representation for the δ functions from Eq. 2.16, giving

$$g(k, \omega, x_0, t_0) = \frac{e^{i(\omega t_0 - kx_0)}}{4\pi^2(c^2k^2 - \omega^2)}. \tag{9.36}$$

The solution for the field is then given by

$$\phi(x,t) = \int dx_0 dt_0 G(x, x_0, t, t_0) S(x_0, t_0), \tag{9.37}$$

and in field theories the Green's function is known as the propagator, since it propagates information from the point x_0, t_0 to the point x, t. Now evaluate the Green's function. We have

$$G(x, x_0, t, t_0) = \frac{1}{4\pi^2} \int \frac{e^{i[k(x-x_0)-\omega(t-t_0)]}}{c^2k^2 - \omega^2} d\omega dk. \tag{9.38}$$

First perform the integration over frequency. There are two poles on the real axis at $\omega = \pm ck$. To determine the integration contour, note that causality implies that there can be no contribution to $\phi(x,t)$ from values of the source at times later than t, i.e. for $t < t_0$ the Green's function must vanish. If $t < t_0$ the factor $e^{-i\omega(t-t_0)}$ is exponentially decreasing in the upper half ω plane, and thus the contour in ω must be taken above the singularities in this plane so that it can be closed in the upper half plane to give $G = 0$. For $t > t_0$ the contour can be closed in the lower half plane, giving

$$G(x, x_0, t, t_0) = \frac{i\Theta(t-t_0)}{2\pi} \int \left[\frac{e^{ik[x-x_0-c(t-t_0)]}}{2ck} - \frac{e^{ik[x-x_0+c(t-t_0)]}}{2ck} \right] dk. \tag{9.39}$$

Again these terms are ambiguous, since the integration contour passes through the singularity at $k = 0$. Now use the fact that $d\Theta(x)/dx = \delta(x)$ to find a representation for the theta function

$$\int \frac{e^{ika}}{k} dk = 2\pi i \Theta(a), \tag{9.40}$$

with the integration contour taken below the singularity at $k = 0$, so that for $a < 0$ the contour can be closed in the lower half plane giving zero. Taking the contour below the singularity we then find

$$\begin{aligned} G(x, x_0, t, t_0) = &-\Theta(t-t_0)\Theta(x - x_0 - c(t-t_0))/2c \\ &+\Theta(t-t_0)\Theta(x - x_0 + c(t-t_0)/2c. \end{aligned} \tag{9.41}$$

But

$$\begin{aligned} \Theta(x - x_0 + c(t-t_0)) &- \Theta(x - x_0 - c(t-t_0)) \\ &= \Theta(c(t-t_0) - |x - x_0|) \end{aligned} \tag{9.42}$$

so the final Green's function is nonzero only in the backward light cone from the point x, t. Using the δ function in S we find

$$\phi(x,t) = \frac{\Theta(t)}{2c} \int \Theta(ct - |x - x_0|) S(x_0) dx_0, \tag{9.43}$$

and the contributions from the source only involve values within the causal domain, where information can reach x, t travelling at less than the speed of light.

9.3 Comments

The result of obtaining a soluble equation for $f(t)$ or for $M(s)$ is largely a matter of trial and error. The outcome in the case of an integral representation depends on the kernel chosen, the variables used, and on a great deal of patience and luck. A standard technique can often lead to simplification if initial attempts fail. It involves extracting the asymptotic behavior of the function as an external factor, writing $y(z) = z^p g(z)$ where z^p is the behavior of $y(z)$ at a regular singular point of the differential equation, or writing $y(z) = e^{S(z)} g(z)$ with $S(z)$ the leading behavior at an irregular singular point (typically $z = \infty$), and then attempting an integral representation of $g(z)$. In addition, if there exist two different possible asymptotic behaviors to choose from, we can obtain in this manner two different integral representations. These methods will be illustrated with a few examples in the following chapters.

9.4 Problems

1. Consider

$$\frac{d^2y}{dx^2} + x\frac{dy}{dx} + y = 0.$$

   a) Find two local series solutions about $x = 0$, $y_1(x)$ and $y_2(x)$.

   b) Find two possible asymptotic behaviors at $x \to +\infty$.

   c) Find two Fourier–Laplace integral solutions

$$y(x) = \int e^{ixt} f(t) dt.$$

   Hint: Use real and imaginary parts, and examine all possible contours.

   d) Identify the solutions in part (a) with those in part (c), i.e. find integral solutions for $y_1(x)$ and $y_2(x)$.

   e) Find the behavior of the two solutions in part (a) at $x \to +\infty$ including normalization.

2. Consider

$$x\frac{d^2y}{dx^2} + \left(\frac{7}{8} - x\right)\frac{dy}{dx} - \frac{1}{4}y = 0.$$

Find integral representations for the two linearly independent solutions.

3. Consider the differential equation

$$x\frac{d^3y}{dx^3} + 2y = 0.$$

Find an integral solution with $y(0) = 1$, $y(\infty) = 0$, assuming a form

$$y(x) = \int e^{-x/\sqrt{t}} f(t) dt.$$

Find the asymptotic behavior for $x \to \infty$ and compare it to that obtained from dominant balance.

4. Consider

$$x\frac{d^2y}{dx^2} - x\frac{dy}{dx} + 2y = 0.$$

Find integral representations for the two linearly independent solutions. Find the asymptotic limits of your solutions for $x \to 0$ and for $x \to \infty$ and verify them using dominant balance.

5. Find integral representations for the two linearly independent solutions of

$$\frac{d^2y}{dx^2} + x\frac{dy}{dx} + xy = 0.$$

Find the asymptotic limits of your solutions for $x \to 0$ and for $x \to \infty$ and verify them using dominant balance.

Chapter 10

Expansion in Basis Functions

After the discovery by Fourier of the possibility of expanding functions in terms of a basis consisting of sine and cosine functions the idea was generalized to representing functions by expansions in other sets, one of the first being the Legendre polynomials. Adrien-Marie Legendre was born in Paris in 1752. He was independently wealthy but taught with Laplace at the École Militaire out of interest in mathematics. He contributed primarily to analysis. In 1782 he won a prize offered by the Berlin Academy for a treatise on the trajectory of a cannonball, including the effects of air resistance. The Legendre functions have many applications in mathematical physics, including the description of heat flow on a spherical surface, and the derivation of the gravitational attraction of a solid ellipsoidal mass at any exterior point. He invented the technique of least squares fitting of a smooth function to a set of data points, and discovered many of the basic facts regarding elliptic integrals. He published a text on geometry which replaced Euclid's *Elements* as a textbook in most of Europe. His attempt to prove the parallel postulate extended over 30 years, and even after the discovery of non-Euclidean geometry he regarded the geometry of flat space as an incontestable truth.

After Legendre, other such special bases, each associated with a differential equation and an inner product, were given by Gegenbauer, Chebychev, Hermite, Jacobi, Laguerre and others. Each of these sets of functions has a particular use, but all suffer from the defect that functions which have strong discontinuities, or structure on many different scale lengths, require a very large number of basis functions to give a reasonable description. This problem has only recently been addressed by the invention of wavelets.

10.1 Legendre Functions

The Legendre differential equation is

$$(1 - z^2)y'' - 2zy' + \nu(\nu + 1)y = 0, \tag{10.1}$$

with ν a real parameter. We note immediately that the equation is unchanged under the replacement of ν by $-\nu - 1$.

10.1.1 *Local analysis*

The domain of interest of the variable z for the Legendre functions is $-1 < z < 1$. Clearly $z = 0$ is an analytic point, but the points $z = \pm 1$ are both regular singular points. Thus for any ν both solutions are given by Taylor's series of the form

$$y(z) = \sum_0^\infty a_k z^k. \tag{10.2}$$

Substituting into the differential equation and renaming summation indices we find

$$a_{k+2} = \frac{(k - \nu)(k + \nu + 1)}{(k + 1)(k + 2)} a_k. \tag{10.3}$$

Choosing $a_0 \neq 0$ gives a series in z^2

$$y \sim \sum_s \frac{(-1)^s 2^{2s} \Gamma(\nu/2 + 1/2 + s)}{(2s)! \Gamma(\nu/2 - s + 1)} z^{2s}, \tag{10.4}$$

and choosing $a_0 = 0, a_1 \neq 0$ gives a series in odd powers of z

$$y \sim \sum_s \frac{(-1)^s 2^{2s+1} \Gamma(\nu/2 + 1 + s)}{(2s + 1)! \Gamma(\nu/2 - s + 1/2)} z^{2s+1}. \tag{10.5}$$

Note that the first series terminates if ν is a positive even integer n, and the second series terminates if ν is a positive odd integer n. Thus for any integer ν one of the solutions is a polynomial.

The two solutions for any ν are easily distinguished by their behavior at $z = \pm 1$. Let $w = z - 1$ and examine the regular singular point at $w = 0$. Attempting a solution of the form $y = \sum a_n w^{\alpha+n}$ we find $\alpha^2 = 0$, a degenerate case, (see section 4.2.2) and the second solution to behave as $y \sim ln(w)$. The solution which is well behaved at $z = 1$ we denote by $P_\nu(z)$. Since $\alpha = 0$ we can normalize these functions by requiring $P_\nu(1) = 1$, and since the second solution is infinite at $z = 1$, this normalization is sufficient to completely determine the solution.

Now consider the special case of $\nu = n$, integer. Letting $2s = n - 2r$ for n even and $2s + 1 = n - 2r$ for n odd we then find for $\nu = n$ the polynomial solution

$$y \sim \sum_r \frac{(-1)^r 2^{n-2r}\Gamma(n + 1/2 - r)}{(n-2r)!r!} z^{n-2r}, \tag{10.6}$$

and the summation is over all integers r giving positive powers of z. Using Eq. 8.55 and normalizing so that $P_n(1) = 1$ we find the Legendre polynomials

$$P_n(z) = \frac{1}{2^n} \sum_{r=0}^{m} \frac{(-1)^r (2n-2r)!}{r!(n-r)!(n-2r)!} z^{n-2r}, \tag{10.7}$$

where $m = n/2$ or $m = (n-1)/2$, whichever is an integer, and

$$P_0(z) = 1, \qquad P_1(z) = z, \qquad P_2(z) = (3z^2 - 1)/2 \ldots. \tag{10.8}$$

10.1.2 *Euler integral representation*

Attempt a solution using the Euler kernel,

$$y(x) = \int_C (t-z)^\mu f(t)dt. \tag{10.9}$$

Substituting we find

$$0 = \int_C dt[(1-z^2)\mu(\mu-1) + 2z\mu(t-z) + \nu(\nu+1)(t-z)^2]f(t-z)^{\mu-2}. \tag{10.10}$$

Choose μ to eliminate the z^2 terms, giving $\mu = \nu, -\nu - 1$. First set $\mu = -(\nu + 1)$. Rearranging terms we find

$$0 = \int_C \left[(1-t^2)f\frac{d}{dt} + 2(\nu+1)tf\right](t-z)^{\mu-1}dt, \tag{10.11}$$

and integrating by parts we obtain a solution provided that

$$f = (t^2-1)^\nu, \qquad \frac{(t^2-1)^\nu}{(t-z)^\nu}\Big|_a^b = 0. \tag{10.12}$$

Consider the term due to integration by parts. In continuing along the integration contour from a to b in the complex t plane there is a factor of

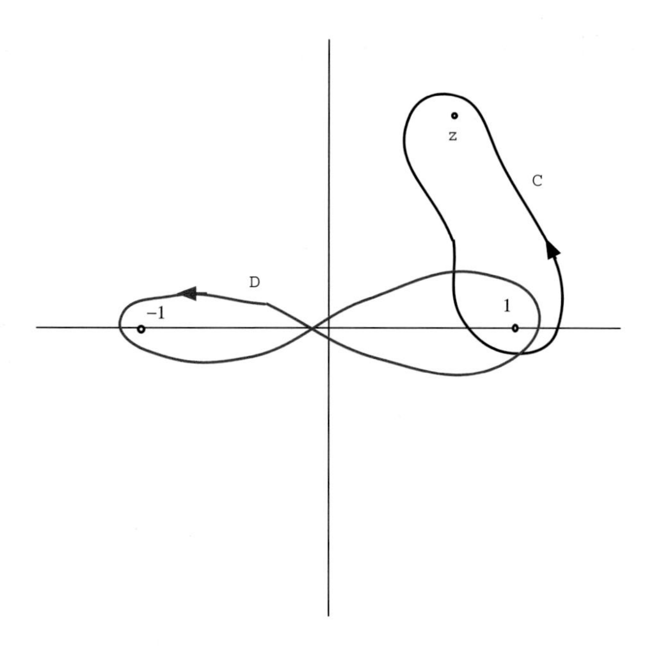

Fig. 10.1 Integration paths for the Legendre function.

$e^{2\pi i\nu}$ upon circling counterclockwise either of the points $t = \pm 1$ or circling clockwise the point $t = z$. This term thus vanishes for any closed contour circling both 1 and z in the same sense, or circling 1 and -1 oppositely. Two such contours are shown in Fig. 10.1, with a solution

$$y(z) = \int \frac{(t^2 - 1)^\nu}{(t - z)^{\nu+1}} dt. \tag{10.13}$$

Consider the contour C circling z and 1. To simplify the task of normalization, take $\nu = n$, integer. The integrand is then analytic at $t = 1$ and the contour can be deformed to become a simple circle about $t = z$, giving

$$y(z) = 2\pi i n! \frac{d^n}{dz^n} (z^2 - 1)^n. \tag{10.14}$$

Setting $y(1) = 1$ gives Rodrigues' formula

$$P_n(z) = \frac{1}{2^n n!} \frac{d^n}{dz^n} (z^2 - 1)^n, \tag{10.15}$$

and results in the integral representation

$$P_\nu(z) = \frac{1}{2\pi i}\int_C \frac{(t^2-1)^\nu}{2^\nu(t-z)^{\nu+1}}dt, \tag{10.16}$$

with the contour C circling both $t = 1$ and $t = z$, which for $\nu = n$, integer, is Schlafli's integral formula.

The second solution Q_ν is defined through

$$Q_\nu(z) = \frac{1}{4isin(\nu\pi)}\int_D \frac{(t^2-1)^\nu}{2^\nu(t-z)^{\nu+1}}dt, \tag{10.17}$$

with branches chosen with $|arg(z)| < \pi$ and $|arg(z-t)| \to arg(z)$ as $t \to 0$ on the contour, and so that $t-1, t+1$ are positive real on the real axis for $t > 1$. If $\nu = n$, a non-negative integer, then beginning to the right of $t = 1$, we have, along the first part of the contour $t-1 \sim e^{-i\pi}, t+1 \sim real$, and after rounding the point $t = -1$ we have $t-1 \sim e^{-i\pi}, t+1 \sim e^{2i\pi}$, the $sin(n\pi)$ factor cancels out, and the integral reduces to

$$Q_n(z) = \frac{1}{2^{n+1}}\int_{-1}^{1} \frac{(1-t^2)^n}{(z-t)^{n+1}}dt. \tag{10.18}$$

The first few functions of the second type are

$$\begin{aligned} Q_0(z) &= \frac{1}{2}ln\frac{1+z}{1-z}, \\ Q_1(z) &= \frac{z}{2}ln\frac{1+z}{1-z} - 1, \\ Q_2(z) &= \frac{1}{4}ln\frac{1+z}{1-z} - \frac{3z}{2} \\ &\quad \ldots . \end{aligned} \tag{10.19}$$

10.1.3 *Recurrence relations*

As with any of the special functions distinguished by an index ν, an integral representation permits the derivation of algebraic relations among the solutions of different ν. Begin with Eq. 10.16 and note that

$$\frac{d}{dt}\frac{(t^2-1)^{\nu+1}}{(t-z)^{\nu+1}} = \frac{2(\nu+1)t(t^2-1)^\nu}{(t-z)^{\nu+1}} - \frac{(\nu+1)(t^2-1)^{\nu+1}}{(t-z)^{\nu+2}}, \tag{10.20}$$

which gives upon integration

$$0 = 2\int_C \frac{t(t^2-1)^\nu}{(t-z)^{\nu+1}}dt - \int_C \frac{(t^2-1)^{\nu+1}}{(t-z)^{\nu+2}}dt, \tag{10.21}$$

and thus

$$\frac{1}{2^{\nu+1}\pi i}\int_C \frac{(t^2-1)^\nu}{(t-z)^\nu}dt = \frac{1}{2^{\nu+2}\pi i}\int_C \frac{(t^2-1)^{\nu+1}}{(t-z)^{\nu+2}}dt - \frac{z}{2^{\nu+1}\pi i}\int_C \frac{(t^2-1)^\nu}{(t-z)^{\nu+1}}dt. \tag{10.22}$$

This gives

$$P_{\nu+1}(z) - zP_\nu(z) = \frac{1}{2^{\nu+1}\pi i}\int_C \frac{(t^2-1)^\nu}{(t-z)^\nu}dt. \tag{10.23}$$

Differentiating this equation gives the first recurrence relation

$$P'_{\nu+1}(z) - zP'_\nu(z) = (\nu+1)P_\nu(z). \tag{10.24}$$

Next expand

$$\int_C \frac{d}{dt}\frac{t(t^2-1)^\nu}{(t-z)^\nu} = 0, \tag{10.25}$$

which gives

$$0 = \int_C \frac{(t^2-1)^\nu}{(t-z)^\nu}dt + 2\nu\int_C \frac{t^2(t^2-1)^{\nu-1}}{(t-z)^\nu}dt - \nu\int_C \frac{t(t^2-1)^\nu}{(t-z)^{\nu+1}}dt. \tag{10.26}$$

Replacing t^2 with t^2-1+1 and t with $t-z+z$ gives

$$0 = (\nu+1)\int_C \frac{(t^2-1)^\nu}{(t-z)^\nu}dt + 2\nu\int_C \frac{t^2(t^2-1)^{\nu-1}}{(t-z)^\nu}dt - \nu z\int_C \frac{(t^2-1)^\nu}{(t-z)^{\nu+1}}dt, \tag{10.27}$$

and using Eq. 10.23

$$(\nu+1)P_{\nu+1}(z) - (2\nu+1)zP_\nu(z) + \nu P_{\nu-1}(z) = 0, \tag{10.28}$$

which is the second recurrence relation. The remaining recurrence relations can be derived from these two. Differentiating Eq. 10.28 and using Eq. 10.24 to eliminate $P'_{\nu+1}(z)$ we find

$$zP'_\nu(z) - P'_{\nu-1}(z) = \nu P_\nu(z), \tag{10.29}$$

adding Eqs. 10.24 and 10.29 gives

$$P'_{\nu+1}(z) - P'_{\nu-1}(z) = (2\nu+1)P_\nu(z), \tag{10.30}$$

and combining Eqs. 10.24 and 10.29 we have

$$(z^2 - 1)P'_\nu(z) = \nu z P_\nu(z) - \nu P_{\nu-1}(z). \tag{10.31}$$

Note that the choice of the contour did not enter into these derivations, and thus the recurrence relations are also satisfied by the $Q_\nu(z)$.

10.1.4 *Laplace integral representation*

In some cases it is possible to find a particularly suitable kernel for the integral representation of a function by deriving a differential equation for the kernel itself. Consider the representation

$$y(x) = \int_C K(z,t)f(t)dt, \tag{10.32}$$

and substitute into the differential equation Eq. 10.1, giving

$$\int_C [(1 - z^2)K_{zz} - 2zK_z + \nu(\nu + 1)K]f(t)dt. \tag{10.33}$$

Now require the kernel K to satisfy the equation

$$(1 - z^2)K_{zz} - 2zK_z + t(tK)_{tt} = 0, \tag{10.34}$$

thus eliminating the derivatives with respect to z in Eq. 10.33 and replacing them with derivatives with respect to t, allowing integration by parts. In addition, the differential equation for the kernel is solvable, giving

$$K(z,t) = \frac{1}{\sqrt{1 - 2zt + t^2}}. \tag{10.35}$$

We then find for $f(t)$

$$t(tf)'' - \nu(\nu + 1)f = 0, \tag{10.36}$$

with solutions $f = t^\nu$ and $f = t^{-\nu-1}$, and the condition that

$$t^2(fK_t - Kf_t)|_a^b = 0. \tag{10.37}$$

This leads to the representations

$$P_\nu(z) = \frac{1}{2\pi i}\int_{C_1} \frac{t^{-\nu-1}}{\sqrt{1 - 2zt + t^2}}dt, \tag{10.38}$$

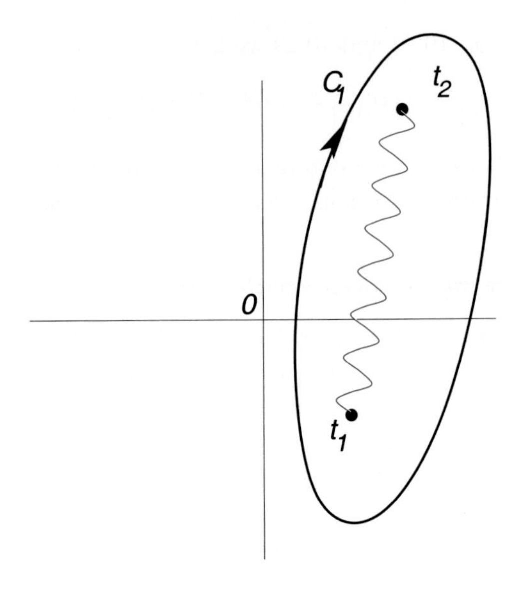

Fig. 10.2 Integration path for the Laplace representation of P_ν.

and

$$Q_\nu(z) = \frac{1}{4isin(\pi\nu)} \int_{C_2} \frac{t^{-\nu-1}}{\sqrt{1-2zt+t^2}} dt, \tag{10.39}$$

with the branch of the square root chosen to be +1 for $t \to 0$. The integration contours must be chosen to return to the initial point on the same Riemann sheet on which they started. The simplest contour is C_1, which is a contour passing clockwise around the two points $t_1 = z - \sqrt{z^2-1}$ and $t_2 = z + \sqrt{z^2-1}$. This contour, shown in Fig. 10.2, gives the Legendre polynomial P_ν. To verify this it is sufficient to show that $P_\nu(1) = 1$. For $z = 1$ we have $t_1 = t_2 = z$ and the integral reduces to

$$P_\nu(1) = \frac{1}{2\pi i} \oint \frac{t^{-\nu-1}}{t-1} dt, \tag{10.40}$$

with $t = 1$ circled in a counterclockwise manner, leaving a residue of 1, giving $P_\nu(1) = 1$.

A more complicated contour is given by C_2, a contour which crosses both of the cuts in such a way to return to the initial Riemann sheet as shown in Fig. 10.3. This contour gives Q_ν. Transformation of integration

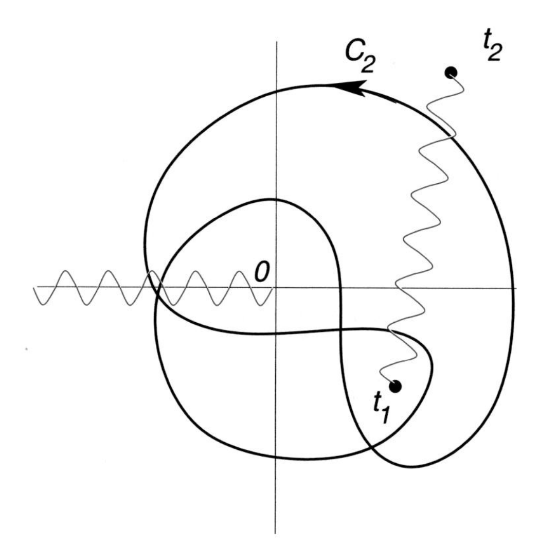

Fig. 10.3 Integration path for the Laplace representation of Q_ν.

variable through

$$t = z + \sqrt{z^2 - 1}cos\phi, \tag{10.41}$$

gives the representations originally given by Laplace

$$P_\nu(z) = \frac{1}{\pi}\int_0^\pi (z + \sqrt{z^2 - 1}cos\phi)^{-\nu-1}d\phi, \tag{10.42}$$

and

$$Q_\nu(z) = \int_0^\infty (z + \sqrt{z^2 - 1}cos\phi)^{-\nu-1}d\phi, \quad (\nu < -1) \tag{10.43}$$

and from the initial differential equation, we can also replace $-\nu - 1$ by ν in these equations.

10.1.5 *Generating function*

For integer ν the kernel Eq. 10.35 immediately leads to a generating function for the Legendre polynomials. The kernel $K(z, t)$ gives a solution to the Legendre equation for any closed contour which begins and ends on the same Riemann sheet, and for integer ν circling the origin provides such a contour. Consider the solution

$$y(z) = \frac{1}{2\pi i} \oint \frac{t^{-n-1}}{\sqrt{1-2zt+t^2}} dt, \tag{10.44}$$

with the integral circling the origin counterclockwise. Making the expansion

$$\frac{1}{\sqrt{1-2zt+t^2}} = \sum_n f_n(z) t^n, \tag{10.45}$$

we find $y(z) = f_n(z)$. But for $z = 1$ we have

$$\frac{1}{\sqrt{1-2t+t^2}} = \sum_n t^n, \tag{10.46}$$

so $f_n(1) = 1$, and thus $f_n(z) = P_n(z)$. (See problem 2 at the end of this chapter.)

10.1.6 *Legendre polynomials*

To prove that the Legendre polynomials are orthogonal, begin with Rodrigues' formula, Eq. 10.15

$$P_n(z) = \frac{1}{2^n n!} \frac{d^n}{dz^n} (z^2 - 1)^n, \tag{10.47}$$

and note that if $s < n$ then $d^s(z^2-1)^n/dz^s = 0$ for $z = \pm 1$. Then define the inner product through $(P_m, P_n) = \int_{-1}^{1} P_m(z) P_n(z) dz$ and note that

$$(P_m, P_n) = \frac{1}{2^n n! 2^m m!} \int_{-1}^{1} \frac{d^m}{dz^m} (z^2-1)^m \frac{d^n}{dz^n} (z^2-1)^n dz. \tag{10.48}$$

Without loss of generality take $m \geq n$ and integrate by parts repeatedly, discarding all terms which have factors of $z^2 - 1$ evaluated at $z = \pm 1$. Removing all derivatives from the first factor we find

$$(P_m, P_n) = \frac{(-1)^m}{2^n n! 2^m m!} \int_{-1}^{1} (z^2-1)^m \frac{d^{m+n}}{dz^{m+n}} (z^2-1)^n dz. \tag{10.49}$$

But if $m > n$ the derivative term in the integral is zero since derivatives of order higher than $2n$ vanish. For $m = n$

$$\begin{aligned}(P_n, P_n) &= \frac{(-1)^n}{2^{2n} n! n!} \int_{-1}^{1} (z^2-1)^n \frac{d^{2n}}{dz^{2n}} (z^2-1)^n dz \\ &= \frac{(2n)!}{2^{2n} n! n!} \int_{-1}^{1} (1-z^2)^n dz. \end{aligned} \tag{10.50}$$

But

$$\int_{-1}^{1}(1-z^2)^n dz = 2\int_0^{\pi/2} sin^{2n+1}(\theta)d\theta = 2\frac{2\cdot 4\cdot\ldots(2n)}{3\cdot 5\ldots(2n+1)}, \quad (10.51)$$

and using Eq. 8.55 we have finally

$$(P_m, P_n) = \frac{2\delta_{mn}}{2n+1}. \quad (10.52)$$

The polynomials can thus be constructed using the Gram–Schmidt process. Choose $P_0(z) = 1$. Then, aside from normalization

$$\begin{aligned} P_1(z) &= z - (z, P_0)P_0 = z \\ P_2(z) &= z^2 - (z^2, P_1)P_1 - (z^2, P_0)P_0 = z^2 - 2/3 \\ &\ldots \end{aligned} \quad (10.53)$$

and the normalization is easily done by requiring that $P_n(1) = 1$, giving

$$\begin{aligned} P_0(z) &= 1, \\ P_1(z) &= z, \\ P_2(z) &= \frac{1}{2}(3z^2 - 1) \\ P_3(z) &= \frac{1}{2}(5z^3 - 3z) \\ &\ldots. \end{aligned} \quad (10.54)$$

10.2 Orthogonal Polynomials

For a given metric $w(x)$ we define the inner product between two complex functions $f(x), g(x)$ as

$$(f, g) = \int_a^b f^*(x)g(x)w(x)dx, \quad (10.55)$$

where the range of integration may be finite or infinite. The function w defines a permissible metric if $(f, f) = 0$ implies $f = 0$, which means that w must be real and positive in the range a, b. The inner product satisfies the Schwartz inequality

$$(f, g)^2 \leq (f, f)(g, g), \quad (10.56)$$

which follows directly from the identity

$$(f,g)^2 = (f,f)(g,g) - \frac{1}{2}\int (f(x)g(y) - f(y)g(x))^2 w(x)w(y)dxdy. \tag{10.57}$$

Orthogonal polynomials can be constructed using the Gram–Schmidt process. Choose $P_0(x) = 1$. Then, aside from normalization

$$\begin{aligned} P_1(x) &= x - (x, P_0)P_0 \\ P_2(x) &= x^2 - (x^2, P_1)P_1 - (x^2, P_0)P_0 \\ &\dots . \end{aligned} \tag{10.58}$$

The norm of a function is defined as $\sqrt{(f,f)}$, and if the orthogonal polynomials are normalized by dividing by their norm, they are said to constitute an orthonormal set. These functions form a useful basis provided they are complete for the set of functions under consideration. Choose the basis P_n to be an orthonormal set. For any given function $f(x)$ the expansion coefficients with respect to the basis P_k are given by

$$c_n = (f, P_n). \tag{10.59}$$

Completeness is the requirement that

$$lim_{N\to\infty}(f - f_N, f - f_N) = 0, \tag{10.60}$$

with $f_N = \sum_0^N c_n P_n(x)$. If this is satisfied the series is said to converge to the function f in the mean. If this series in fact converges uniformly the limit can be taken inside the integral and we find

$$lim_{N\to\infty} f_N = f(x). \tag{10.61}$$

From the relation

$$(f - f_N, f - f_N) \geq 0, \tag{10.62}$$

we find

$$(f,f) - 2\sum_0^N c_n(f, P_n) + \sum_0^N c_n^2 \geq 0, \tag{10.63}$$

and thus

$$\sum_0^N c_n^2 \leq (f,f), \tag{10.64}$$

which is known as Bessel's inequality. This proves that the sum of the squares of the expansion coefficients always converges if f has a finite norm.

The simplest example of a set of orthogonal polynomials is found by taking the weight $w(x) = 1$ and the domain to be $-1 < x < 1$, with normalization $(P_n, P_m) = \delta_{mn}/(n+1/2)$, giving the Legendre polynomials. Other examples are given in the following table:

Table 10.1 Weights and Domains for Orthogonal Polynomials

Polynomial	*notation*	*weight*	*domain*	*norm*
Legendre	P_n	1	$-1, 1$	$(n+1/2)^{-1}$
Gegenbauer	C_n^λ	$(1-x^2)^{\lambda-1/2}$	$-1, 1$	$\frac{2\pi\Gamma(2\lambda+n)}{(n+\lambda)n!4^\lambda\Gamma^2(\lambda)}$
Chebychev	T_n	$(1-x^2)^{-1/2}$	$-1, 1$	$\pi/(2-\delta_{n0})$
Hermite	H_n	e^{-x^2}	$-\infty, \infty$	$2^n\sqrt{\pi}n!$
Jacobi	$P_n^{\alpha\beta}$	$(1-x)^\alpha(1-x)^\beta$	$-1, 1$	$\frac{\Gamma(\alpha+1+n)\Gamma(\beta+1+n)2^{\alpha+\beta+1}}{\Gamma(n+1)\Gamma(\alpha+\beta+1+n)(\alpha+\beta+1+2n)}$
Laguerre	L_n^α	$x^\alpha e^{-x}$	$0, \infty$	$\frac{\Gamma(\alpha+n+1)}{\Gamma(n+1)}$

These functions are solutions to the following differential equations:

Table 10.2 Differential Equations for Orthogonal Polynomials

Polynomial	*equation*
Legendre	$(1-z^2)f" - 2zf' + n(n+1)f = 0$
Gegenbauer	$(z^2-1)f'' + (2\lambda+1)zf' - n(2\lambda+n)f = 0$
Chebychev	$(1-z^2)f'' - zf' + n^2 f = 0$
Hermite	$f'' - 2zf' + 2nf = 0$
Jacobi	$(1-z^2)f'' + [\beta - \alpha - (\alpha+\beta+2)z]f' + n(n+\alpha+\beta+1)f = 0$
Laguerre	$zf'' + (\alpha - z + 1)f' + nf = 0$

The methods used above for the Legendre polynomials can be used to obtain integral representations, orthogonality relations using the given norm, etc., for each of these examples.

10.3 Wavelets

10.3.1 *Introduction*

Representation of functions through a sum of appropriate basis functions is a technique which began with the trigonometric series of Fourier, and was extended to the use of special polynomials, some of which are described in the previous section. The method is particularly useful only if the functions under consideration are faithfully represented with only a few terms. If however the functions to be represented possess detail at many different length scales, or even discontinuities, it is easy to see that trigonometric or polynomial expansions must have very many terms to give a good representation, making them of limited use.

Wavelets solve this problem by providing a two-parameter family of basis functions, each of which has significant magnitude only in a finite domain. One parameter describes the location of the basis function, and the second parameter the width scale. The family of basis functions is constructed to be self-similar under scaling, and also sometimes continuously differentiable, but this latter requirement is not necessary. The construction begins by defining nested sets of functions of higher and higher resolution, orthogonal to the wavelets. These scaling functions are then used to define the wavelets. The wavelet functions can be very complicated, but the great power of the wavelet analysis is that they are easily constructed and also one in fact seldom has to use the form of the functions themselves. Some useful references are Daubechies ([1992]), Mallat ([1998]), Van den Berg ([1999]), Antione ([2004]), and Welland ([2003]).

10.3.2 *Scaling function*

We wish to construct a prototype scaling function $\phi(t)$ which will be used to construct a mutually orthogonal basis, with the inner product[1] given by

$$(\phi, \alpha) = \int_{-\infty}^{\infty} dt \phi(t) \alpha(t). \tag{10.65}$$

This function must be reasonably localized, and in particular it must, under translation by integers, generate a set of functions, the elements of which are mutually orthogonal. A two-parameter family of functions is constructed through

$$\phi_{k,n}(t) = 2^{k/2} \phi(2^k t - n), \tag{10.66}$$

[1] We consider only real functions.

with n giving translation and k scaling, and the normalization chosen so that $(\phi_{k,n}, \phi_{k,n})$ is independent of k, n. These functions are clearly self-similar under both the scaling operation and the translation operation. Now we require that the scaling function satisfy the following conditions:

(a) Orthogonality under translation by an integer:

$$(\phi(t-n), \phi(t-m)) = E\delta_{nm}, \tag{10.67}$$

which implies

$$(\phi_{k,n}, \phi_{k,m}) = E\delta_{nm}. \tag{10.68}$$

Normalization is, however, not defined through the inner product. It is much more convenient to require that $\phi(t)$ have nonzero integral and normalize the functions through

$$\int_{-\infty}^{\infty} dt\phi(t) = 1, \tag{10.69}$$

and this condition determines the constant E.

(b) The space spanned by a given resolution k must contain the space spanned by coarser resolutions, i.e. if

$$\begin{aligned} V_0 &= [\phi(t-n)]_{n=-\infty,\infty} \\ V_k &= [\phi(2^k t-n)]_{n=-\infty,\infty} \end{aligned} \tag{10.70}$$

then $V_0 \subset V_1 \subset V_2 \ldots$.

(c) The function $\phi = 0$ is the only function common to all the V_k, i.e.

$$\bigcap_k V_k = 0. \tag{10.71}$$

(d) The basis is complete, i.e. all square integrable functions on the real line R can be expressed in terms of the finest resolution,

$$lim_{k\to\infty} V_k = L^2(R). \tag{10.72}$$

Any scaling function which satisfies these conditions permits the construction of a wavelet basis, which satisfies many important properties leading to an efficient means of expanding functions. Condition (b) implies that a scaling function of resolution k can be expressed in terms of basis functions of resolution $k+1$. Because of the self-similarity of the scaling functions this follows by requiring it for $k=0$. That is from

$$\phi(t) = \sum_{n=-\infty}^{\infty} a_n \sqrt{2}\phi(2t-n), \tag{10.73}$$

we find

$$\phi(2^k t - n) = \sum_{m=-\infty}^{\infty} a_{m-2n} \sqrt{2}\phi(2^{k+1}t - m), \tag{10.74}$$

or

$$\phi_{k,n}(t) = \sum_{m=-\infty}^{\infty} a_{m-2n} \phi_{k+1,m}(t). \tag{10.75}$$

Now use the orthogonality relation Eq. 10.67 to find that

$$\sum_n a_n a_{n-2k} = \delta_{k,0}. \tag{10.76}$$

The normalization $\int_{-\infty}^{\infty} \phi(t)dt = 1$ gives

$$\sum_n a_n = \sqrt{2}, \tag{10.77}$$

but we make the more stringent requirement that

$$\sum_{n=even} a_n = \sum_{n=odd} a_n = \frac{1}{\sqrt{2}}. \tag{10.78}$$

We will see that the a_n operate as a filter for processing a given function $f(t)$. Any set of a_n which satisfy the condition given by Eq. 10.76 is called a quadratic mirror filter, terminology originating in the theory of signal processing, which predated wavelet theory.

The function ϕ can be constructed from the a_n by regarding Eq. 10.73 as a fixed point equation, making an initial guess for ϕ, and applying it iteratively through

$$\phi^{k+1}(t) = \sum_{n=-\infty}^{\infty} a_n \sqrt{2}\phi^k(2t-n). \tag{10.79}$$

If the a_n are such that this recursion relation converges, it gives the function $\phi(t)$. We thus have the amazing result that the filter coefficients completely determine the functional form of the scaling function. However, it is certainly not *a priori* obvious or trivial that such wavelet systems satisfying all the requirements given above exist. The formulation of these requirements and the demonstration of a large number of useful solutions is due to Daubechies [1992; 1988]. See also Hubbard [1996] for a history and overview of wavelet theory.

The complexity of the wavelet representation increases with the length of the filter a_n. If the number of a_n is finite the scaling functions have compact support, and, in fact, if there are N filter coefficients, $a_0, \ldots, a_{N-1}$ then the support of $\phi(t)$ is the range $0 < t < N-1$. This is easily seen by considering Eq. 10.79 and noting that a_n shifts $\phi(2t)$, which has half the support size of $\phi(t)$, by an amount $(n-1)/2$, so the largest shift in the iteration is by $(N-1)/2$. The range $0 < t < N-1$ is seen to be invariant under these operations, that is if $\phi(t)$ is zero outside this range it remains so under Eq. 10.79. If only two of the a_n are nonzero (length two) there is a unique solution to Eqs. 10.76 and 10.78, giving

$$a_M = \frac{1}{\sqrt{2}}, \qquad a_{M+1} = \frac{1}{\sqrt{2}}, \tag{10.80}$$

for some M, the value of which only shifts the support range. If $M = 0$ then the support is $0 < t < 1$ and $\phi(t) = 1$ in this domain and zero outside. For representations given by a_n of longer length there is some freedom in the choice of the values. Daubechies [1992] used conditions of the vanishing of a number of moments of the wavelets, and smoothness criterion, to construct useful wavelet systems for higher length vectors a_n. A solution of length four given by Daubechies is

$$a_n = \frac{(1+\sqrt{3}, 3+\sqrt{3}, 3-\sqrt{3}, 1-\sqrt{3})}{4\sqrt{2}}, \tag{10.81}$$

which is easily seen to satisfy Eqs. 10.76 and 10.78. Solutions exist for any even length, and there are more degrees of freedom in their form as the length increases. An example of a solution with infinite length is given by

$$a_n = \frac{2}{\pi n} sin\left(\frac{\pi n}{2}\right), \tag{10.82}$$

with

$$\phi(t) = \frac{sin(\pi t)}{\pi t}, \tag{10.83}$$

which has infinite support but is also infinitely differentiable.

10.3.3 *Wavelet basis construction*

We now use the scaling function and the filter coefficients to construct a basis which is orthogonal, and consists of a sequence of functions giving higher and higher resolution of detail. Let W_0 be the space of functions orthogonal to V_0, but contained in V_1. By construction we have $V_0 \subset V_1 \subset V_2 \ldots$. Consider a wavelet ψ, belonging to W_0. Then

$$(\phi(t-j), \psi(t)) = 0 \quad for \;\; all \;\; j. \tag{10.84}$$

The expansion of $\phi(t-j)$ in terms of the basis of V_1 is

$$\phi(t-j) = \sum_n a_n \sqrt{2}\phi(2t - 2j - n), \tag{10.85}$$

and let that for ψ be

$$\psi(t) = \sum_n b_n \sqrt{2}\phi(2t - n). \tag{10.86}$$

Then the orthogonality condition gives

$$\sum_{n=-\infty}^{\infty} a_n b_{n-2j} = 0 \quad all \;\; j, \tag{10.87}$$

with a solution

$$b_n = \pm(-1)^n a_{L-n}, \tag{10.88}$$

with L an arbitrary odd integer chosen to position ψ. That this is a solution is easily seen by noting that the terms in the sum given by n and $L+2j-n$ cancel one another. In fact this can be shown to be the only solution (Burrus *et al.* [1998]). This result is essential, since it implies that the functions given above span the whole space V_1, i.e. there are no other functions in this space orthogonal both to the ϕ and to the ψ. The basic wavelet is then given by Eq. 10.86.

Now continue in the same manner to levels of higher resolution. The topology of the nested set of vector spaces is shown in Fig. 10.4. Let W_1 be the space of functions orthogonal to V_1 but contained in V_2. Orthogonality to functions in V_1 naturally implies orthogonality to functions in W_0, so these wavelets are orthogonal to the wavelets previously constructed. In this manner we construct the orthogonal decomposition in the form

$$V_0 \bigoplus W_0 \bigoplus W_1 \bigoplus W_2 \bigoplus W_3 \ldots \tag{10.89}$$

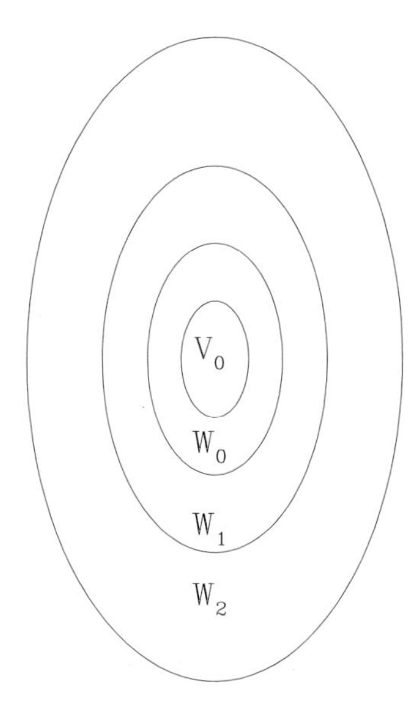

Fig. 10.4 Nested function spaces, with $V_{k+1} = W_k \bigoplus V_k$.

and all of these spaces are orthogonal to each other. Since $W_k \subset V_{k+1}$ and the $\phi_{k,j}$ are more and more localized with increasing k, the wavelets are also more and more localized. At level k, the $\phi_{k,j}$ spanning V_k are given by

$$\phi_{k,j}(t) = 2^{k/2}\phi(2^k t - j), \qquad j = -\infty, \infty \tag{10.90}$$

and the basis for W_k is given by

$$\psi_{k,j}(t) = 2^{k/2}\psi(2^k t - j), \qquad j = -\infty, \infty. \tag{10.91}$$

Note that

$$\int_{-\infty}^{\infty} dt\psi(t) = \sum_n (-1)^n a_{L-n} = -\sum_{j=even} a_j + \sum_{j=odd} a_j = 0, \tag{10.92}$$

and it follows from the construction that

$$(\psi_{k,j}, \psi_{l,i}) = \delta_{k,l}\delta_{j,i}\sum_n (a_{L-n})^2 = E\delta_{k,l}\delta_{j,i}. \tag{10.93}$$

By assumption (Eq. 10.72), any square integrable function can be expanded[2] in the orthogonal basis as

$$f(t) = \sum_{j=-\infty}^{\infty} c(j)\phi(t-j) + \sum_{k=0}^{\infty}\sum_{j=-\infty}^{\infty} d_k(j)\psi_{k,j}(t), \tag{10.94}$$

with $\phi \in V_0$, and $\psi_{k,j} \in W_k$. Furthermore we have

$$Ec(j) = \int_{-\infty}^{\infty} dt\phi(t-j)f(t), \qquad Ed_k(j) = \int_{-\infty}^{\infty} dt\psi_{k,j}(t)f(t). \tag{10.95}$$

The series is truncated according to the magnitude of the coefficients, which typically falls off rapidly with both j and k, but keeping all coefficients above a certain threshold can produce isolated bands of j and k. The advantage over a usual expansion in terms of orthogonal polynomials is that the range of the summations is localized around the location and the resolution scale of the initial function $f(t)$.

It is not necessary that the wavelets be made mutually orthogonal, there are also methods for using generalized basis systems (Burrus *et al.* [1998]).

10.3.4 *Determining the expansion coefficients*

All information about the function is contained in the expansion at the finest scale, and thus it is not surprising that expansion coefficients for the coarser scales are determined by the fine scale coefficients. Expand $f(t) \in V_{k+1}$ using the complete basis at level k

$$f(t) = \sum_j c_k(j)2^{k/2}\phi(2^k t - j) + \sum_j d_k(j)2^{k/2}\psi(2^k t - j). \tag{10.96}$$

Now calculate the coefficients by taking the inner product

$$\begin{aligned} c_k(j) &= \int_{-\infty}^{\infty} dt2^{k/2}\phi(2^k t - j)f(t) \\ &= \sum_m a_{m-2j}\int_{-\infty}^{\infty} dt2^{(k+1)/2}\phi(2^{k+1}t - m)f(t). \end{aligned} \tag{10.97}$$

From these equations we see that the support of the $c_k(j)$ is given by the support of $\phi(t)$, which is $0 < t < N-1$, giving $c_k(j) \neq 0$ for the range

[2]That this is true for the basis constructed must of course be proven.

$-(N-1) < j < (N-1)2^k$. We then find

$$c_k(j) = \sum_m a_{m-2j} c_{k+1}(m) = \sum_n a_n c_{k+1}(n+2j), \tag{10.98}$$

and a similar analysis for $d_k(j)$ gives

$$d_k(j) = \sum_m b_{m-2j} c_{k+1}(m) = \sum_n b_n c_{k+1}(n+2j), \tag{10.99}$$

and thus once the expansion coefficients are determined at the finest scale they are easily calculated at all coarser scales using these filtering relations, so called because the coefficients at level $k+1$ are filtered by the coefficients a_m, b_m. At scale k there are only half as many nonzero coefficients as at scale $k+1$, hence the $2j$ appearing in the argument of c_{k+1}. This result makes the construction of the wavelet coefficients extremely simple. Consider the determination of the coefficients at the finest scale to be used

$$c_k(j) = \int_{-\infty}^{\infty} dt \phi_{k,j}(t) f(t), \qquad d_k(j) = \int_{-\infty}^{\infty} dt \psi_{k,j}(t) f(t). \tag{10.100}$$

At this scale the function $f(t)$ may be considered to be constant over the domain of the scaling and wavelet functions. Removing $f(t)$ from the integrals gives the result that $d_{k,j} = 0$, and using $\int_{-\infty}^{\infty} dt \phi_{k,j}(t) = 2^{-k/2}$ the $c_{k,j}$ are simply equal to $2^{-k/2} f(t_{k,j})$ with $t_{k,j}$ the point of support of $\phi_{k,j}$. The center of the support domain is given by

$$t_{k,j} = 2^{-k} \left[\frac{N-1}{2} + j \right]. \tag{10.101}$$

Thus the wavelet coefficients are determined simply by sampling values of the function; the forms of the scaling function and wavelet function are not used.

If the domain of $f(t)$ is finite, the domain of $\phi(t)$ is chosen to match it. If a final resolution of $1/2^K$ of this domain is required, then the total number of wavelets in the sum is

$$(N-1) \sum_0^K 2^k \simeq (N-1) 2^{K+1}, \tag{10.102}$$

but many of the coefficients can be very small, so the number of terms needed for the representation can be much smaller.

The algorithm for computing the wavelet coefficients of a function in the case of filters of finite length, due to Mallat [1998], is known as the

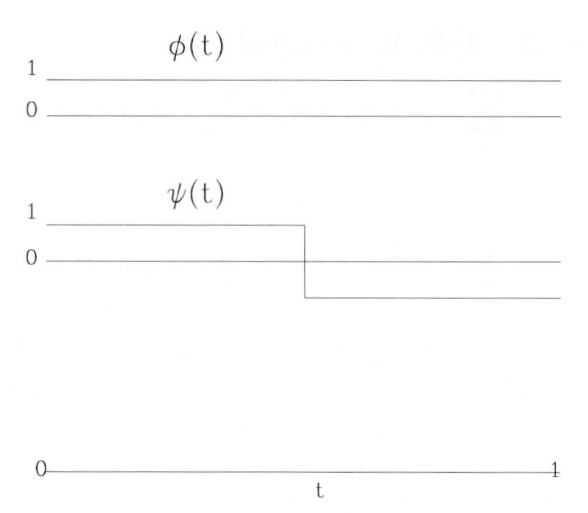

Fig. 10.5 The functions given by the length two filter, Eqs. 10.104, 10.105.

fast wavelet transform. In fact it is more efficient than the fast Fourier transform. A signal of length L requires $Lln(L)$ calculations for the fast Fourier transform, but the number of calculations is only of order L for the fast wavelet transform.

10.3.5 *Examples*

The simplest example of a wavelet system is the Haar system [1910], given by the length two filter

$$a_0 = \frac{1}{\sqrt{2}}, \qquad a_1 = \frac{1}{\sqrt{2}}, \tag{10.103}$$

and the scaling function $\phi(t)$ is given by

$$\phi(t) = \begin{cases} 1 & 0 < t < 1 \\ 0 & otherwise \end{cases}, \tag{10.104}$$

and we see that $E = 1$, $b_n = (1/\sqrt{2}, -1/\sqrt{2})$ and

$$\psi(t) = \begin{cases} 1 & 0 < t < 1/2 \\ -1 & 1/2 < t < 1 \\ 0 & otherwise. \end{cases} \tag{10.105}$$

Figure 10.5 shows the scaling function and the wavelet for the Haar system. For this set of functions the orthogonality of the set is easily seen, but this is not true for longer length filters. Figures 10.6 and 10.7 shows

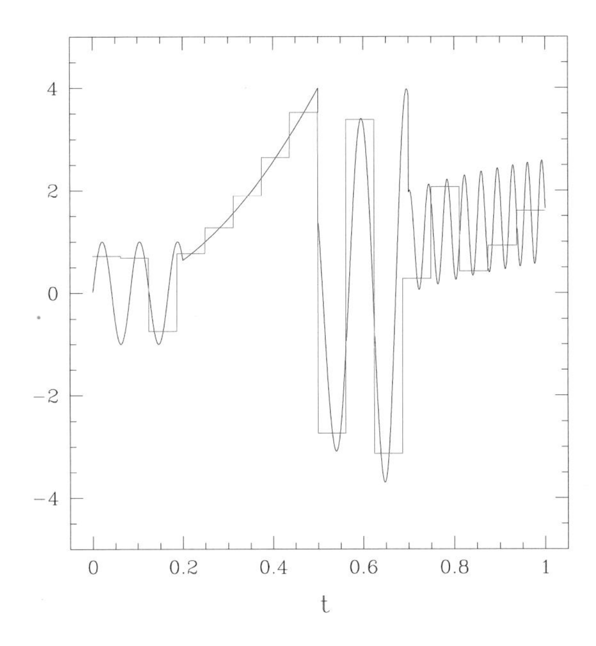

Fig. 10.6 A wavelet approximation using the Haar wavelet system with $0 \leq k \leq 4$.

the wavelet approximations to a function with $0 \leq k \leq 4$ and $0 \leq k \leq 8$ terms present in the Haar expansion. With $0 \leq k \leq 4$ the wavelet of greatest resolution has a width of $1/16$, so the short wave length oscillations cannot be resolved. With $0 \leq k \leq 8$ the wavelet of greatest resolution has a width of $1/256$, sufficient to resolve the short wave length oscillations in this function. This second representation matches the function everywhere within an error of 0.2, with a maximum of eight terms in the wavelet sum for any value of t, but all wavelets are used in the decomposition. Restricting the sum to terms of magnitude greater than 0.02 reduces the number of wavelets to 156 and still gives a good representation of the function. Figure 10.8 shows the distribution of the magnitudes of the wavelet amplitudes for the approximation shown in Fig. 10.7.

An example of length four is given by the Daubechies set, Eq. 10.81. An iteration of Eq. 10.79 produces the functions shown in Fig. 10.9. Various schemes are given in the literature for this construction, but if the number of points in $f(t)$ is free (and interpolation can be used on the initial data set to produce this) we find rapid convergence from an initial function of $\phi(t) = 1$ by simply taking the number of points to be $2^k(N-1)$ with $t = j*(N-1)/2^k$,

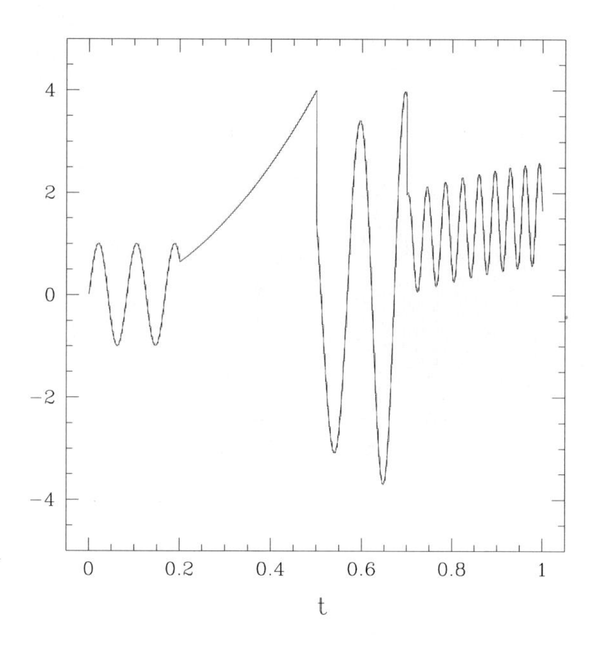

Fig. 10.7 A wavelet approximation using the Haar wavelet system with $0 \leq k \leq 8$.

so that translations given in the iterations take data points exactly into data points. Note that the scaling function and the wavelet appear to have fractal character, but orthogonality of $\phi(t)$ to $\phi(t-1)$ is not obvious, and neither is the orthogonality of $\phi(t)$ and $\psi(t)$. The Daubechies wavelets are much more effective than the Haar wavelets in compressing data. Choosing a filter of length N allows one to impose that the associated wavelets have a certain number of continuous derivatives. If the wavelets are chosen to have m continuous derivatives, the coefficient $c_k(j)$ resulting in a decomposition in the Daubechies basis will be of order $2^{-(m+1/2)k}$ whereas it would be of order $2^{-3k/2}$ in the Haar system.

The use of wavelets has proven to be the most efficient method of compressing data yet devised, and is used for many standard procedures in signal processing. The techniques are readily extended to more than one dimension, and, for example, the Daubechies basis is used for the encoding of fingerprints for numerical storage and transmission. A wavelet analysis of a two-dimensional image with noise present can, by truncating small amplitude wavelet coefficients, result in a much clearer image than the original, as well as giving a characterization of the noise if that is desired.

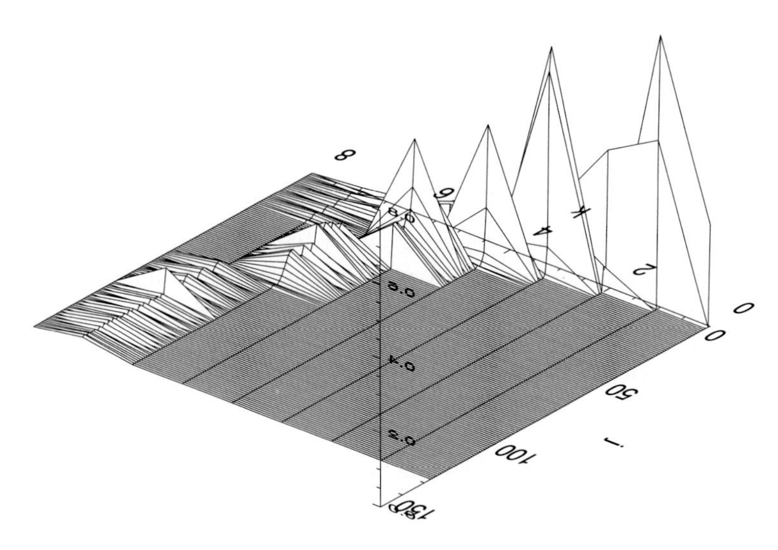

Fig. 10.8 Distribution of the wavelet amplitudes for the expansion given in Fig. 10.7.

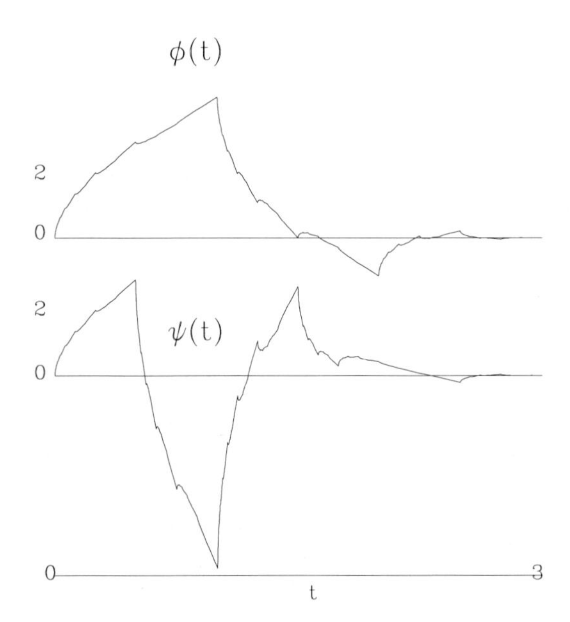

Fig. 10.9 The functions given by the length four filter, Eq. 10.81.

It is instructive to program a few simple wavelet routines to familiarize oneself with the techniques. In addition there is available a library of MATLAB routines for wavelets and related time–frequency transforms (Wavelab [2005]), which can be retrieved over the web.

10.3.6 *Time–frequency analysis*

Representing speech or music immediately presents a fundamental problem also underlying quantum mechanics, namely the fact that, unlike the way music is written, with a definite note appearing at a definite time, it is impossible to define frequency instantaneously. There is an uncertainty relating the length of time the note sounds to the frequency of the note, given by

$$\Delta t \Delta \omega \simeq 2\pi, \tag{10.106}$$

equivalent to the Heisenberg uncertainty principle. Thus the signal must be coded by time–frequency atoms which each occupy an area in the time–frequency plane of at least 2π (Jaffard and Meyer [1996]). See also Cohen [1995] and Flandrin [1999]. Wavelets are thus constructed to consist of a wave packet made of a sine or cosine function with a fixed frequency, multiplied by a slowly varying envelope function, with adjustable lengths of attack, duration, and attenuation. An example of this construction is given by the Malvar–Wilson basis. Partition the real line with the intervals $[a_j, a_{j+1}]$. Then $l_j = a_{j+1} - a_j$ is the length of this interval. Also choose numbers $\alpha_j > 0$ such that $\alpha_j + \alpha_{j+1} \leq l_j$. The lengths α_j give the domain in which the signal is increasing and decaying, and the l_j gives the length of the note. Choose the envelope function $w_j(t)$ through

$$\begin{aligned}
&0 \leq w_j(t) \leq 1 \quad all \;\; t, \\
&w_j(t) = 1 \quad for \;\; a_j + \alpha_j \leq t \leq a_{j+1} - \alpha_{j+1}, \\
&w_j(t) = 0 \quad for \;\; t \leq a_j - \alpha_j \;\; or \;\; t > a_{j+1} + \alpha_{j+1}, \\
&w_j^2(a_j + \tau) + w_j^2(a_j - \tau) = 1 \quad for \;\; |\tau| < \alpha_j, \\
&w_{j-1}(a_j + \tau) = w_j(a_j - \tau) \quad for \;\; |\tau| < \alpha_j.
\end{aligned} \tag{10.107}$$

We then note that $\sum_j w_j^2(t) = 1$ for all t. Wavelets are then constructed to be

$$u_{j,k}(t) = \sqrt{\frac{2}{l_j}} w_j(t) cos\left(\frac{\pi(k+1/2)}{l_j}(t - a_j)\right), \tag{10.108}$$

and these functions form an orthonormal basis for square integrable functions on the real line. Figure 10.10 shows an example of a Malvar–Wilson time–frequency wavelet, with $a_j = 1/2$, $a_{j+1} = 7$, $k = 8$, $\alpha_j = 1$, and $\alpha_{j+1} = 2$. The envelope is shown in black and the wavelet itself in red.

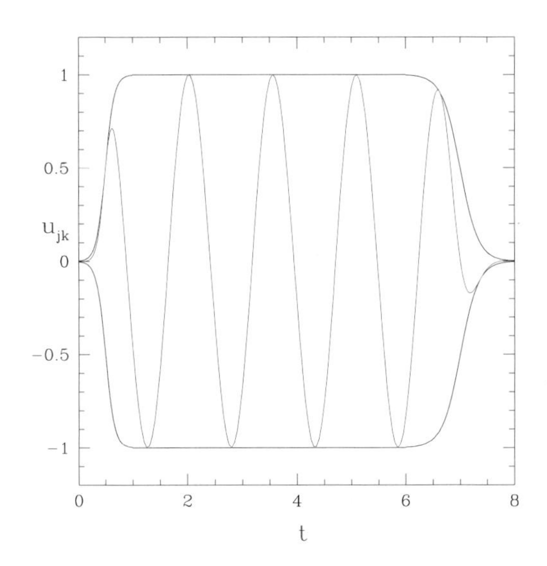

Fig. 10.10 An example of a Malvar–Wilson basis function.

In quantum mechanics there is a function which comes close to giving the local weight of a distribution in the time–frequency plane. Given a function $f(t)$ the Wigner–Ville transform is defined to be

$$W(t,\omega) = \int_{-\infty}^{\infty} f(t+\tau/2)f^*(t-\tau/2)e^{-i\omega\tau}d\tau. \tag{10.109}$$

This function gives the local probability density in t upon integration over frequency

$$\int_{-\infty}^{\infty} W(t,\omega)\frac{d\omega}{2\pi} = |f(t)|^2, \tag{10.110}$$

and the local probability density in ω upon integrating over time

$$\int_{-\infty}^{\infty} W(t,\omega)dt = |f^*(\omega)|^2. \tag{10.111}$$

If the signal $f(t)$ is square integrable then $W(t, \omega)$ is a real valued continuous function in the time–frequency plane, but it can assume negative values so cannot be interpreted as a local weight in the plane. It also can take on nonzero values in domains in which the function $f(t)$ has no support. For example, if the Fourier transform of $f(t)$ has support in the domain $\omega_0 \leq |\omega| \leq \omega_1$ then $W(t, \omega)$ may have support in the whole domain $|\omega| \leq \omega_1$.

10.4 Problems

1. Find the radius of convergence of the series

$$y \sim \sum_s \frac{(-1)^s 2^{2s} \Gamma(\nu/2 + 1/2 + s)}{(2s)! \Gamma(\nu/2 - s + 1)} z^{2s}.$$

2. Show that $z = \infty$ is an ordinary point of the differential equation for the Legendre function, and thus that for integer $\nu = n$ the contour circling the axis counterclockwise, which was shown to give $P_n(z)$, can be deformed through infinity to become the contour C_1 of Fig. 10.2.

3. Evaluate

$$\int_{-1}^{1} z P_n(z) P_{n+1}(z) dz.$$

4. Evaluate

$$\int_{-1}^{1} z^2 P_n(z) P_{n+1}(z) dz.$$

5. Show that

$$\int_{-1}^{1} z(1 - z^2) \frac{dP_n}{dz} \frac{dP_m}{dz} dz,$$

is zero unless $m - n = \pm 1$ and evaluate it in these cases. (Math Trip [1896])

6. Directly verify Laplace's representation Eq. 10.42 by differentiation.

7. Set up a numerical iteration scheme and use it to find the form of the Daubechies length four scaling functions shown in Fig. 10.9.

8. Show that the Haar system and the infinite support system given by $\phi(t) = sin(\pi t)/(\pi t)$ are Fourier transforms of each other.

9. Show that the wavelet for the infinite support system given by the scaling function $\phi(t) = sin(\pi t)/(\pi t)$ is given by $\psi(t) = 2\phi(2t) - \phi(t)$.

Chapter 11

Airy

George Biddell Airy was born in Alnwick, Northumberland, England, in 1801. He had an impressive early education, including the Latin and Greek classics, mathematics, and an early introduction to Newton's *Principia*, and was then sent to Trinity College, Cambridge. He kept copious notes throughout his life, and his autobiography (Airy [1896]) is a rich portrait of the topics covered in classes at Cambridge in the 1800s. While he was a student the transition from the fluxion notation of Newton to the differential calculus of Leibnitz was initiated. Shortly after obtaining his degree he took charge of the Cambridge observatory as professor. At this time he calculated a correction to the planetary motion of Venus, one of the last such corrections to be calculated by hand. In 1835 he took over the Greenwich observatory as Astronomer Royal, greatly improving the quality of the observations there, and making a series of observations of Jupiter's fourth satellite to determine the mass of Jupiter. He also conducted pendulum experiments in a mine, to determine the mean density of the earth. His interest in mathematics was very much connected with real physical problems, and he had a continual debate with mathematicians at Cambridge, objecting to examination questions of "purely idle algebra, arbitrary combinations of symbols, applicable to no further purpose". In addition to mechanics, he was much interested in optics, designing special lenses for his own glasses and contributing to the elimination of chromatic aberration. In a paper by Airy [1838] "On the Intensity of Light in the Neighborhood of a Caustic" there appears the integral[1] $\int_0^\infty cos(t^3/3 \pm xt)dt$, which Airy laboriously integrated by hand. Subsequently the integral (with the plus

[1] Airy actually used the form $\int_0^\infty cos[\frac{\pi}{2}(w^3 - mw)]dw$, which can be put in what is now the standard form.

sign) was shown by Stokes to satisfy the differential equation

$$y'' = xy, \tag{11.1}$$

the standard solutions to which are now known as Airy functions. The modern notations of $Ai(z)$ and $Bi(z)$ were introduced by Jeffreys [1942].

11.1 WKB Analysis

The Stokes diagram for the Airy function is shown in Fig. 11.1, and the analytic continuation is carried out in section 5.3. The Stokes constant is $T = i$, and $Q(z) = -z$. Using WKB we can relate the asymptotic behavior of $Ai(z)$ at $+\infty$ to that at $-\infty$. For large z write the solution as $y = e^S$, and substitute, giving $S'' + (S')^2 = z$. Dominant balance gives $S' = \pm z^{1/2}$, and finding the next order gives

$$S = \pm\frac{2}{3}z^{3/2} - \frac{ln(z)}{4}. \tag{11.2}$$

On the positive real axis $Ai(z)$ is defined to be the subdominant solution, which aside from normalization has the form

$$(0, z)_s \sim \frac{e^{-\frac{2}{3}z^{3/2}}}{z^{1/4}}. \tag{11.3}$$

Note that the sign of the square root is determined by taking $\sqrt{Q}$ positive on the negative real axis and by our placement of the branch cut, making this solution subdominant, and that we have multiplied by a constant phase to choose the solution to be real. Writing $z = re^{-i\pi}$, a phase choice determined by our placement of the cut in Fig. 11.1, we find on the negative real axis

$$(0, z) = \frac{e^{i(-\frac{2}{3}r^{3/2}+\frac{\pi}{4})}}{r^{1/4}}, \tag{11.4}$$

and

$$(z, 0) = \frac{e^{i(\frac{2}{3}r^{3/2}+\frac{\pi}{4})}}{r^{1/4}}. \tag{11.5}$$

Placing the cut in another position will modify some expressions, but not final results. The solution in domain 5 is

$$(0, z)_d - i(z, 0)_s = \frac{2sin(\frac{2}{3}r^{3/2} + \frac{\pi}{4})}{r^{1/4}}. \tag{11.6}$$

Choose overall normalization for the Airy function for positive x to be $Ai(x) \simeq e^{-\frac{2}{3}x^{3/2}}/(2\sqrt{\pi}x^{1/4})$, giving for large negative x

$$Ai(x) \simeq \frac{sin(\frac{2}{3}r^{3/2} + \frac{\pi}{4})}{\sqrt{\pi}r^{1/4}}, \tag{11.7}$$

with $r = |x|$, which is the leading order asymptotic expression for the Airy function for x negative.

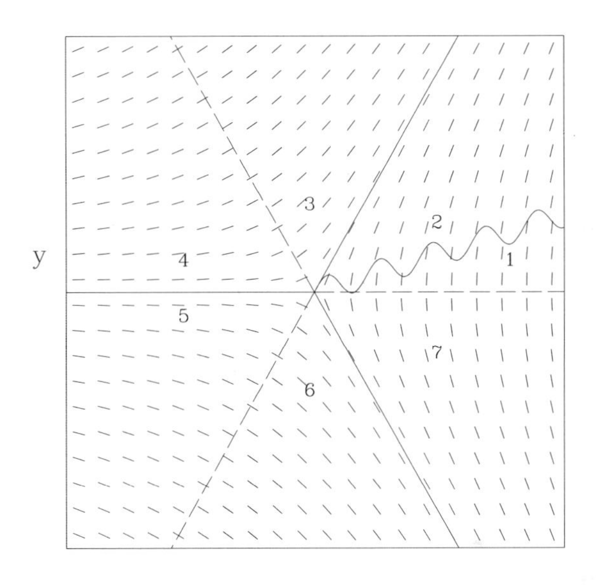

Fig. 11.1 Stokes diagram for the Airy function.

The continuation of the second function, $Bi(x)$, is slightly more subtle, since requiring the solution to be dominant at large positive x does not uniquely define it within WKB theory. Choose instead for large negative x the asymptotic form

$$Bi(x) \simeq \frac{cos(\frac{2}{3}r^{3/2} + \frac{\pi}{4})}{\sqrt{\pi}r^{1/4}}, \tag{11.8}$$

which complements $Ai(x)$. First write this as

$$Bi(x) \simeq \frac{e^{i(\frac{2}{3}r^{3/2}+\frac{\pi}{4})} + e^{-i(\frac{2}{3}r^{3/2}+\frac{\pi}{4})}}{2\sqrt{\pi}r^{1/4}}, \tag{11.9}$$

which is in domain 5 of Fig. 11.1, aside from the $2\sqrt{\pi}$ normalization equal to $(z,0)_s - i(0,z)_d$. In domain 6 this becomes $2(z,0)_s - i(0,z)_d$ and in domain 7 it becomes

$$2(z,0)_d - i(0,z)_s. \tag{11.10}$$

Above the Stokes line in domain 1 this becomes $2(z,0)_d + i(0,z)_s$ and exactly on the line it is $2(z,0)_d$ or, restoring normalization

$$Bi(x) \simeq \frac{e^{\frac{2}{3}x^{3/2}}}{\sqrt{\pi}x^{1/4}}, \tag{11.11}$$

for large positive x.

Note that by using the special rule concerning solutions exactly on Stokes lines, one may in fact begin with a purely dominant solution at large positive x and carry out the continuation to negative x, giving the same result. This is because this rule guarantees flux conservation, a condition which is exact, not merely a WKB approximation.

11.2 Argand Plot of the Airy Function

The WKBJ approximation $\psi_-(z)$ for the Airy function $Ai(z)$ reads:

$$\psi_- = \frac{1}{2\sqrt{\pi}z^{1/4}}e^{-\frac{2}{3}z^{3/2}}.$$

Continued in the upper complex plane Heading's uniformly valid approximation reads

$$Ai_{\text{unif}}(z) = \frac{1}{2\sqrt{\pi}z^{1/4}}\, e^{-\frac{2}{3}z^{3/2}} + C(\theta)\frac{1}{2\sqrt{\pi}z^{1/4}}\, e^{\frac{2}{3}z^{3/2}}$$

where

$$C(\theta) = \begin{cases} 0 & 0 < \theta < \frac{2}{3}\pi \\ i/2 & \theta = \frac{2}{3}\pi \\ i & \frac{2}{3}\pi < \theta < \pi \end{cases}.$$

We write $z = re^{i\theta}$. We choose $r = 1.7171$ and plot the Airy function and its approximations in the range $0 < \theta < \pi$. The first plot, Fig. 11.2, shows the exact Airy function $Ai(z)$ and its WKBJ approximation $\psi_-(z)$. The approximation is good for $\theta < 2\pi/3$ but as we approach $\theta = \pi$ the approximation breaks down.

The second plot, Fig. 11.3, compares the Airy function to Heading's uniformly valid approximation $Ai_{\text{unif}}(z)$. Obviously it remains useful when θ approaches π. The Stokes jump is almost invisible.

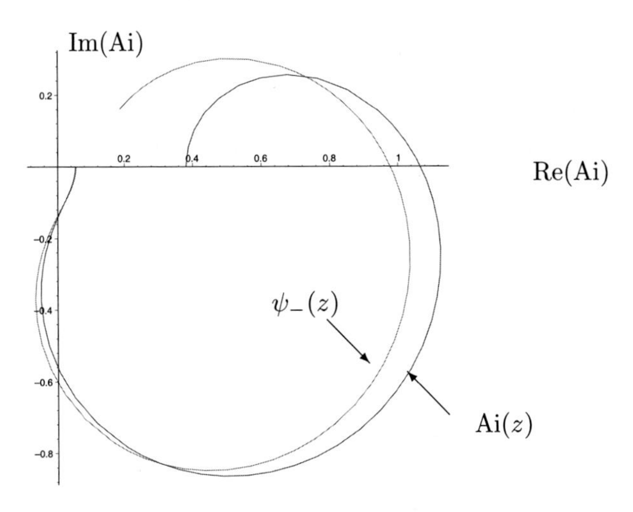

Fig. 11.2 Argand diagram of the Airy function $\text{Ai}(z)$ and its WKBJ approximation ψ_- continued in the complex plane.

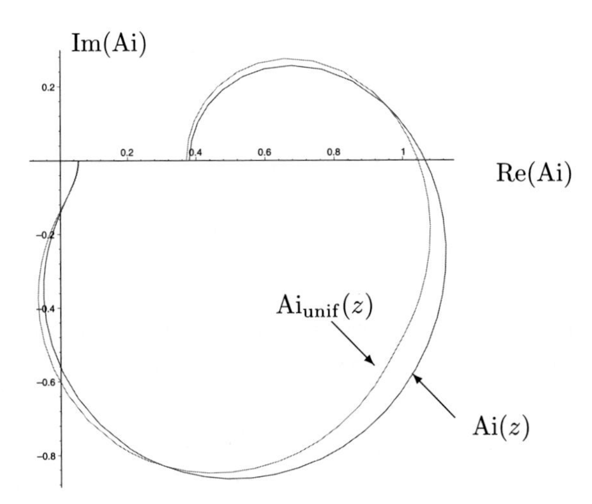

Fig. 11.3 Argand diagram of the Airy function $\text{Ai}(z)$ and its uniformly valid approximation $\text{Ai}_{\text{unif}}(z)$.

11.3 Fourier–Laplace Integral Representation

Use the Fourier–Laplace kernel, i.e. write

$$y(z) = \int_C e^{zt} f(t) dt. \tag{11.12}$$

Writing $ze^{zt} = \frac{d}{dt} e^{zt}$ and integrating by parts we find a solution to the original equation provided that

$$t^2 f + \frac{df}{dt} = 0, \qquad f(t) e^{zt}|_a^b = 0, \tag{11.13}$$

where the contour C has end points at $t = a, b$. The resulting equation for $f(t)$ is only first order, so the solution is easily obtained

$$f(t) = e^{-t^3/3}, \tag{11.14}$$

and the possible contours, which make the end point contributions vanish for all complex z, are shown in Fig. 11.4, with the contours beginning and ending at the points $t = \infty e^{i2\pi k/3}$ with $k = -1, 0, 1$. Note that there are only two independent contours, since $C_3 = C_1 + C_2$. The information that there are only two solutions to this equation is embedded in the analytic structure of the complex plane, i.e. in the asymptotic form of the integrand at large z and in the location of points where the term due to integration by parts vanishes.

Define the functions $E_1(z), E_2(z)$ given by

$$E_k(z) = \frac{1}{\pi i} \int_{C_k} e^{zt - t^3/3} dt, \tag{11.15}$$

which are analogous to the Hankel functions in the analysis of the Bessel equation, section 13.3.2. We will verify that

$$\begin{aligned} E_1(z) &= Ai(z) + iBi(z), \\ E_2(z) &= Ai(z) - iBi(z). \end{aligned} \tag{11.16}$$

Making use of the WKB analysis, to verify these relations it is sufficient to confirm the asymptotic behavior of the function $Ai(z)$ for $z \to +\infty$ or $z \to -\infty$, and $Bi(z)$ for $z \to -\infty$.

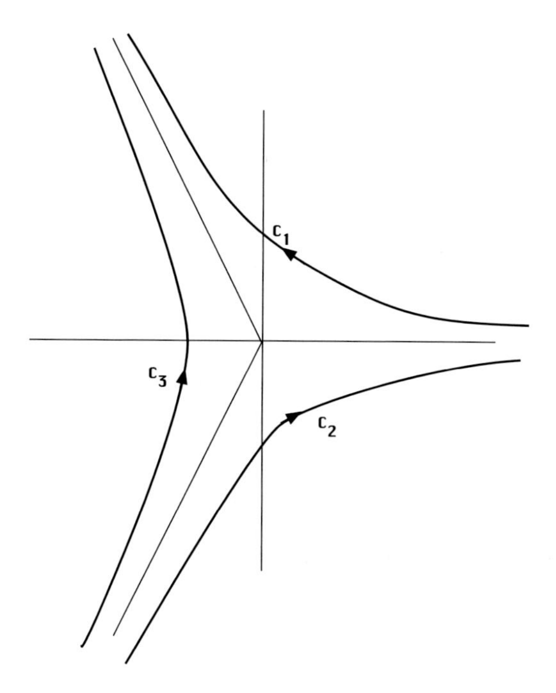

Fig. 11.4 Integration paths for the Airy function.

11.4 Asymptotic Limits

11.4.1 *Large negative z*

Examine $z \to -\infty$. Writing $e^{zt-t^3/3} = e^{\phi}$ we find $\phi' = 0$, giving saddle points, at $t = \pm\sqrt{z}$. Take $z = re^{-i\theta}$. The location and orientation of the saddle points for $\theta = \pi$ is shown in Fig. 11.5. The contours C_1 and C_2 are taken to pass through the saddle points as shown, allowing evaluation of the integrals for large negative real $z = re^{-i\pi}$. The lowest order saddle point evaluation, Eq. 7.16, gives

$$
\begin{aligned}
E_1(z) &\sim \frac{ie^{-i(\frac{2}{3}r^{3/2}+\frac{\pi}{4})}}{\sqrt{\pi}r^{1/4}} = \frac{e^{-\frac{2}{3}z^{3/2}}}{\sqrt{\pi}z^{1/4}}, \\
E_2(z) &\sim \frac{e^{i(\frac{2}{3}r^{3/2}+\frac{\pi}{4})}}{i\sqrt{\pi}r^{1/4}} = \frac{e^{\frac{2}{3}z^{3/2}}}{i\sqrt{\pi}z^{1/4}},
\end{aligned}
\tag{11.17}
$$

proving Eqs. 11.16.

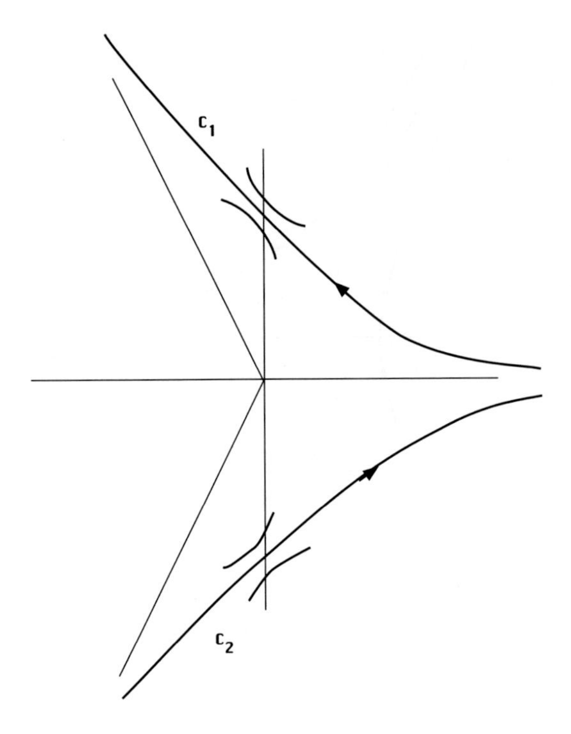

Fig. 11.5 Integration paths showing saddle points for $Re z < 0$.

Using the integral representation we can find the higher order corrections to these asymptotic expressions. There are two means of doing this. The most reliable way is to use the dominant asymptotic form of the solution $y \sim e^S$, write $y(z) \sim e^S W(z)$, find the differential equation for $W(z)$ and generate its asymptotic series. In some cases it is instead possible to directly generate the series from the integral. This happens to be the case for the Airy function, the method being described by Erdelyi [1955]. To directly evaluate the series from the integral one writes the integral in the form $\int e^{-u}(dt/du)du$ with u real. For this method to be successful one must invert the expression $\phi(z,t) = -u$ to give $t(u)$ and thus dt/du as a function of u. For the Airy function this is possible using Lagrange's theorem (Appendix A). However it is often not possible to find the inversion and in general it is easier in any case to generate the asymptotic expansion for $W(z)$ using the differential equation.

Thus write

$$y = \frac{1}{z^{1/4}} e^{\pm\frac{2}{3}z^{3/2}} W_\pm(z), \tag{11.18}$$

and find the differential equation for $W_\pm(z)$

$$W''_\pm - \frac{W'_\pm}{2z} \pm 2z^{1/2} W'_\pm + \frac{5}{16z^2} W_\pm = 0. \tag{11.19}$$

We then have the asymptotic series

$$W_\pm = \sum_0^\infty b_n z^{-3n/2}, \qquad b_{n+1} = \pm \frac{3(n+\frac{5}{6})(n+\frac{1}{6})}{4(n+1)} b_n. \tag{11.20}$$

Using Eq. 8.51 the expression for b_n can be simplified to

$$b_n = \frac{(\pm 1)^n 3^n \Gamma(n+\frac{1}{6})\Gamma(n+\frac{5}{6})}{2\pi 4^n n!} b_0 = \frac{(\mp 1)^n 3^n \Gamma(n+\frac{1}{6})}{4^n n! \Gamma(\frac{1}{6}-n)} b_0 \tag{11.21}$$

and ± 1 is associated with $e^{\pm 2z^{3/2}/3}$, respectively. Substituting we find

$$E_1(z) = \frac{ie^{-i(\frac{2}{3}r^{3/2}+\frac{\pi}{4})}}{\sqrt{\pi} r^{1/4}} \sum_n \frac{(-3)^n \Gamma(n+\frac{1}{6})\Gamma(n+\frac{5}{6}) r^{-3n/2} e^{3in\pi/2}}{2\pi 4^n n!}. \tag{11.22}$$

Separate this into two sums with $n = 2k, 2k+1$ giving

$$E_1(z) = \frac{e^{-i(\frac{2}{3}r^{3/2}+\frac{\pi}{4})}}{\sqrt{\pi} r^{1/4}} [i\Sigma_1 - \Sigma_2], \tag{11.23}$$

with

$$\Sigma_1 = \sum_k \frac{3^{2k}\Gamma(2k+\frac{1}{6})\Gamma(2k+\frac{5}{6}) r^{-3k}(-1)^k}{2\pi 4^{2k}(2k)!} = 1 - \frac{385}{4608} r^{-3} + \ldots,$$

$$\Sigma_2 = \sum_k \frac{3^{2k+1}\Gamma(2k+\frac{7}{6})\Gamma(2k+\frac{11}{6}) r^{-3k-3/2}(-1)^k}{2\pi 4^{2k+1}(2k+1)!} = \frac{5}{48} r^{-3/2} - \ldots. \tag{11.24}$$

Similarly we find

$$E_2(z) = \frac{e^{i(\frac{2}{3}r^{3/2}+\frac{\pi}{4})}}{i\sqrt{\pi} r^{1/4}} \sum_n \frac{3^n \Gamma(n+\frac{1}{6})\Gamma(n+\frac{5}{6}) r^{-3n/2} e^{3ni\pi/2}}{2\pi 4^n n!}, \tag{11.25}$$

and again separating n even and n odd terms

$$E_2(z) = \frac{e^{i(\frac{2}{3}r^{3/2}+\frac{\pi}{4})}}{\sqrt{\pi} r^{1/4}} [-i\Sigma_1 - \Sigma_2]. \tag{11.26}$$

We then find for $Ai(z)$ the asymptotic series

$$Ai(z) = \frac{E_1 + E_2}{2} = \frac{sin(\frac{2}{3}r^{3/2} + \frac{\pi}{4})}{\sqrt{\pi}r^{1/4}}\Sigma_1 - \frac{cos(\frac{2}{3}r^{3/2} + \frac{\pi}{4})}{\sqrt{\pi}r^{1/4}}\Sigma_2. \quad (11.27)$$

The asymptotic series for $Bi(z)$ is

$$Bi(z) = \frac{E_1 - E_2}{2i} = \frac{cos(\frac{2}{3}r^{3/2} + \frac{\pi}{4})}{\sqrt{\pi}r^{1/4}}\Sigma_1 + \frac{sin(\frac{2}{3}r^{3/2} + \frac{\pi}{4})}{\sqrt{\pi}r^{1/4}}\Sigma_2. \quad (11.28)$$

Both Ai and Bi contain combinations of dominant and subdominant terms for z near the negative real axis. From the Stokes plot, we know that continuation to $arg(z) = \pm 2\pi/3$ will contribute a new subdominant term, which will remain subdominant only up to $arg(z) = \pm\pi/3$, after which it becomes dominant, so these expansions for Ai and Bi are limited to $|arg(z) - \pi| < 2\pi/3$, or equivalently $\pi/3 < arg(z) < 5\pi/3$.

11.4.2 *Large positive z*

Examine $z \to +\infty$. The saddle points are shown in Fig. 11.6, with the direction of steepest descent indicated, given by $\phi''(t - t_0)^2 = -real$, showing also the contours taken through the saddle leading to the asymptotic evaluation of the integrals for large z. We then find from Eq. 7.16

$$E_1(z) \simeq -\frac{e^{\frac{2}{3}z^{3/2}}}{i\sqrt{\pi}z^{1/4}},$$
$$E_2(z) \simeq \frac{e^{\frac{2}{3}z^{3/2}}}{i\sqrt{\pi}z^{1/4}}. \quad (11.29)$$

To this order, the sum of these two functions is zero, however, for the sum of the two functions the contour $C_1 + C_2$ can be deformed to path C_3 as is apparent from Fig. 11.6, and in this case we find

$$\frac{E_1(z) + E_2(z)}{2} \simeq \frac{e^{-\frac{2}{3}z^{3/2}}}{2\sqrt{\pi}z^{1/4}}, \quad (11.30)$$

and this combination is $Ai(z)$, and we have, for all z

$$Ai(z) = \frac{1}{2\pi i}\int_{C_3} e^{zt - t^3/3}dt. \quad (11.31)$$

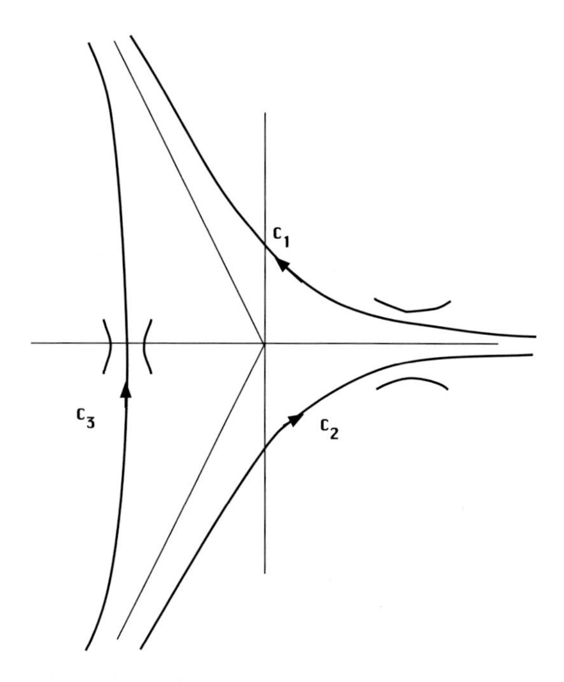

Fig. 11.6 Integration paths showing saddle points for $Re z > 0$.

The series expansion given by W_- is valid also for large positive z, so using Eq. 11.30 we then find

$$Ai(z) \simeq \frac{e^{-\frac{2}{3}z^{3/2}}}{2\sqrt{\pi}z^{1/4}} \sum_n \frac{3^n \Gamma(n+\frac{1}{6}) z^{-3n/2}}{4^n n! \Gamma(\frac{1}{6}-n)}, \tag{11.32}$$

and the first few terms are

$$Ai(z) \simeq \frac{e^{-\frac{2}{3}z^{3/2}}}{2\sqrt{\pi}z^{1/4}} \left[1 - \frac{5}{48} z^{-3/2} + \frac{385}{4608} z^{-3} - \ldots\right]. \tag{11.33}$$

From the Stokes plot, we know that a subdominant solution for positive real z, when continued around the plane, becomes dominant for $arg(z) = \pm\pi/3$, and picks up a subdominant term at $arg(z) = \pm 2\pi/3$. This new term remains subdominant only up to $arg(z) = \pm\pi$, after which it becomes important, so this expansion is limited to $|arg(z)| < \pi$.

Similarly we find

$$\frac{E_1(z) - E_2(z)}{2i} = Bi(z) \simeq \frac{e^{\frac{2}{3}z^{3/2}}}{2\sqrt{\pi}z^{1/4}} \sum_n \frac{(-3)^n \Gamma(n+\frac{1}{6}) z^{-3n/2}}{4^n n! \Gamma(\frac{1}{6}-n)}, \tag{11.34}$$

and the first few terms are

$$Bi(z) \simeq \frac{e^{\frac{2}{3}z^{3/2}}}{2\sqrt{\pi}z^{1/4}}\left[1+\frac{5}{48}z^{-3/2}+\frac{385}{4608}z^{-3}-\dots\right]. \tag{11.35}$$

From the Stokes plot, we know that a dominant solution for real positive z, with error contributing some subdominant component, when continued around the plane, becomes subdominant for $arg(z) = \pm\pi/3$, and the error becomes dominant, so this expansion is limited to $|arg(z)| < \pi/3$.

11.4.3 *Small* $|z|$

For $|z|$ small we can directly generate the series from the integral. Choose contours along the paths $t = \infty e^{\pm i2\pi/3}$ and expand the integrand in powers of z. Using the integration variable $u = t^3/3$ we find

$$\begin{aligned} E_1(z) = & \frac{1}{\pi i}\int_0^\infty e^{-u}\sum_k \frac{(3u)^{(k-2)/3}z^k e^{(2\pi i(k+1)/3)}}{k!}du \\ & -\frac{1}{\pi i}\int_0^\infty e^{-u}\sum_k \frac{(3u)^{(k-2)/3}z^k}{k!}du. \end{aligned} \tag{11.36}$$

Integrate and separate into terms with $k = (0,1,2) + 3n$ giving

$$E_1(z) = \frac{3^{1/2}e^{i\pi/3}}{\pi}\sum_n \frac{z^{3n}\Gamma(n+\frac{1}{3})3^{n-2/3}}{(3n)!} - \frac{3^{1/2}e^{-i\pi/3}}{\pi}\sum_n \frac{z^{3n+1}\Gamma(n+\frac{2}{3})3^{n-\frac{1}{3}}}{(3n+1)!} \tag{11.37}$$

but $(3n)! = 3^{3n}n!\Gamma(n+\frac{2}{3})\Gamma(n+\frac{1}{3})/(\Gamma(\frac{1}{3})\Gamma(\frac{2}{3}))$ and using Eq 8.51 we find

$$E_1(z) = 3^{-2/3}2e^{i\pi/3}\sum_n \frac{z^{3n}}{9^n n!\Gamma(n+\frac{2}{3})} - 3^{-4/3}2e^{-i\pi/3}\sum_n \frac{z^{3n+1}}{9^n n!\Gamma(n+\frac{4}{3})}, \tag{11.38}$$

which of course has an infinite radius of convergence, solutions being analytic in the plane except for the point ∞.

In the same way we find

$$E_2(z) = 3^{-2/3}2e^{-i\pi/3}\sum_n \frac{z^{3n}}{9^n n!\Gamma(n+\frac{2}{3})} - 3^{-4/3}2e^{i\pi/3}\sum_n \frac{z^{3n+1}}{9^n n!\Gamma(n+\frac{4}{3})}. \tag{11.39}$$

We then find

$$Ai(z) = 3^{-2/3}\sum_n \frac{z^{3n}}{9^n n!\Gamma(n+\frac{2}{3})} - 3^{-4/3}\sum_n \frac{z^{3n+1}}{9^n n!\Gamma(n+\frac{4}{3})}. \quad (11.40)$$

Similarly the expansion for $Bi(z)$ is

$$Bi(z) = 3^{-1/6}\sum_0^\infty \frac{z^{3n}}{9^n n!\Gamma(n+\frac{2}{3})} + 3^{-5/6}\sum_0^\infty \frac{z^{3n+1}}{9^n n!\Gamma(n+\frac{4}{3})}. \quad (11.41)$$

11.5 Mellin Integral Representation

A straightforward attempt to find a Mellin representation for the Airy function leads to a three term recursion relation for $M(s)$, which is not useful. Instead, extract the leading asymptotic behavior at large positive z, writing $Ai(z) = e^{-2z^{3/2}/3}f$, giving

$$f'' - 2z^{1/2}f' - \frac{1}{2z^{1/2}}f = 0, \quad (11.42)$$

and write

$$f = \frac{1}{2\pi i}\int_C M_s z^{-s} ds. \quad (11.43)$$

This gives for $M(s)$

$$s(s+1)M(s)z^{-s-2} + (2s-1/2)z^{-s-1/2}M(s) = 0, \quad (11.44)$$

or

$$s(s+1)M(s) = -2(s+5/4)M(s+3/2). \quad (11.45)$$

A solution is given by (see Eq. 9.26)

$$M(s) = N\left(\frac{3}{2}\right)^{2s/3}\frac{\Gamma(2s/3)\Gamma(2s/3+2/3)}{(-2)^{2s/3}\Gamma(2s/3+5/6)}. \quad (11.46)$$

Changing integration variable to $2s/3 = t + 1/6$ we have

$$M(t) = N\left(\frac{3}{4}\right)^{t}\frac{\Gamma(t+1/6)\Gamma(t+5/6)}{(-1)^t\Gamma(t+1)}. \quad (11.47)$$

To find an asymptotic series for large z we need a form of $M(t)$ with singularities for positive t, so take the alternate form (see the discussion following Eq. 9.26)

$$M(t) = N\left(\frac{3}{4}\right)^t \frac{\Gamma(t+1/6)\Gamma(-t)}{(-1)^t\Gamma(1/6-t)}, \tag{11.48}$$

and fix the normalization using the leading asymptotic form coming from the pole at $t=0$ to give the Mellin representation

$$Ai(z) = \frac{e^{-2z^{3/2}/3}}{2\pi i z^{1/4}} \int_{\gamma-i\infty}^{\gamma+i\infty} \left(\frac{3}{4}\right)^t \frac{\Gamma(t+1/6)\Gamma(-t)}{(-1)^t\Gamma(1/6-t)} z^{-3t/2} dt, \tag{11.49}$$

with $0 < \gamma < 1$. Moving the integration contour to the right and using the fact that the residue of $\Gamma(-t)$ at integer t is $(-1)^t/\Gamma(1+t)$ we find the asymptotic expansion

$$Ai(z) \simeq \frac{e^{-\frac{2}{3}z^{3/2}}}{2\sqrt{\pi}z^{1/4}} \sum_n \frac{3^n\Gamma(n+\frac{1}{6})z^{-3n/2}}{4^n n!\Gamma(\frac{1}{6}-n)}, \tag{11.50}$$

agreeing with Eq. 11.32. The Stirling approximation can be used to discover the domain of z in which this asymptotic expansion is applicable. Writing $t = v + iu$ and $z = re^{i\phi}$ the magnitude of the integrand in Eq. 11.49 takes the form for large u

$$e^{-3\pi|u|/2+3\phi u/2}, \tag{11.51}$$

where the ambiguity of the factor $(-1)^{iu}$ has been resolved by continuing from the positive real axis, i.e. take $-1 = e^{i\pi}$ for $u > 0$ and take $-1 = e^{-i\pi}$ for $u < 0$. We thus find this representation valid for $|\phi| < \pi$, agreeing with the onset of the Stokes phenomenon.

11.6 Matching Local Solutions

It is not always possible to find an integral representation for the solution of a given differential equation. In this case a global representation of a solution can be constructed by numerically patching together local approximations. We illustrate this for the Airy equation by matching the local solutions about $x = \pm\infty$ and $x = 0$. To obtain a unique solution, choose the exponentially decaying solution for $x \to +\infty$, with normalization $Ai(x) \simeq e^{-\frac{2}{3}x^{3/2}}/(2\sqrt{\pi}x^{1/4})$. Constructing an optimum asymptotic expansion for large positive x, $y_+(x)$ we find that the error $|dy_+/y_+| < .005$ for $x > 2$. From this expression find the value of the function and its derivative at $x = 2$, and use these values to determine the two constants a_0, a_1

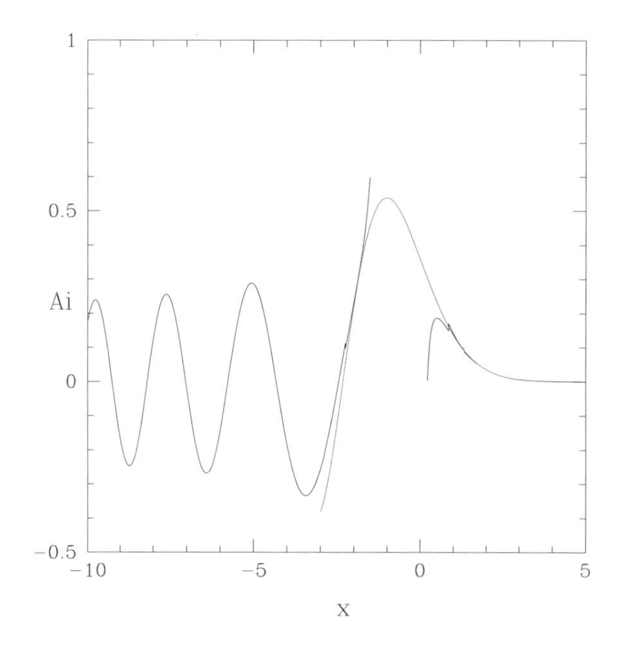

Fig. 11.7 Airy function by matching local expansions.

in the local expansion about $x = 0$, $y_0(x)$. Fifteen terms were used in the Taylor series about $x = 0$. The asymptotic expression for large negative x has similar accuracy for $x < -2$. Then use the value and derivative of $y_0(x)$ at $x = -2$ to determine two constants A, B in the asymptotic expansion for $x \to -\infty$, $y(r) \sim Asin(2r^{3/2}/3)/r^{1/4} + Bcos(2r^{3/2}/3)/r^{1/4}$. The final result is shown in Fig. 11.7, showing the two asymptotic solutions and the Taylor series solution. Each solution has been continued past the matching points, $x = \pm 2$ to show how they diverge from the correct values.

Figure 11.8 shows the number of terms retained in the asymptotic expansions as a function of x to give optimum accuracy. Naturally for large $|x|$ it is not necessary to retain nearly so many terms to obtain accuracy of $|dy/y| < .005$. It is then possible to numerically examine the solution at large negative x and determine that $y_-(x) \to 1/(\sqrt{\pi}r^{1/4})sin(\frac{2}{3}r^{3/2} + \pi/4)$ and some analyses assume this phase shift *ab initio*. But without a full numerical solution or a matching as carried out above, there is no way to know this phase, so it is really beyond the scope of local analysis, depending on a WKB analysis or an integral representation. The behavior near $x = 0$ cannot be determined by WKB analysis, which is invalid in this domain,

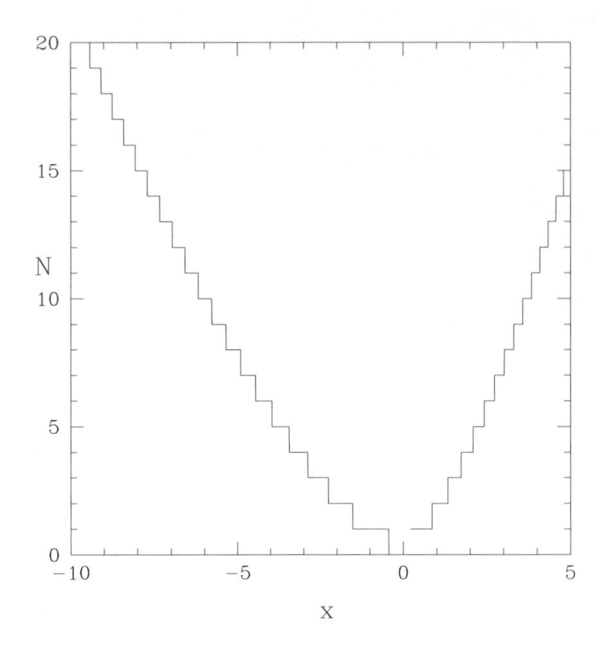

Fig. 11.8 Optimal number of terms in asymptotic expansions for Airy.

and requires a matching procedure as shown above, or the use of an integral representation.

11.7 The Wronskian

To find the Wronskian, note from Abel's formula Eq. 2.10 that it is a constant, and use the expansions near $z = 0$ to evaluate it at zero, giving

$$W(z) = Ai(0)Bi'(0) - Ai'(0)Bi(0) = \frac{2}{3^{3/2}\Gamma(2/3)\Gamma(4/3)} = \frac{1}{\pi}. \quad (11.52)$$

11.8 Problems

1. Show that

$$\frac{2}{3^{3/2}\Gamma(2/3)\Gamma(4/3)} = \frac{1}{\pi}.$$

2. Show that

$$y(z) = \int_0^\infty cos(t^3/3 - xt)dt$$

satisfies the Airy equation. Hint: Change variable to $s = t^3/3$ and first show that expressions for y, y', y'' converge. Then show that the Airy equation is satisfied except for an ambiguous term at $s = \infty$. Evaluate this term by averaging expressions over a small range of x, i.e. $[x, x + \delta]$ with δ small but fixed, before taking the limit $s = \infty$.

3. Extract the dominant behavior $e^{2z^{3/2}/2}$ and find the Mellin representation for large positive z for the Airy function $Bi(z)$. Examine convergence of the integral and confirm the domain of validity in the variable z.

4. Find the number of terms to be kept in the asymptotic expansion of the error function, Eq. 4.68, for large positive z.

5. Use the integral representation for the Airy function to show that

$$\int_0^\infty Ai(x)dx = \frac{1}{3},$$

and

$$\int_{-\infty}^0 Ai(x)dx = \frac{2}{3}.$$

6. Show that $f = Ai^2(z)$ satisfies

$$f''' - 4zf' - 2f = 0.$$

7. Use the Wronskian to show that

$$\frac{d}{dx}\frac{Bi(x)}{Ai(x)} = \frac{1}{\pi Ai^2(x)}.$$

8. Show that

$$\int Ai^2(x)dx = xAi^2(x) - (Ai'(x))^2.$$

9. Show that

$$\int Ai(x)Bi(x)dx = xAi(x)Bi(x) - Ai'(x)Bi'(x).$$

Chapter 12

Phase-Integral Methods II

12.1 Introduction

In Chapter 5 we saw how the analytic continuation of a solution of a differential equation through the complex plane is accomplished by the rules derived by Heading, using the Stokes constants associated with the singularities of $Q^{1/2}$. The Stokes constants and the rules for continuation can be more clearly understood with the use of the integral representations obtained for the solutions.

In addition, a major gap in the applicability of the rules derived by Heading is the fact that the Stokes constants are derived only for isolated singularities, whereas many of the routinely treated problems have Stokes diagrams consisting of two or more connected singularities. It is commonly assumed that using these Stokes constants one obtains asymptotic expressions valid in the limit of large separation between the singularities. However it is often the case that arbitrary separation is of interest physically. In sections 5.8 and 5.9 we demonstrated that for scattering problems the Stokes constant can be derived for arbitrary singularity separation by using flux conservation, but this method is not applicable to all problems. In this chapter we examine this problem with a few examples allowing analytic solution and find a generalization of Stokes constants that gives the correct asymptotic behaviors of the solutions for arbitrary separation between two singularities. The analytic solutions are obtained by finding integral solutions and performing analytic continuation. Unfortunately there are not a large number of problems for which this is possible.

In section 12.2 we show how integral representations can be used to obtain the Stokes constants, and demonstrate the origin of the $T/2$ rule for the case when z is exactly on a Stokes line. In section 12.3 we examine scattering from an overdense barrier, in section 12.4 scattering from an underdense barrier, section 12.5 consists of an analysis of the Budden problem, and section 12.6 that of the instability or bound state problem. Section 12.7 concerns the accuracy of WKB approximations. Finally in section 12.8 we present some conclusions.

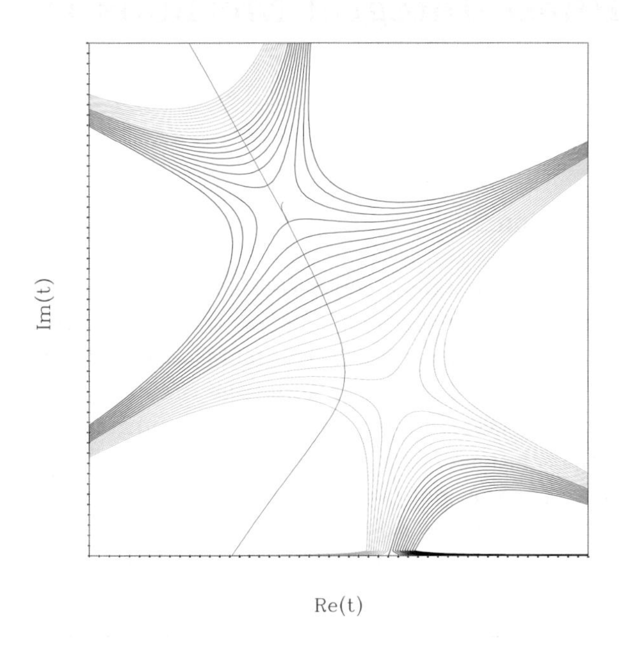

Fig. 12.1 Integration path showing contours of $Re\phi$ just before z arrives at the Stokes line. Positive values of $Re\phi$ are black, and negative values green. Following the line of steepest descent from the upper saddle point leads to the asymptote at $t \sim e^{-i2\pi/3}$ without encountering the lower saddle point.

12.2 Stokes Phenomena and Integral Representations

The Stokes phenomenon for the Airy function manifests itself in the integral representation Eq. 11.31 $f(z) \sim \int e^{\phi(t,z)}dt$ by forcing a change in the integration contour when z is continued past the Stokes line. The saddle points located at $t = \pm\sqrt{z}$, are shown for large positive z in Fig. 11.6. As z is continued downward in the complex plane the left saddle moves upward. Shortly before arriving at the Stokes line at $z \sim e^{-2i\pi/3}$ the contours of $Re\phi$ are as shown in Fig. 12.1, and following the line of steepest descent for

$Re\phi$ from the upper saddle point one arrives at the asymptote $t \sim e^{-i2\pi/3}$, as shown. There is thus no contribution from the lower saddle point. When z is exactly at the Stokes line the line of steepest descent from the upper saddle leads to the lower saddle, as shown in Fig. 12.2. Note that at this point the contribution from the second saddle point is one half of the full saddle point integral. This explains why, when z is exactly on the Stokes line, one half of the Stokes constant must be used.

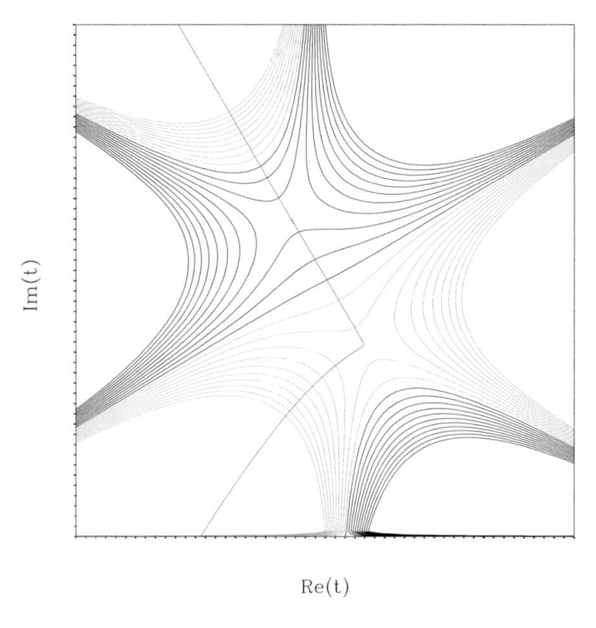

Fig. 12.2 Integration path showing contours of $Re\phi$ with z at the Stokes line. Positive values of $Re\phi$ are black, and negative values green. Following the line of steepest descent from the upper saddle point leads to the the lower saddle point.

However when z passes the Stokes line contours are as shown in Fig. 12.3 and by following the line of steepest descent for $Re\phi$ the contour is forced to the asymptote $t \sim real$. Then to complete the integration path it is necessary to pass through the second saddle in order to arrive at the asymptote $t \sim e^{-i2\pi/3}$. Thus one must add the full contribution from the second saddle point, i.e. use the full Stokes constant. At this point the contribution from the lower saddle is maximally subdominant, so the lower saddle adds a negligible part to the integral for large z. This small addition corresponds to the modification of the asymptotic representation by the addition of the subdominant term multiplied by the Stokes constant. Note that the analysis of the integration contours is valid for all z, but if z is not large the contributions to the integrals come from all along the path, not just the saddle points. Finally as z is continued all the way around to become negative real the saddles are located on the imaginary axis and the integration path is as shown in Fig. 11.5, and the contributions from the two saddle points are equal in magnitude.

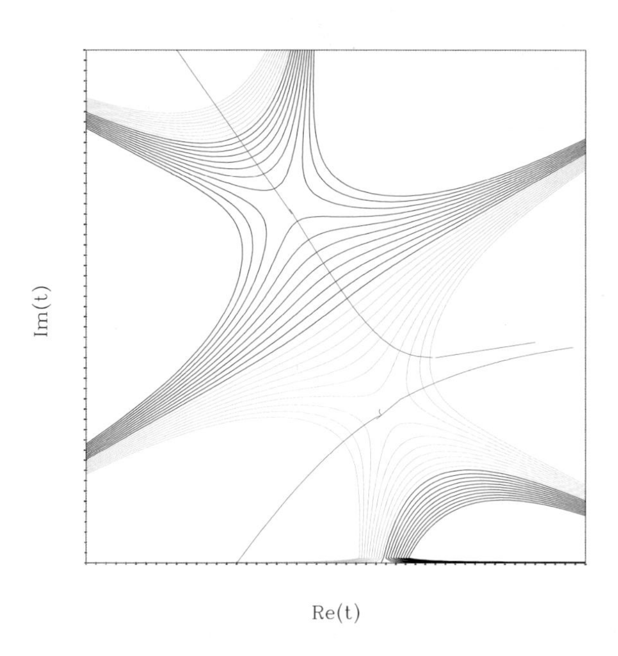

Fig. 12.3 Integration path showing contours of $Re\phi$ just after z passes the Stokes line. Positive values of $Re\phi$ are black, and negative values green. Now to complete the integration one must pass also through the lower saddle point.

Furthermore, this analysis allows us to see why the Stokes constant is $T = i$, namely the contribution from the subdominant saddle point is not exactly equal to the subdominant asymptotic solution, it differs by a phase. When $z = re^{-i2\pi/3}$, exactly on the Stokes line between domains 5 and 6 of Fig. 11.1, the function $(0, z) = e^{r^{3/2}+i\pi/6}/r^{1/4}$ is dominant and $(z, 0) = e^{-r^{3/2}+i\pi/6}/r^{1/4}$ is subdominant. At this point $z^{3/2}$ is real, the phase of $\pi/6$ comes from the denominator $z^{1/4}$, common to each of them.

The contribution to

$$y(z) = \frac{1}{\pi i} \int e^{zt - t^3/3} dt \qquad (12.1)$$

from the upper saddle point, except for normalization, is $y_{s,u} = e^{r^{3/2} + i\pi/6}/r^{1/4}$, agreeing with the dominant part of the approximate solution, but the contribution from the lower saddle point is $y_{s,l} = -e^{r^{3/2} + i2\pi/3}/r^{1/4}$, differing from $(z, 0)$ by a factor of $-i$, so $(z, 0)$ must be multiplied by $-i$, the Stokes constant (clockwise), to give the correct result. This factor comes from the phase of the saddle point orientation, the lower saddle differing from the upper by a rotation of $\pi/2$.

Now consider a zero of order n

$$y''(z) + z^n y = 0. \qquad (12.2)$$

The WKB solutions have the form

$$(0, z) = \frac{e^{iz^p/p}}{z^{n/4}}, \qquad (z, 0) = \frac{e^{-iz^p/p}}{z^{n/4}} \qquad (12.3)$$

with $p = n/2 + 1$, and the real axis is an anti-Stokes line. The first Stokes line to be encountered moving in a clockwise direction from the real axis is at $z \sim e^{-i\pi/(n+2)}$. Begin on the positive real axis with $(0, z)$, which becomes dominant upon continuation below the axis. At the Stokes line the solution becomes

$$(0, z) - T(z, 0) \qquad (12.4)$$

where from Eq. 5.22 $T = 2icos[\pi/(n + 2)]$.

Now construct an integral representation. Change independent variable to w with $w = (z/2)^{n+2}$ giving

$$(n + 2)^2 wy'' + (n + 2)(n + 1)y' + 2^{n+2} y = 0 \qquad (12.5)$$

with a Fourier–Laplace integral representation

$$y(w) = A \int t^a e^{\phi} dt, \qquad (12.6)$$

with $\phi = wt - b^2/t$, $a = -(n + 3)/(n + 2)$ and $b = 2^p/(n + 2)$, and the condition that $t^{a+2} e^{wt - b^2/t}$ vanish at the end points of the integration.

Saddle points are located at points with $\phi' = 0$, given by $t_s = \pm ib/\sqrt{w}$, and an integration path for positive real z is shown in Fig. 12.4 that gives the solution $(0, z)$. End points are at $t = 0$, approached with t positive

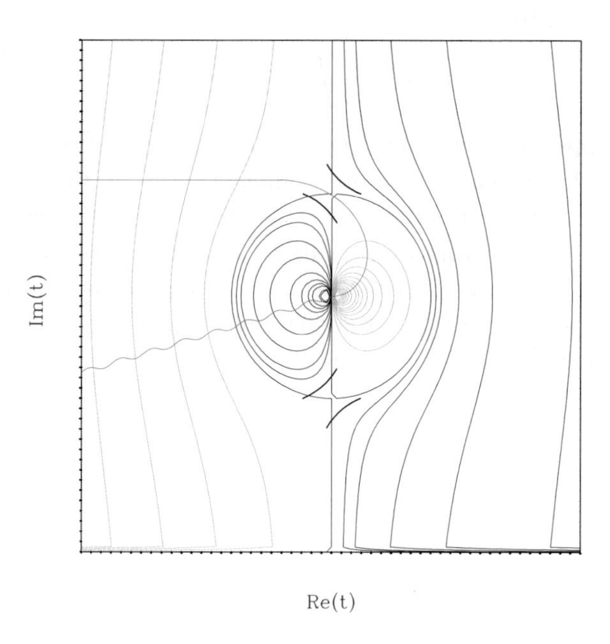

Fig. 12.4 Integration path for $Q = z^n$ and positive real z, showing some level contours, the saddle points, the cut, and the integration path. Negative levels of $Re\phi$ are shown in green.

real, and at $t = \infty$, with phase such that $wt < 0$. There is a cut from zero to infinity, but it is not relevant for the path, the end point condition is satisfied on all Riemann surfaces. At the saddle point $\phi_s'' = -2b^2/t_s^3$ and $\phi_s = wt_s - b^2/t_s$.

Now analytically continue z toward the Stokes line at $w = re^{-i\pi}$. The variable w rotates clockwise and the integration contour must be rotated counterclockwise to maintain convergence. The end point of the contour rotates as $t \sim 1/w$, while the saddle point positions move at half this rate, $t_s \sim 1/\sqrt{w}$. Because the initial part of the path must lie in the region $Re\, t > 0$ the path is wrapped around the axis as z rotates. Just before arriving at the Stokes line the integration path is as shown in Fig. 12.5. The saddle point near the positive real axis does not yet contribute to the integration.

For z lying on the Stokes line, the line of steepest ascent from $t = 0$ leads to the saddle on the positive real axis. From the midpoint of the saddle a line of steepest ascent moves counterclockwise in a circle to the saddle on the negative real axis. The line of steepest descent from the

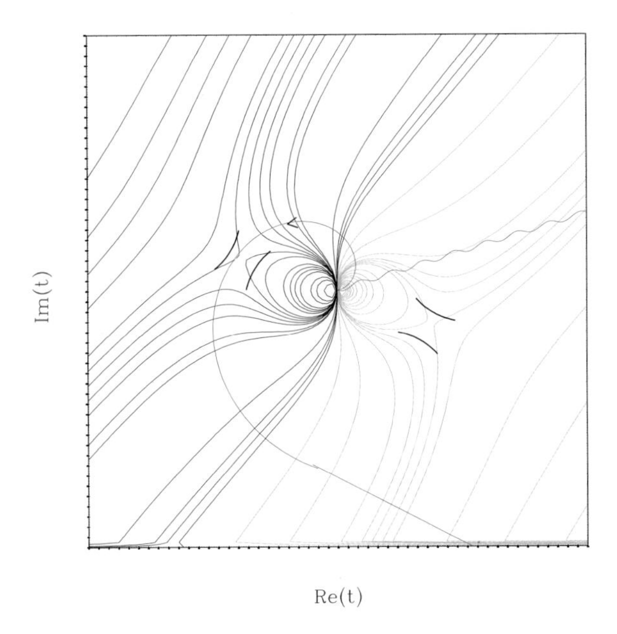

Fig. 12.5 Integration path for $Q = z^n$ with z above the Stokes line at $z \sim e^{-\pi i/(n+2)}$ showing some level contours, the saddle points, the cut, and the integration path. Negative levels of $Re\phi$ are shown in green.

saddle located at negative real t leads again to the saddle point on the real axis, as shown in Fig. 12.6. Thus the saddle on the positive real axis gives two half contributions, one from $t \sim real$ and one from $t \sim e^{2\pi i}$, leading to the rule of using only half the Stokes constant when z lies exactly on the Stokes line.

Figure 12.7 shows the level contours and the integration path after z has moved past the Stokes line. Now there are two full contributions from the saddle in the right half plane, with phases in t differing by 2π.

Now consider the contributions from the saddle points. The saddle point contribution to y from the saddle located at t_s is given by

$$y(z) = A\frac{\sqrt{2\pi}t_s^{a-2}e^{wt_s - b^2/t_s}}{\sqrt{-\phi''(t_s)}}. \tag{12.7}$$

For z positive real, substituting the values of t_s and ϕ'' we take $A = e^{-in\pi/(4n+8)}/(\sqrt{2\pi}2^{n/4}b^{n/(2n+4)-1})$ so that for positive real z, we have $y(z) = (0, z)$.

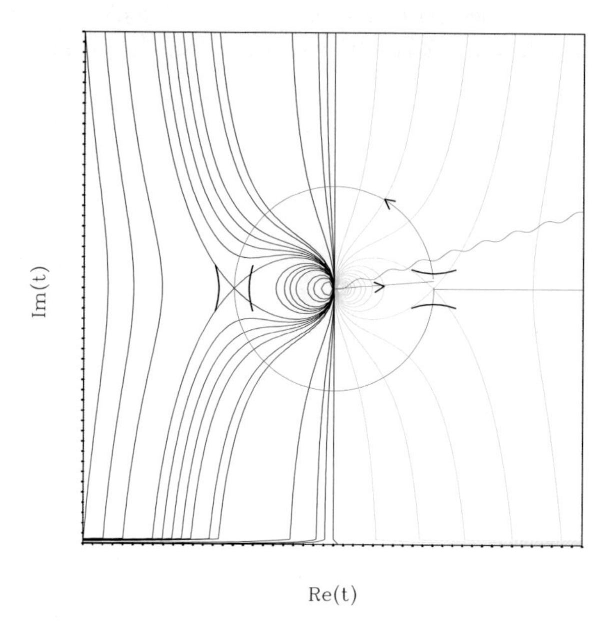

Fig. 12.6 Integration path for $Q = z^n$ and z on the Stokes line at $z \sim e^{-\pi i/(n+2)}$ showing some level contours, the saddle points, the cut, and the integration path. Negative levels of $Re\phi$ are shown in green.

The magnitude of the contribution from any saddle point is then seen to be

$$y(z) = \frac{e^{\phi_s}}{z^{n/4}} \tag{12.8}$$

with the phase given by $e^{-in\pi/(4n+8)} t_s^{-(n+3)/(n+2)} dt$ where dt is the infinitesimal part of the integration path crossing the saddle point, where we use the fact that the phase of $\sqrt{-\phi''}dt$ is positive real at the saddle. (See the discussion leading to the derivation of Eq. 7.16.) Finally, to compare with the WKB expressions $(0, z)$ and $(z, 0)$ one must take account of the phase of $z^{n/4}$ in the denominator of these functions.

Now evaluate the contribution for z on the Stokes line, Fig. 12.6. The saddle in the left half plane gives $(0, z)$. There are two half-saddle contributions from the saddle in the right half plane, one with $t_s = real$ and one with $t_s = 2\pi i$. Adding these two terms we find from this saddle

$$y(z) = -icos\left(\frac{\pi}{n+2}\right)(z, 0). \tag{12.9}$$

As soon as z is continued past the Stokes line this saddle contributes two full terms, as seen in Fig. 12.7, so this expression gives one half the Stokes constant.

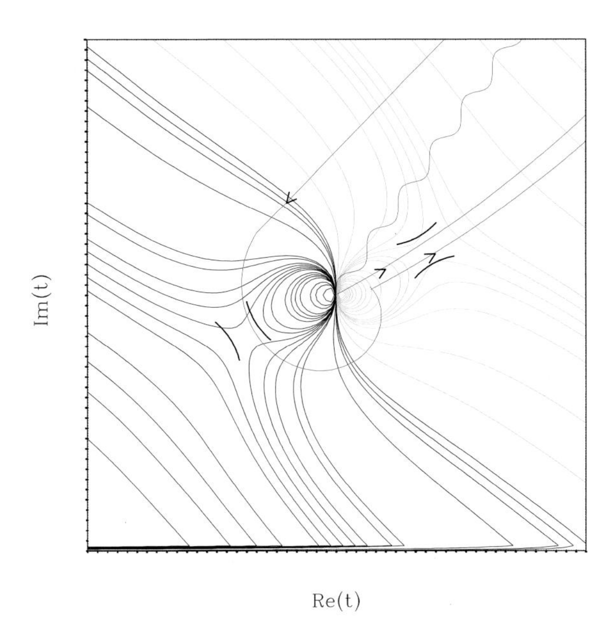

Fig. 12.7 Integration path for $Q = z^n$ and z rotated past the Stokes line at $z \sim e^{-\pi i/(n+2)}$ showing some level contours, the saddle points, the cut, and the integration path. Negative levels of $Re\phi$ are shown in green.

We find again that the Stokes constant is

$$T = 2icos\left(\frac{\pi}{n+2}\right) \tag{12.10}$$

as given by Eq. 5.22, and the angle $\pi/(n+2)$ is the angle between adjacent Stokes and anti-Stokes lines in the Stokes diagram.

12.3 Scattering, Overdense Barrier

The overdense barrier is given by a function $Q(z)$ which is real on the real axis, with two first order zeros at real points a and b, and Q is negative between a and b and positive otherwise. An analytically tractable example of scattering from a barrier is given by

$$y'' + Q(z)y = 0, \qquad Q = z^2 - b^2. \tag{12.11}$$

The solution is asymptotically propagating. The physical problem to which it is associated is that of an incident wave arriving from the left, being partly transmitted through the barrier and partly reflected. Thus to understand the boundary conditions it is necessary to introduce the time dependence of the solution, which we take as $e^{-i\omega t}$. Thus solutions can be determined to be right or left moving according to the asymptotic dependence of the phase on x.

12.3.1 *WKB analysis*

The Stokes diagram for scattering from an overdense barrier is shown in Fig. 12.8, and propagating oscillatory solutions exist for large positive and negative x. We consider an incident wave from the left, the problem

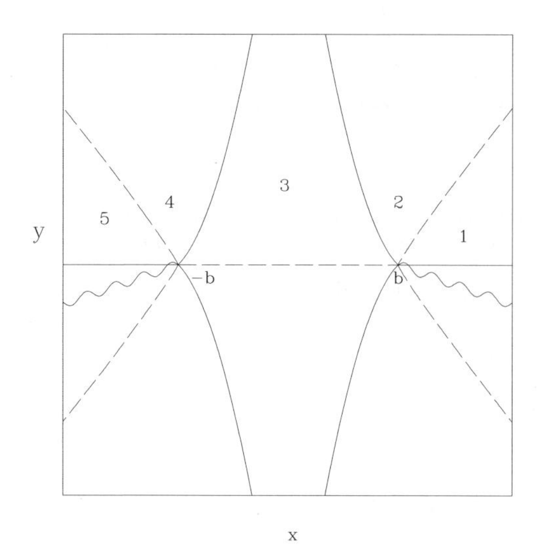

Fig. 12.8 Stokes plot for the overdense barrier.

being to determine the reflected and transmitted waves. Define W through $[-b, b] = e^{-W}$. For convenience we choose the cuts so that the WKB functions are analytic in the upper half plane. Take Q to be real and positive for $b < x$, then with the choice of the cuts as shown in Fig. 12.8

along the real axis between $-b$ and b Q has phase $Q \sim e^{i\pi}$ and W is real and positive, $W = \pi b^2/2$, and for $x < -b$ Q has phase $e^{2i\pi}$. Requiring outgoing wave conditions for large positive x gives boundary conditions of a subdominant solution in domain (1). Continuation through the upper half plane and using $Q^{1/2} \simeq z - b^2/(2z)$ gives

(1) $(b, z)_s = \frac{e^{iz^2/2}}{\sqrt{z}}(z)^{-ib^2/2}$
(2) $(b, z)_s$
(3) $(b, z)_d = e^W(-b, z)_s$
(4) $e^W(-b, z)_d$
(5) $e^W(-b, z)_d + Se^W(z, -b)_s$,

with S the Stokes constant associated with the zero at $z = -b$, and subscripts indicate subdominance and dominance. Making use of the analyticity of these expressions in the upper half plane it is convenient to write the solution in domain 5 as

$$\psi \simeq (b, z) + Se^{2W}(z, b). \tag{12.12}$$

We will also need an explicit expression for the solution for large negative x to compare with the result obtained using an integral representation. We have $Q^{1/2} \simeq z - b^2/(2z)$, so in domain 5, using Eq. 12.12 we have

$$y \simeq \frac{e^{iz^2/2}(z)^{-ib^2/2}}{\sqrt{z}} + S\frac{e^{\pi b^2}e^{-iz^2/2}(z)^{ib^2/2}}{\sqrt{z}}, \tag{12.13}$$

with the first term the reflected wave and the second the incoming wave. Note that to obtain the amplitudes of these waves one must substitute $z = re^{i\pi}$. We have left them in this form to stress their analytic continuation through the upper half plane. In terms of normalized waves the second term becomes

$$-iSe^{\pi b^2/2}\frac{e^{-iz^2/2}(-z)^{ib^2/2}}{\sqrt{-z}}. \tag{12.14}$$

12.3.2 *Integral representation*

Extract the asymptotic behavior by writing $y = e^{-iz^2/2}f(z)$, giving a solution with

$$f(z) = \frac{2^{ib^2/2}e^{i\pi/4}}{\sqrt{2\pi}}\int_{\infty e^{3i\pi/4}}^{\infty e^{-i\pi/4}} e^{-it^2/4+itz}t^{-ib^2/2-1/2}dt \tag{12.15}$$

with normalization chosen to match the WKB solution. For positive real z there is a saddle point at $t = 2z$, as shown in Fig. 12.9. Evaluation of this contribution gives

$$y(z) \simeq \frac{e^{iz^2/2}}{\sqrt{z}}(z)^{-ib^2/2}, \tag{12.16}$$

showing that this is an outgoing wave, satisfying the correct boundary condition for the scattering problem.

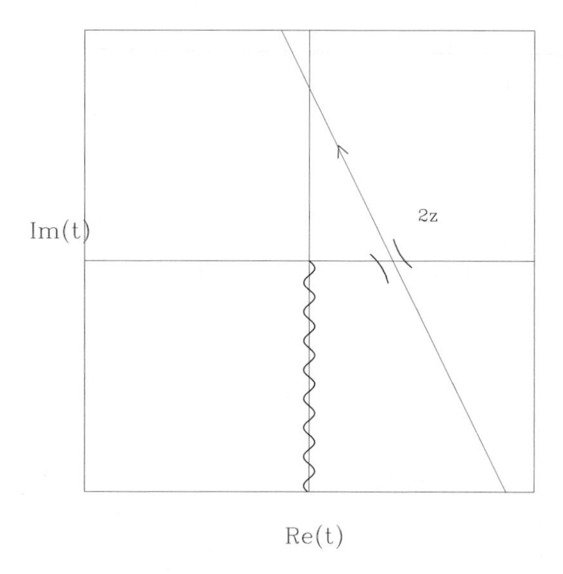

Fig. 12.9 Integration contour for the overdense barrier, z real positive.

For negative real $z = re^{i\pi}$ the saddle point moves to the negative real axis, deforming the contour as shown in Fig. 12.10. We then find a contribution from the cut as well as from the saddle, giving

$$y(z) \simeq \frac{e^{iz^2/2}}{\sqrt{z}} z^{-ib^2/2}$$

$$+\frac{e^{-iz^2/2}}{\sqrt{-z}}(-2z)^{ib^2/2} e^{\pi b^2/4} \frac{\Gamma(1/2 - ib^2/2)}{\sqrt{2\pi}}(e^{-\pi b^2/2} + e^{\pi b^2/2}). \tag{12.17}$$

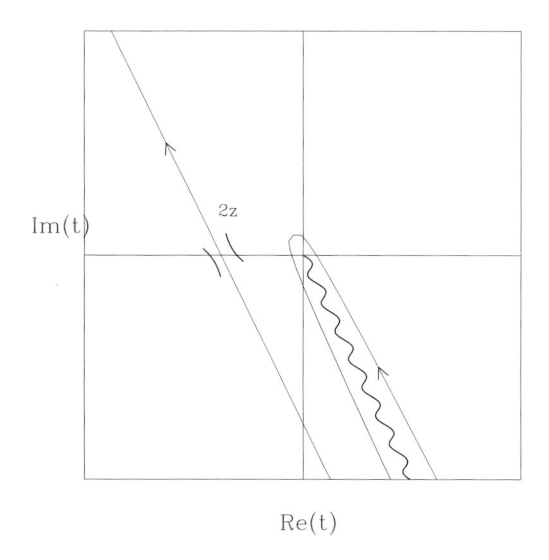

Fig. 12.10 Integration path for the overdense barrier, z real negative.

Using $\Gamma(1/2 - ib^2/2)\Gamma(1/2 + ib^2/2) = \pi/cosh(\pi b^2/2)$ we then find the amplitude of the incoming wave to be

$$\frac{\sqrt{2\pi}e^{\pi b^2/4}2^{ib^2/2}}{\Gamma(1/2 + ib^2/2)} \tag{12.18}$$

and that of the the reflected wave $e^{\pi b^2/2}$, and the transmitted wave has normalization equal to 1. We thus have for this problem the reflection and transmission values

$$R = \frac{\Gamma(1/2 + ib^2/2)e^{\pi b^2/4}}{\sqrt{2\pi}}2^{-ib^2/2}, \tag{12.19}$$

$$T = \frac{\Gamma(1/2 + ib^2/2)e^{-\pi b^2/4}}{\sqrt{2\pi}}2^{-ib^2/2}. \tag{12.20}$$

From $\Gamma(z^*) = \Gamma^*(z)$ we find $|\Gamma(1/2 - ib^2/2)| = \sqrt{\pi/cosh(\pi b^2/2)}$. In the limit $b \to \infty$ we have $|\Gamma| \to \sqrt{2\pi}e^{-\pi b^2/4}$ and the wave is entirely reflected. For $b \to 0$ the incoming amplitude is $\sqrt{2}$ and the reflected and transmitted amplitudes are both 1, giving the well-known result that the wave is half reflected and half transmitted.

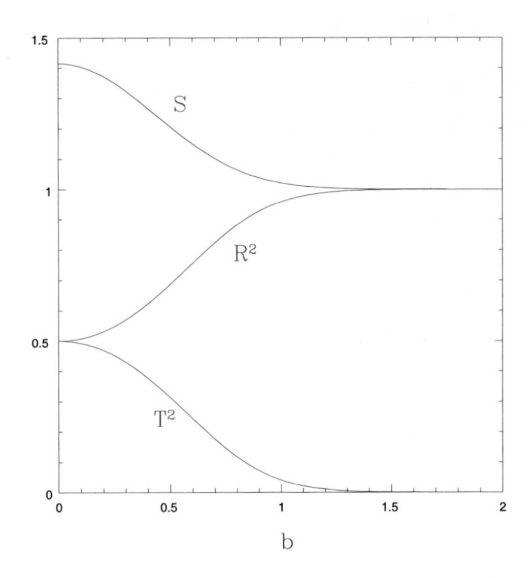

Fig. 12.11 Magnitude of the Stokes constant for the overdense barrier, and absolute squares of reflection and transmission amplitudes vs b.

Comparing this result with the WKB analysis, Eq. 12.14, we find the Stokes constant

$$S = i(2)^{ib^2/2} \frac{\sqrt{2\pi} e^{-\pi b^2/4}}{\Gamma(1/2 + ib^2/2)}. \tag{12.21}$$

Note that the limit of large b does not produce the Stokes constant given by Heading, even though the reflection and transmission amplitudes are equal in this limit to those given by his analysis using $S = i$. Thus the limit of large separation is in fact a singular limit so far as the Stokes constants are concerned. This is because the Stokes constant given by Eq. 12.21 contains correct phase information for the reflected and transmitted waves, whereas the Stokes constants calculated using isolated singularities cannot contain this information.

Figure 12.11 shows the magnitude of the Stokes constant S as a function of b, as well as the absolute squares of the reflection and transmission amplitudes R and T.

12.4 Scattering, Underdense Barrier

A solvable underdense barrier problem is given by $Q(z) = z^2 + b^2$, real on the real axis, with two first order zeros at points $\pm ib$ and Q is positive everywhere on the real axis, as analyzed in Chapter 5. The Stokes diagram is shown in Fig. 12.12, where the cut has been deformed to follow the Stokes line, making the WKB solutions analytic in the upper half plane above ib. Choose the sheet with $\sqrt{Q}$ positive for $x > 0$, then continuing through the upper half plane to $x < 0$ we have $\sqrt{Q} \sim e^{i\pi}$. The rotation of the plot has been exaggerated; it can be infinitesimal. Define W through $[-ib, ib]_r = e^{-W}$ where the subscript refers to the right-hand side of the cut, and $W = \pi b^2/2$. Then $[-ib, ib]_l = e^W$. Consider an incident wave from the left, giving again outgoing boundary conditions at the far right.

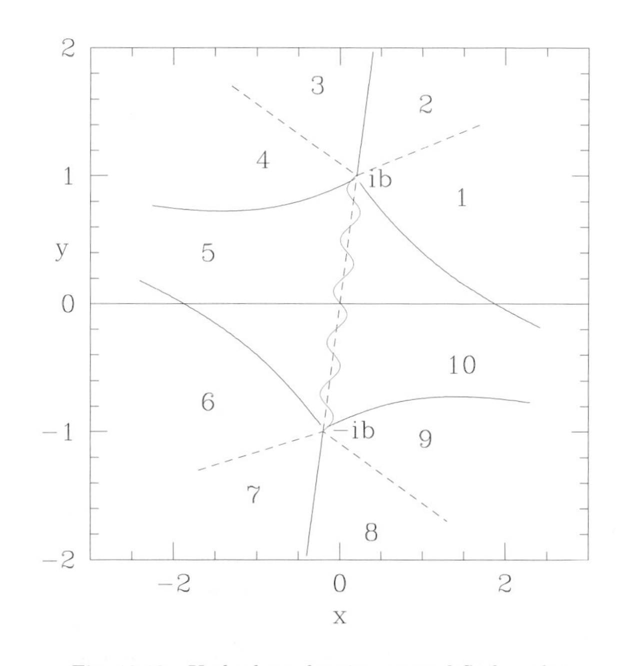

Fig. 12.12 Underdense barrier, rotated Stokes plot.

Continuation then gives:

(1) $(ib, z)_s$
(2) $(ib, z)_s$
(3) $(ib, z)_d$

(4) $(ib, z)_d + S(z, ib)_s$
(5) $(ib, z)_s + S(z, ib)_d$
(5) $[ib, -ib](-ib, z)_d + S(z, -ib)_s[-ib, ib]_l$
(6) $[ib, -ib](-ib, z)_s + S(z, -ib)_d[-ib, ib]_l$

Take the reference point for phase to be $z = 0$. Then we have in (1) $[ib, 0]_r(0, z)$ and in (6) $[ib, 0]_l(0, z) + S(z, 0)[0, ib]_l$. Because of the choice of the cut locations the incoming wave is $(z, 0)$ and the reflected wave is $(0, z)$ giving reflection of $R = e^{-W}/S$ and transmission of $T = 1/S$. Normalizing the solution to the transmitted wave we have for large negative x

$$\psi \simeq e^{-W}(0, z) + S(z, 0). \tag{12.22}$$

12.4.1 *Integral representation*

The integral representation proceeds exactly following the overdense case, with the substitution $b^2 \to -b^2$, giving a solution with

$$f(z) = \frac{2^{-ib^2/2} e^{i\pi/4}}{\sqrt{2\pi}} \int_{\infty e^{3i\pi/4}}^{\infty e^{-i\pi/4}} e^{-it^2/4 + itz} t^{ib^2/2 - 1/2} dt \tag{12.23}$$

with normalization chosen to match the WKB solution. For positive real z there is a saddle point at $t = 2z$. Evaluation of this contribution gives

$$y(z) \simeq \frac{e^{iz^2/2}}{\sqrt{z}} z^{ib^2/2}, \tag{12.24}$$

showing that this is an outgoing wave, satisfying the correct boundary condition for the scattering problem.

For negative real $z = re^{i\pi}$ the saddle point moves to the negative real axis, with a contribution from the cut as well as from the saddle, giving

$$y(z) \simeq \frac{e^{iz^2/2}}{\sqrt{z}} z^{ib^2/2}$$
$$+ \frac{e^{-iz^2/2}}{\sqrt{-z}} (-2z)^{-ib^2/2} e^{-\pi b^2/4} \frac{\Gamma(1/2 + ib^2/2)}{\sqrt{2\pi}} (e^{-\pi b^2/2} + e^{\pi b^2/2}). \tag{12.25}$$

Using $\Gamma(1/2 - ib^2/2)\Gamma(1/2 + ib^2/2) = \pi/cosh(\pi b^2/2)$ we then find the amplitude of the incoming wave to be

$$\frac{\sqrt{2\pi} e^{-\pi b^2/4} 2^{ib^2/2}}{\Gamma(1/2 - ib^2/2)} \tag{12.26}$$

and that of the the reflected wave $e^{-\pi b^2/2}$, and the transmitted wave has normalization equal to 1. We thus have for this problem the reflection and transmission values.

$$R = \frac{\Gamma(1/2 - ib^2/2)e^{-\pi b^2/4}}{\sqrt{2\pi}} 2^{ib^2/2}. \tag{12.27}$$

$$T = \frac{\Gamma(1/2 - ib^2/2)e^{\pi b^2/4}}{\sqrt{2\pi}} 2^{ib^2/2}. \tag{12.28}$$

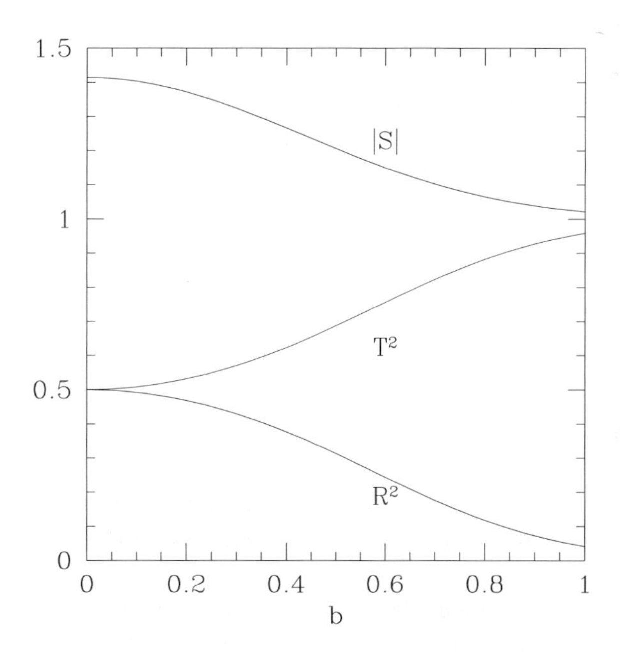

Fig. 12.13 Underdense barrier, reflection, transmission and Stokes constant.

From $\Gamma(z^*) = \Gamma^*(z)$ we find $|\Gamma(1/2 - ib^2/2)| = \sqrt{\pi/cosh(\pi b^2/2)}$. In the limit $b \to \infty$ we have $|\Gamma| \to \sqrt{2\pi}e^{-\pi b^2/4}$ and the wave is entirely transmitted. For $b \to 0$ the incoming amplitude is $\sqrt{2}$ and the reflected and transmitted amplitudes are both 1, giving the well-known result that the wave is half reflected and half transmitted.

Comparing this result with the WKB analysis we find the Stokes constant

$$S = i(2)^{-ib^2/2} \frac{\sqrt{2\pi} e^{-\pi b^2/4}}{\Gamma(1/2 - ib^2/2)}. \tag{12.29}$$

Note that this is not simply the analytic continuation of Eq. 12.21 because the Stokes diagrams are not analytic continuations of one another due to the cut choices. Figure 12.13 shows the magnitude of the Stokes constant S as a function of b, as well as the absolute squares of the reflection and transmission amplitudes R and T.

12.5 The Budden Problem

We consider the standard problem treating the penetration and resonant absorption of an electromagnetic wave, analyzed by Budden [1979]. The differential equation is $Ly = [d^2/dx^2 + Q(x)]y = 0$. The function $Q(x)$ is taken to be $Q(x) = 1 + c/x$, having a first order zero or cutoff at $x = -c$ and a resonance or first order pole at $x = 0$, and $Q(x) > 0$ for $0 < x$ and $x < -c$.

12.5.1 *WKB analysis*

The Stokes diagram for this equation is shown in Fig. 12.14. The cut at the pole, $z = 0$, must be taken upwards as shown. This condition is equivalent to the Landau prescription for the pole to describe Landau damping (Stix [1992]). Note that the singularity at $z = -c$ is an artifact of the WKB analysis. The differential equation has only a pole at $z = 0$, and thus a single cut extending upward from this point. The asymptotic form of the two solutions are $y \sim e^{iz} z^{ic/2}$ and $y \sim e^{-iz} z^{-ic/2}$.

With the choice of cuts as shown, and taking $Q^{1/2}$ to be real and positive in the interval $0 < x$, we have $arg(Q^{1/2}) = \pi/2$ on the real axis for $-c < x < 0$ and $arg(Q^{1/2}) = \pi$ for $x < -c$. Now perform the standard analytic continuation, as done by Heading, but making no assumption about the Stokes constants. Begin at large positive x in domain 1, with an outgoing

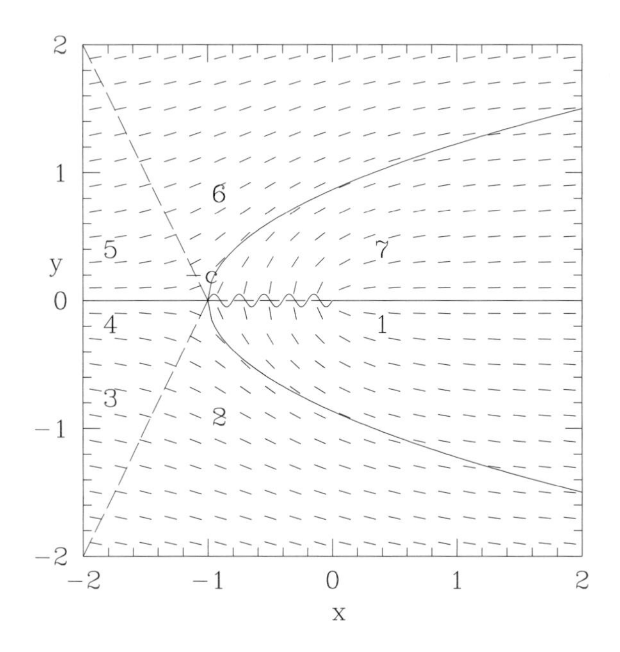

Fig. 12.14 Stokes diagram for Budden problem.

solution $y \sim e^{ix}$. In domain 1 this is dominant. The continuation is then

$$
\begin{array}{ll}
1 & (0,z)_d = z^{ic/2}e^{iz}, \\
1 & [0,-c]_b(-c,z)_s, \\
2 & [0,-c]_b(-c,z)_d, \\
3 & [0,-c]_b(-c,z)_d - S_2(z,-c)_s, \\
4 & [0,-c]_b(-c,z)_s - S_2(z,-c)_d, \\
5 & -S_2(z,-c)_d + [[0,-c]_b + S_5S_2](-c,z)_s, \\
6 & -S_2(z,-c)_s + [[0,-c]_b + S_5S_2](-c,z)_d, \\
6 & -S_2(z,0)_s[0,-c]_a + [[0,-c]_b + S_5S_2][-c,0]_a(0,z)_d,
\end{array}
\tag{12.30}
$$

with S_2, S_5 the unknown Stokes constants, $[0,-c]_{a,b} = e^{i\int_0^{-c} Q^{1/2}dx}$ and the subscript indicates whether it is evaluated above or below the cut. Now note that $[0,-c]_b = e^{i\int_0^{-c} Q^{1/2}dx} = e^W$ with $W = c\pi/2$, and $[0,-c]_a = e^{-W}$. Notice that with the cut chosen as shown the function $(0,z)$ can be analytically continued through the whole complex z plane.

12.5.2 *Integral representation*

Assuming an integral representation of the Fourier–Laplace form

$$y = \frac{\Gamma(1+ic/2)2^{-ic/2}e^{\pi c/4}}{\pi} \int_C e^{zt} f(t)dt \tag{12.31}$$

we find a solution provided

$$f(t) = \left(\frac{t-i}{t+i}\right)^{c/2i} \frac{1}{t^2+1}, \qquad \left(\frac{t-i}{t+i}\right)^{c/2i} e^{xt}|_a^b = 0 \tag{12.32}$$

where a, b are the end points of the contour C. For $z > 0$ the only possible end point is $t = -\infty$. Two contours giving independent solutions are shown in Fig. 12.15. The contour C_1 gives a right moving wave, and C_2 a left moving wave. For large $z > 0$ the dominant contribution to the integral from contour C_1 comes from near $t = i$, and using $t = e^{i\pi/2} + \rho e^{i\pi}$ above the cut and $t = e^{i\pi/2} + \rho e^{-i\pi}$ below the cut we find

$$y(x) \simeq e^{iz} z^{ic/2}, \tag{12.33}$$

equal to the asymptotic WKB expression.

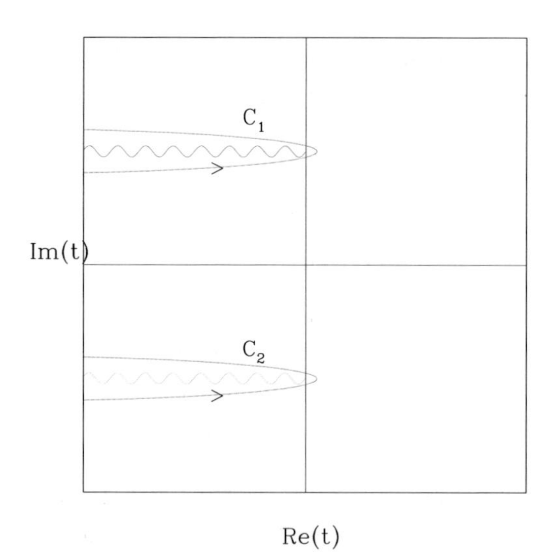

Fig. 12.15 Contour for the Budden problem, $z = r$.

Now analytically continue to negative z. Because of the cut due to the singularity at $z = 0$, (see Fig. 12.14) in order to compare expressions with the WKB results we must continue clockwise through the negative half plane in z. Thus, for the integral representation to remain valid, the contour must be continued counterclockwise through the negative half t plane as z is continued, keeping $zt < 0$ for large t.

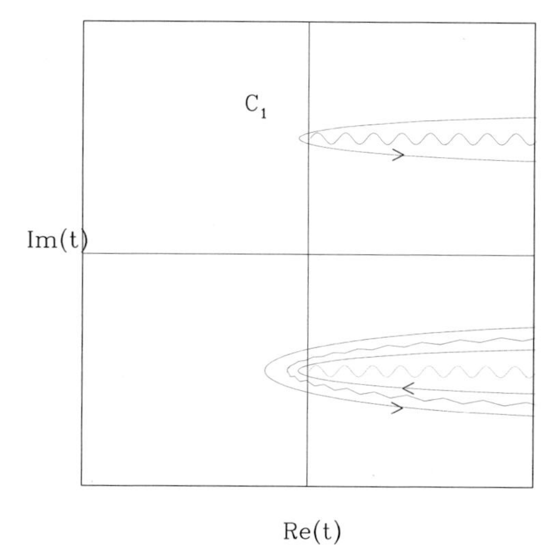

Fig. 12.16 Contour for the Budden problem, $z = re^{-i\pi}$.

Pulling the contour C_1 out to positive t it deforms as shown in Fig. 12.16. The dominant contributions to this contour now come from near $t = i$, simply the analytic continuation of the expression for positive z, and from near $t = -i$, where we use $t = e^{-i\pi/2}+\rho$ above the cut and $t = e^{-i\pi/2}+\rho e^{2\pi i}$ below the cut, and noting that clockwise rotation about the point $t = i$ produces a factor of $e^{\pi c}$ we find

$$y(z) \simeq e^{iz} z^{ic/2} + \frac{e^{-iz}2^{-ic}\Gamma(1 + ic/2)}{(-z)^{ic/2}\Gamma(1 - ic/2)}[1 - e^{-\pi c}] \tag{12.34}$$

and the first term is the incoming wave, with amplitude $y \sim e^{iz}(re^{-i\pi})^{ic/2} \sim e^{iz}r^{ic/2}e^{\pi c/2}$, the second term the reflected wave. Comparing this to the WKB expression we find

$$S_2 = \frac{\Gamma(1 + ic/2)}{\Gamma(1 - ic/2)}2^{-ic}[1 - e^{-\pi c}]. \tag{12.35}$$

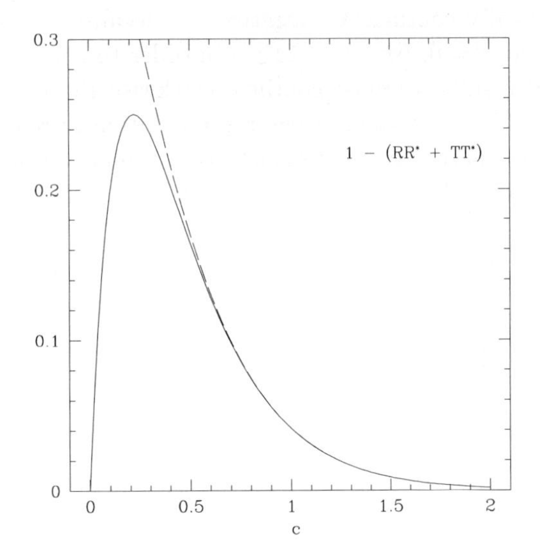

Fig. 12.17 Power absorption at the Budden singularity. The dashed line shows the result using the Stokes constants for isolated singularities.

The transmission coefficient is

$$T = e^{-\pi c/2} \tag{12.36}$$

and the reflection coefficient is

$$R = \frac{\Gamma(1 + ic/2)}{\Gamma(1 - ic/2)} 2^{-ic}(1 - e^{-\pi c}). \tag{12.37}$$

Note that for $c = 0$ there is complete transmission, $R = 0$, and in the limit of $c = \infty$ there is complete reflection. The absorption is given by

$$1 - RR^* - TT^* \tag{12.38}$$

and is shown as a function of c in Fig. 12.17. Maximum absorption at the singularity of $1/4$ of the incident power is obtained for $c = ln(2)/\pi \simeq 0.22$. Also shown as a dashed line is the result obtained using the Stokes constants for isolated singularities, as derived by Heading, and given in Chapter 5. The isolated singularity result from Eq. 5.36 gives maximum absorption of $4/9$ at $c = 0$.

These results are independent of the Stokes constant at the first order pole at $z = 0$. This constant can be determined by continuing all the way around the complex plane to $z = re^{-2\pi i}$. Since $z = 0$ is a regular singular

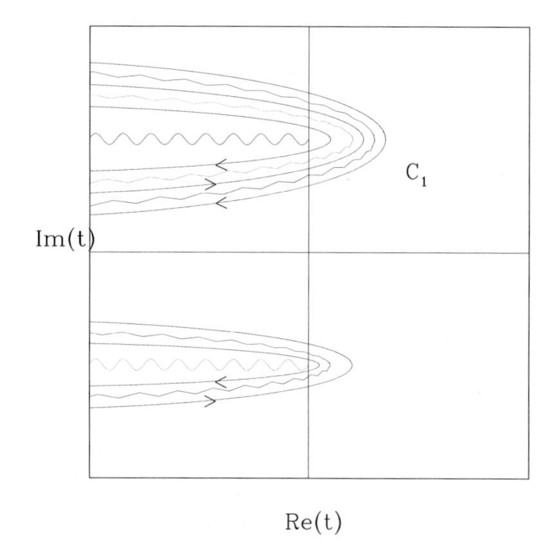

Fig. 12.18 Contour for the Budden problem, $z = re^{-2i\pi}$.

point of the differential equation, the solution is not single valued in the plane, and we arrive at $z = r$ on a second nonphysical sheet. The integration contour becomes as shown in Fig. 12.18. The dominant contributions to this contour again come from near $t = i$ and near $t = -i$, giving

$$y(x) \simeq e^{iz} z^{ic/2} \frac{\Gamma(1+ic/2)}{\Gamma(1-ic/2)}(1-e^{-c\pi})$$

$$+e^{-iz} z^{-ic/2} 2^{-ic} \frac{\Gamma(1+ic/2)}{\Gamma(1-ic/2)} sinh(\pi c/2) \tag{12.39}$$

and the first term is the transmitted wave. The second term is a left moving wave for $x > 0$, not present on the physical sheet. Comparing this to the WKB expression we find the Stokes constant at the first order pole

$$S_5 = \frac{\Gamma(1-ic/2)}{\Gamma(1+ic/2)} e^{\pi c} 2^{ic}, \tag{12.40}$$

almost the inverse of the Stokes constant at the zero.

12.6 Bound state

The bound state or instability problem is unusual in that one is not interested in arbitrary values of the separation of the singularities, but rather only those isolated values giving a normalizable solution, i.e. which require that the dominant part of the solution vanish for $x = \pm\infty$. This requirement alone gives the most important result for such systems, the Bohr–Sommerfeld condition (Bohr [1913], Schiff [1955]). An analytically tractable example of a bound state is given by taking

$$Q = p + \frac{1}{2} - \frac{z^2}{4}, \tag{12.41}$$

giving the Weber, or parabolic cylinder equation (Whittaker and Watson [1962]). Generalizations of this case, and corresponding generalizations of the parabolic cylinder function have been given by Olver [1974] Define $b^2 = 4p + 2$ so that the zeros of Q occur at $z = \pm b$.

12.6.1 *WKB analysis*

The Stokes diagram is shown in Fig. 12.19. This is the standard quantum mechanics bound state diagram, but also occurs in slightly modified form in the examination of instabilities in general, in that the zeros of Q may be located at complex z values. The solution of all such problems requires that the function approach zero at plus and minus infinity, and thus the asymptotic real axis must lie in domains which contain Stokes lines. In this example the real axis for large x coincides with a Stokes line, but this is not necessary in general.

Thus we require that the asymptotic solution for real z be entirely given by a subdominant term. Begin at large positive x with a subdominant solution. Define $[-b, b] = e^{iW}$. Along the real axis between $-b$ and b, Q is real and positive. Choose the branch of the square root making also $\sqrt{Q}$ real and positive, and thus W is real and positive with $W = \pi b^2/4$. $Q(z)$ is a given function, but we must determine at each location the phase of $Q^{1/2}$ and $Q^{1/4}$. With the choice of cuts as shown, and choosing $Q^{1/2}$ positive and real for $-b < x < b$ we have for $x > b$ and for $x < -b$ that $Q^{1/2} \sim e^{-i\pi/2}$. Thus for $z > b$ and real the solution (b, z) is dominant. Begin with a subdominant solution at large positive x and continue, giving

(1) $(z, b)_s$
(2) $(z, b)_d$

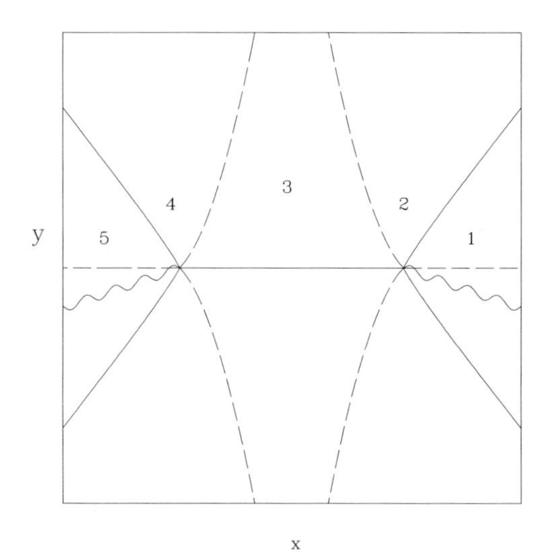

Fig. 12.19 Stokes diagram for the parabolic cylinder function.

(3) $(z,b)_d + S_2(b,z)_s = e^{iW}(z,-b)_d + e^{-iW}S_2(-b,z)_s$
(4) $e^{iW}(z,-b)_d + [e^{iW}S_4 + e^{-iW}S_2](-b,z)_s$
(5) $e^{iW}(z,-b)_s + [e^{iW}S_4 + e^{-iW}S_2](-b,z)_d$
(6) $-ie^{iW}(-b,z)_s - i[e^{iW}S_4 + e^{-iW}S_2](z,-b)_d$

Consider the subdominant solution for large positive z.

$$(z,b)_s = \frac{\sqrt{2}e^{i\pi/4}}{(z^2-b^2)^{1/4}}e^{-\int_b^z \sqrt{z^2-b^2}/2dz}. \tag{12.42}$$

For large z we have $(z^2-b^2)^{1/2} \simeq z - b^2/(2z)$, giving in domain 1

$$(z,b)_s = \sqrt{2}e^{i\pi/4}z^p e^{-z^2/4}, \qquad (b,z)_d \simeq \sqrt{2}e^{i\pi/4}z^{-p-1}e^{z^2/4}. \tag{12.43}$$

Now evaluate $(z,-b)_s$ and $(-b,z)_d$ for negative z. Writing $(z,-b) = (z,b)e^{-iW}$ and $(-b,z) = (b,z)e^{iW}$ and noting that the expressions given by Eq. 12.43 are analytic in the upper half plane, we find the asymptotic phase of $(-b,z)_d \simeq e^{-i\pi/4}$ for negative $z = re^{i\pi}$.

The WKB continuation gives for large negative z in domain 6

$$y(z) \simeq -ie^{iW}(-b,z)_s - i[e^{iW}S_4 + e^{-iW}S_2](z,-b)_d. \tag{12.44}$$

Looking only at phases and requiring the dominant part of the solution to have within a sign the phase $e^{i\pi/4}$ we then find

$$[e^{iW} S_4 + e^{-iW} S_2] \simeq e^{-i\pi/2 + in\pi} \tag{12.45}$$

for all W, n integer. Setting the real part equal to zero gives $S_4 = re^{-iW}$, $S_2 = -re^{iW}$ with r real. If the desired boundary conditions are imposed, namely that the solution be subdominant at $x = \pm\infty$ then the subdominant solution must also have the phase[1] $e^{i\pi/4}$, giving the Bohr–Sommerfeld condition

$$W = (n + 1/2)\pi. \tag{12.46}$$

We then find that the two Stokes constants are equal and pure imaginary, but their magnitude is undetermined.

12.6.2 *Integral representation*

A Fourier–Laplace integral representation can be found by extracting the growing asymptotic form, $y = e^{z^2/4} \int_C e^{2izt} g(t)dt$, giving

$$y(z) = e^{z^2/4} \int_{-\infty}^{\infty} e^{2izt - 2t^2} t^p dt. \tag{12.47}$$

The asymptotic behavior for large positive z comes from a saddle point, shown in Fig. 12.20. We then find

$$y(z) \to \frac{\sqrt{\pi}}{2^{p+1/2}} e^{i\pi p/2} z^p e^{-z^2/4}, \tag{12.48}$$

giving the integral representation

$$D_p(z) = \frac{2^{p+1/2}}{\sqrt{\pi}} e^{-i\pi p/2} e^{z^2/4} \int_{-\infty}^{\infty} e^{2izt - 2t^2} t^p dt. \tag{12.49}$$

In addition, since there is a cut attached to $t = 0$ we have for $z > 0$ the prescription that in the integral $t \sim e^{i\pi}$ for $t < 0$, to allow analytic continuation to large z, with the contour passing through the saddle.

For $z = ir$ the saddle has moved so that the integration contour just touches the real line, and integration past the tip of the cut contributes to the solution, as shown in Fig. 12.21. This is equivalent in WKB terms to the

[1] Note that if the dominant solution is present one cannot domand that the phase of the subdominant part agree with the phase of the solution for positive x, it can in fact be complex.

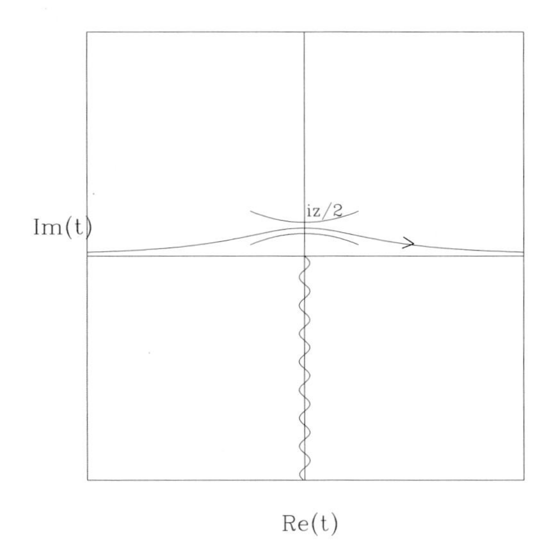

Fig. 12.20 Saddle point for large positive z evaluation of the parabolic cylinder function.

appearance of the subdominant solution upon entering domain 3, Fig. 12.19. However, at this point the new contribution is subdominant, proportional to $e^{z^2/4}$, and is still lost in the error of the WKB approximation.

Integrating along the cut we find the subdominant part of the solution comes from t near zero, giving for $z = re^{i\pi/2}$

$$D_p(z) \simeq r^p e^{i\pi p/2} e^{r^2/4} + \frac{e^{-i\pi p/2}}{\sqrt{2\pi}} e^{-r^2/4} r^{-p-1} \Gamma(1+p). \tag{12.50}$$

Comparing this solution with the WKB expression in domain 3 we find

$$S_2 = \frac{i}{\sqrt{2\pi}} \Gamma(p+1). \tag{12.51}$$

Continuing analytically to negative $z = re^{i\pi}$ causes the saddle point to move around into the cut, distorting the integration contour as shown in Fig. 12.22. The contribution due to the cut for large negative z comes from the end point of the integral,

$$y_{cut} \simeq \frac{\sqrt{2}}{\sqrt{\pi}} e^{z^2/4} z^{-p-1} sin(\pi p) e^{i\pi p} \Gamma(p+1). \tag{12.52}$$

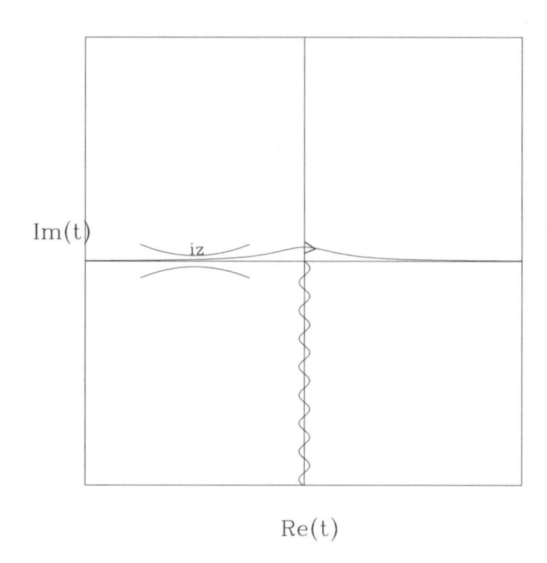

Fig. 12.21 Saddle point for large imaginary z evaluation of the parabolic cylinder function.

Adding the saddle contribution to the contribution from the cut then gives

$$D_p(z) \simeq z^p e^{-z^2/4} + \sqrt{\frac{2}{\pi}} e^{z^2/4} z^{-p-1} \Gamma(1+p) sin(p\pi) e^{ip\pi}. \quad (12.53)$$

Compare Eq. 12.53 with the expression from the WKBJ analysis for large negative real z Eq. 12.44,

$$y(z) \sim z^p e^{-z^2/4} + (e^{iW} S_4 + e^{-iW} S_2) e^{2iW} z^{-p-1} e^{z^2/4}, \quad (12.54)$$

with $W = \int_{-b}^{b} \sqrt{Q} dz = \pi(p + 1/2)$. Substituting the value of S_2 we find

$$S_4 = ie^{-2ip\pi} \frac{2sin(p\pi) + 1}{\sqrt{2\pi}} \Gamma(1+p). \quad (12.55)$$

Setting the dominant part of Eq. 12.53 to zero gives $p = n$ integer so $S_4 = S_2 = i\Gamma(1+n)/\sqrt{2\pi}$. The limit of large separation or equivalently large n does not reproduce the Heading value of $S = i$, but the Bohr–Sommerfeld condition, which is the physically relevant result, requires discrete values of the separation. However if it is desired to analytically continue the solution into the complex plane for arbitrary separation the correct continuation is Eq. 12.53, which corresponds to the use of the Stokes value given by Eq. 12.51.

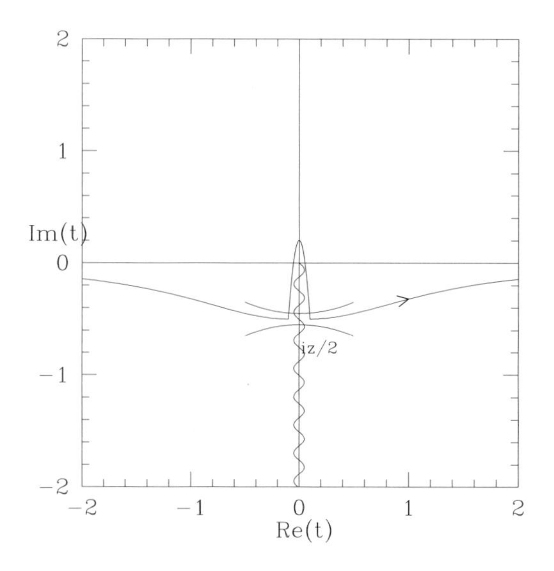

Fig. 12.22 Saddle point for large negative z evaluation of the parabolic cylinder function.

12.7 Concerning Accuracy

In the problems examined in the previous sections the solutions given by the WKB analysis have been exact. The Stokes constants determined for the continuation give exact agreement with the integral representations, the asymptotic phases of the solutions are correct, and the Bohr–Sommerfeld condition determines the eigenvalues exactly. The reason for this is that the analytic function $Q(z)$ in each case possesses precisely the singularities required to represent the function $Q(x)$ on the real axis, and no others. Consideration of a simple example where this is not true will illustrate the importance of this. Take

$$Q(z) = E - z^4. \tag{12.56}$$

Considered along the real axis this again represents a simple potential well with two turning points. It differs from the quadratic well considered in section 12.6 only by its shape. But considered in the complex plane it has a wholly different aspect, namely in addition to the turning points along the real axis there are also two complex turning points at $z = \pm iE^{1/4}$.

The Stokes structure is shown in Fig. 12.23. In the vicinity of the real axis this diagram is equivalent to the canonical Stokes diagram for a bound

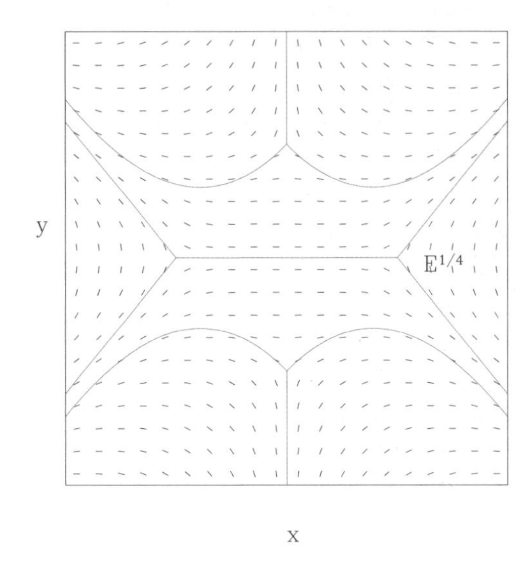

Fig. 12.23 Stokes plot for $Q = E - z^4$, showing anti-Stokes lines.

state, but away from the real axis it is modified by the presence of the additional singularities.

The bound state eigenvalues given by the Bohr–Sommerfeld condition and by a numerical integration for the first few states are given in Table 12.1. As is well known, the error in the eigenvalue decreases with n. The reason for this is simply understood by considering the WKB continuation process. The WKB asymptotic solutions fail only in the vicinity of the turning points. For large n there is ample space to carry out the continuation remaining in the vicinity of the real axis and thus very far from the complex turning points. When this can be done the additional complex turning points do not enter into the solution of the problem. For small n there is insufficient space between the real axis and the complex turning points in which the asymptotic solutions are valid, and thus these additional turning points interfere with the analytic continuation. For the case $Q = E - z^4$ the analytic continuation from the positive real axis must pass between the singularity at $z = E^{1/4}$ and that at $z = iE^{1/4}$. The point midway between these singularities is $z = E^{1/4}e^{i\pi/4}/\sqrt{2}$, and at this point $Q'/Q^{3/2} \simeq 2/E^{3/4}$.

Table 12.1 Eigenvalues for $Q(z) = E - z^4$.

n	E_{WKB}	E_{num}	$\delta E/E$	$Q'/Q^{3/2}$
0	0.867	1.06	.182	2
1	3.752	3.787	.009	.7
2	7.413	7.456	.006	.5
3	11.61	11.62	.001	.3
4	16.23	16.26	.001	.25

This value is shown in Table 12.1. Clearly for small n values the continuation passes through a region in which the WKB asymptotic solution is not a good approximation.

A physics problem, and indeed the solution of the differential equation, is defined by the values of $Q(x)$ on the real axis. However, there is an unique analytic continuation of Q into the complex plane, and this continuation gives rise to aditional zeros and singularities for complex z. These singularities modify the exact solution of the differential equation through the shape of the potential on the real axis, and affect the WKB solution according to how close they are to the real axis.

12.8 Conclusion

In conclusion we have explored the possibility of extending the work of Heading, namely the determination of Stokes constants for singularities of the form x^n, to include diagrams involving multiple singularities. Using analytically tractable examples we compare exact solutions with WKB approximations and find the Stokes constants for diagrams with two singularities. These constants are found to be given by Euler gamma functions of the separation. This transcendental function is ubiquitous in the theory of differential equations, in spite of satisfying none itself. We show that it is possible to consistently define the Stokes constants in this manner to obtain exact asymptotic expressions for amplitudes, and compare these results to those obtained using the isolated singularity expressions derived by Heading. The Stokes constants thus obtained do not approach the values given by Heading in the limit of large separation of the singularities, but this is not a defect, it reflects the fact that the new constants contain correct phase information for the solutions, something missing in the values given by Heading. Since this phase information has no well defined limit for large separation neither do the Stokes constants. In spite of the Stokes constants having no well-defined limit for large separation, the physical results, which for scattering problems means the reflection and transmission constants, and for bound states and instabilities the eigenvalues resulting from the Bohr–Sommerfeld rule, do have asymptotic values agreeing with those obtained using the Stokes constants for isolated singularities.

The first rule of Heading should thus be modified to read:

(a) If a_d and a_s are respectively the coefficients of the dominant and subdominant terms of a solution, then upon crossing a Stokes line in a counterclockwise sense a_s must be replaced by a_s+Sa_d where S is called the Stokes constant. When the Stokes line originates at an isolated zero of order n, $S = 2icos(\pi/(n+2))$. For Stokes diagrams consisting of more than one singularity the use of this Stokes constant produces results that are correct only in the limit of large singularity separation, and does not give correct phase information for reflected and transmitted waves. If arbitrary separation or phase information of the asymptotic solution is required the Stokes constants must be modified as follows:
For two connected first order zeros with $Q = (z-a)(z-b)/c^2$ and boundary conditions for scattering of an incident wave the Stokes constant is

$$S = i\left(\frac{2}{c}\right)^P \frac{\sqrt{2\pi}e^{i\pi P}}{\Gamma(1/2+P)} \tag{12.57}$$

where $P = \int_a^b Q^{1/2}dz/\pi$. Both a and b can be complex, leading to subcritical scattering.
For two connected first order zeros located at a, b with $Q = (z-a)(b-z)/4$ and boundary conditions for a bound state or instability the Stokes constant at the zeros is

$$S = i\frac{\Gamma(1/2+P)}{\sqrt{2\pi}} \tag{12.58}$$

where $P = \int_a^b \sqrt{Q}dz/\pi$. Note that this value of S is practically the inverse of that for the scattering problem.
For a first order zero at $z = a$ connected to a first order pole located at $z = b$, with the cut taken above the line joining these points, the Stokes constant at the zero is

$$S = \frac{\Gamma(P)}{\Gamma(-P)}e^{i\pi P}2^{-2P} \tag{12.59}$$

where $P = \int_a^b \sqrt{Q}dz/\pi$.

For a first order zero at $z = a$ connected to a first order pole located at $z = b$ the Stokes constant at the pole is

$$S = \frac{\Gamma(-P)}{\Gamma(P)} e^{-2i\pi P} 2^{2P} \tag{12.60}$$

where $P = \int_a^b \sqrt{Q} dz / \pi$.

This amendment to the rules of Heading allows analysis of problems with arbitrary separation between the singularities. Interestingly, the limit of large separation is singular, the Stokes constants of Heading are not recovered because the phase information regarding the solution is modified by the presence of the second singularity, but the physical results, the reflection and transmission amplitudes, and the Bohr–Sommerfeld rule, are.

For the problems considered these Stokes constants produce WKB solutions that have exact asymptotic behavior and eigenvalues. Similar functions $Q(x)$ having the same number of zeros and singularities on the real axis, but different functional shape will result only in approximate solutions, with the degree of error given by the proximity of additional complex zeros and singularities resulting from the continuation of the function $Q(x)$ into the complex plane.

12.9 Problems

1. Consider the equation $y'' + Qy = 0$ with $Q = c^2(a^4/4 - a^2x^2 + x^4)$.

a) Find transmission and reflection coefficients, and a condition that the potential is transparent, i.e. that the reflection is zero.

b) Now subtract a small constant from Q, so that $Q = c^2(a^4/4 - a^2x^2 + x^4) - \delta$. Perform a perturbation theory analysis of the change in the reflection and transmission coefficients.

2. Consider the equation $y'' + Qy = 0$ with $Q = c^2x^2(a^2 - x^2)$.

a) Find values of c so that the solution tends to zero for $x \to \pm\infty$.

b) Now subtract a small constant from Q, so that $Q = c^2x^2(a^2 - x^2) - \delta$. Perform a perturbation theory analysis of the change in the eigenvalues.

3. Consider underdense scattering $y'' + Qy = 0$ with $Q = x^2 + b^2$.

a) Find an alternate integral representation by using the variable $z = x^2$ and writing $y(z) = \int e^{zt} f(t)dt$. Find integration contours to give outgoing and incoming waves for $x \to \infty$. Normalize your integral solutions to give the standard WKB solutions $(0, z)$ and $(z, 0)$.

b) Choose an outgoing solution for $x \to \infty$. How is the contour deformed during analytic continuation to $x \to -\infty$? How would you calculate the reflection and transmission coefficients?

4. Consider the Budden problem $\psi'' + Q\psi = 0$ with $Q = 1 + c/x$.

a) Find an alternate integral representation by writing $\psi(x) = e^{ix}y$ and taking $y(x) = \int e^{xt} f(t)dt$. Find the integration contours giving right and left moving waves for $x \to \infty$. Normalize your integral solutions to give the standard WKB solutions $(0, z)$ and $(z, 0)$.

b) How is the contour deformed during analytic continuation to $x \to -\infty$? How would you calculate the reflection and transmission coefficients?

5. For the Stokes caluculation for the pole of order n, $\psi'' + z^n\psi = 0$ of section 12.2, continue the solution to the first anti-Stokes line at $z = e^{-2\pi i/(n+2)}$ and the second Stokes line at $z = e^{-3\pi i/(n+2)}$ and verify that the solution agrees with the WKB expression.

Chapter 13

Bessel

Friedrich Wilhelm Bessel was born in 1784 in Minden, Brandenburg, now Germany. At the age of fourteen he was apprenticed as a clerk for a merchant in the seaport of Bremen. Fascinated by navigation, he began to study mathematics and astronomy as a means of determining longitude. His calculation of the orbit of Halley's Comet obtained for him a position at the observatory at Lilienthal in 1806. In 1809 he became the director of an astronomical observatory in Konigsberg, and calculated 3,222 stellar positions using data from the eighteenth century. In these calculations he explored least-squares estimation and probabilistic error theory, developed by Carl Friedrick Gauss, and coined the term "probable error". He was the first to use parallax to calculate the distance to a star. In 1838 there was a competition to be the first to measure parallax accurately, which Bessel won, announcing that 61 Cygni had a parallax of .314 arcseconds, which, given the diameter of the earth's orbit, indicated that the star was 9.8 light years away. His precise measurements of deviations in the motion of Sirius allowed him to predict the existence of a companion in 1844, eventually leading to the discovery of Sirius B. He was elected a fellow of the Royal Society, and the largest crater in the moon's Mare Serenitatis is named after him

He was primarily an astronomer, and introduced the cylinder functions to study the perturbation of the elliptic motion of a planet caused by a second planet. The Bessel equation was first examined by Euler in 1764, studying the vibrations of a stretched membrane. Then in 1770 Lagrange showed that these functions appeared in the analysis of the motion of a planet about the sun. Only in 1824 Bessel did publish a memoir with a detailed examination of the functions which now bear his name.

The equation is

$$y'' + \frac{y'}{z} + \left(1 - \frac{\nu^2}{z^2}\right) y = 0. \tag{13.1}$$

13.1 Local Analysis

13.1.1 *Local at zero*

Note $z = 0$ is a regular singular point. Thus we attempt a solution of the form

$$y_\alpha(z) = \sum_0^\infty a_n z^{n+\alpha}. \tag{13.2}$$

Substituting into the differential equation and renaming summation indices we find, with $L = d^2/dz^2 + 1/z + 1 - \nu^2/z^2$,

$$\begin{aligned} Ly &= (\alpha^2 - \nu^2) a_0 z^{\alpha-2} + [(1+\alpha)^2 - \nu^2] a_1 z^{\alpha-1} \\ &+ \sum_2^\infty \left([(n+\alpha)^2 - \nu^2] a_n + a_{n-2}\right) z^{n+\alpha-2} = 0. \end{aligned} \tag{13.3}$$

The indicial equation gives $\alpha = \pm\nu$, and we also find that $a_1 = 0$. Thus

$$4\left(k + \frac{\alpha+\nu}{2}\right)\left(k + \frac{\alpha-\nu}{2}\right) a_{2k} = -a_{2k-2} \tag{13.4}$$

and $a_k = 0$ for k odd. This gives the solution

$$y_\alpha(z) = \sum_{k=0}^\infty \frac{a_0(-1)^k \Gamma(1+(\alpha+\nu)/2)\Gamma(1+(\alpha-\nu)/2)}{4^k \Gamma(k+1+(\alpha+\nu)/2)\Gamma(k+1+(\alpha-\nu)/2)} z^{2k+\alpha}. \tag{13.5}$$

Take $Re\nu \geq 0$ with no loss of generality. Two independent solutions for ν non integer are given by taking $\alpha = \pm\nu$ and choosing the normalization a_0. Set $\alpha = \nu$ and take the standard normalization of $a_0 = 1/[2^\nu \Gamma(1+\nu)]$ giving for the function which is well-behaved at zero

$$J_\nu(z) = \left(\frac{z}{2}\right)^\nu \sum_0^\infty (-1)^k \frac{z^{2k}}{2^{2k} k! \Gamma(\nu+k+1)}, \tag{13.6}$$

with an infinite radius of convergence. Rather than using $J_{-\nu}$, the second function, infinite at $z = 0$, is taken to be the Neumann function, which is given[1] for $\nu \neq integer$ by

$$N_\nu(z) = \frac{1}{sin\nu\pi}[cos\nu\pi J_\nu(z) - J_{-\nu}(z)]. \tag{13.7}$$

The normalizations and the combination of $\pm\nu$ terms for N_ν are dictated by behavior at infinity.

13.1.2 *Analytic continuation in ν*

If ν is an integer the second solution is not given directly by $J_{-\nu}$. It is clear from the equation for a_{2k} that setting $\alpha = -\nu$, the two solutions are equal if ν is integer. As seen in section 4.2.2 in this case the second solution often involves terms in $ln(z)$. In fact Eq. 13.7 is analytic considered as a function of ν, so the second solution for $\nu = integer$ can be obtained by analytic continuation. Take the limit of Eq. 13.7 as $\epsilon \to 0$, with $\nu = n + \epsilon$, using

$$\left(\frac{z}{2}\right)^{n+2k+\epsilon} = \left(\frac{z}{2}\right)^{n+2k} (1 + \epsilon ln(z/2) + O(\epsilon^2)), \tag{13.8}$$

and for $j \geq 0$

$$\begin{aligned}\frac{1}{\Gamma(j+1+\epsilon)} &= \frac{1}{\Gamma(j+1)}\left(1 - \epsilon\frac{\Gamma'(j+1)}{\Gamma(j+1)} + O(\epsilon^2)\right) \\ &= \frac{1}{\Gamma(j+1)}\left(1 + \epsilon\gamma - \epsilon\sum_1^j \frac{1}{m} + O(\epsilon^2)\right),\end{aligned} \tag{13.9}$$

with γ the Euler–Mascheroni constant, section 8.3, and

$$\frac{1}{\Gamma(1-j-\epsilon)} = \frac{sin(\pi(j+\epsilon))\Gamma(j+\epsilon)}{\pi} \to (-1)^j \epsilon\Gamma(j), \tag{13.10}$$

for $j > 0$. We then find for n non-negative integer

$$\begin{aligned}N_n(z) &= \frac{2}{\pi} J_n(z)[ln(z/2) + \gamma] - \frac{1}{\pi}\sum_{k=0}^{n-1} \frac{(\frac{z}{2})^{-n+2k}\Gamma(n-k)}{k!} \\ &- \frac{1}{\pi}\sum_{k=1}^{\infty} \frac{(-1)^k(\frac{z}{2})^{n+2k}}{k!(n+k)!}\left[\sum_{m=1}^{k}\frac{1}{m} + \sum_{m=1}^{n+k}\frac{1}{m}\right] - \frac{1}{\pi n!}\left(\frac{z}{2}\right)^n \sum_{k=1}^{n}\frac{1}{k}.\end{aligned} \tag{13.11}$$

[1]Sometimes the Neumann function is denoted by $Y_\nu(z)$.

13.1.3 *Local at infinity*

The point $z = \infty$ is an irregular singular point, as can be checked with the substitution $z = 1/t$. Attempt a solution of the form $y = e^S$. Substitute and find

$$S'' + (S')^2 + \frac{S'}{z} + 1 - \frac{\nu^2}{z^2} = 0. \tag{13.12}$$

Dominant balance gives to leading order $S' = \pm i$, and the next order $S = \pm iz + g$ gives $g = -ln(z)/2$ or

$$y \simeq \frac{1}{z^{1/2}} e^{\pm iz}, \tag{13.13}$$

real solutions being of course sine and cosine functions. Note that at this point there is no means of relating a particular solution near $z = 0$ to a particular combination of sine and cosine at large z. Write

$$y = \frac{1}{z^{1/2}} e^{\pm iz} W_{\pm}(z), \tag{13.14}$$

and find the differential equation for $W_{\pm}(z)$

$$W'' \pm 2iW' + \frac{1/4 - \nu^2}{z^2} W = 0, \tag{13.15}$$

with a solution given by the divergent series

$$W_{\pm} = \sum_{0}^{\infty} a_n z^{-n}, \tag{13.16}$$

and

$$a_{n+1} = \pm \frac{(n + 1/2 + \nu)(n + 1/2 - \nu)}{2i(n+1)} a_n, \tag{13.17}$$

or

$$a_n = (\pm 1)^n \frac{\Gamma(n + 1/2 + \nu)\Gamma(n + 1/2 - \nu)}{n!(2i)^n \Gamma(1/2 + \nu)\Gamma(1/2 - \nu)} a_0. \tag{13.18}$$

13.2 WKB Analysis

To perform a WKB analysis of the Bessel equation, make the substitution $y = \psi/\sqrt{z}$, giving $\psi'' + Q(z)\psi = 0$ with $Q(z) = 1 + (1/4 - \nu^2)/z^2$. For large z the solutions have the form $y \simeq e^{\pm iz}/\sqrt{z}$. Figure 13.1 shows the Stokes plot for the Bessel equation. (We assume $\nu > 1/2$.) Note that this figure, viewed on a very large scale, is not distinguishable from Fig. 5.3. It is not possible to use Stokes constants associated with isolated singularities; the structure must be treated as a whole.

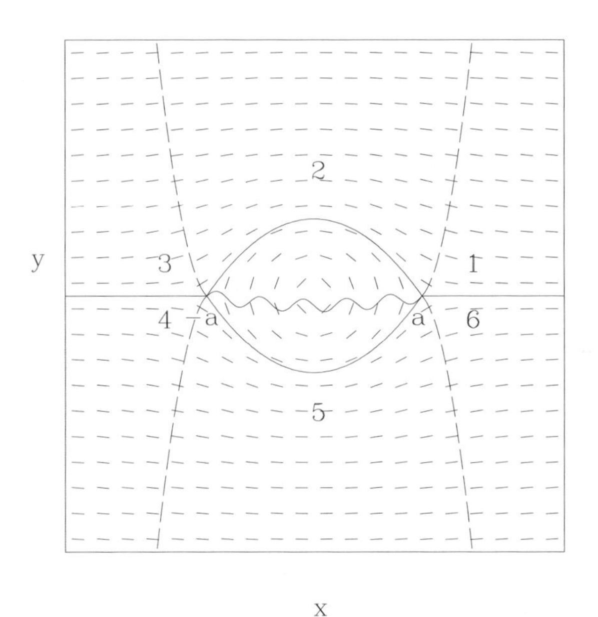

Fig. 13.1 Stokes plot for the Bessel equation.

Write $Q = (z-a)(z+a)/z^2$ with $a = \sqrt{\nu^2 - 1/4}$, and place the cut from $-a$ to a so that along the real line above the cut we have $Q \sim 1$ for $|z| > a$, $Q \sim e^{i\pi}$ for $0 < z < a$ and $Q \sim e^{-i\pi}$ for $-a < z < 0$.

Now carry out the continuation. First notice that $\sqrt{Q}$ is odd, so that $[-a,a]_u = e^{i\pi a}$ where the subscript refers to integration above the cut. Similarly $[-a,a]_l = e^{i\pi a}$. We wish to relate the solution to expansions about $z = 0$ so begin in domain 1 with $(0,z)_s = e^{i\pi a/2}(a,z)$. Continuing to domain 2 there is no change. Then reconnect, giving $e^{-i\pi a/2}(-a,z)_s = (0,z)_s$, with no change on moving to domain 3. In domain 4 this becomes $e^{-i\pi a/2}(-a,z)_d$. In domain 5 we have $(-a,z)_d e^{-i\pi a/2} + T_1 e^{-i\pi a/2}(z,-a)_s$ which upon reconnection becomes $(a,z)_d e^{i\pi a/2} + T_1 e^{-i\pi a}(z,a)_s$ and in domain 6 we have $(a,z)_d e^{i\pi a/2} + [T_1 e^{-i\pi a} + T_2](z,a)_s$, which is $(0,z)_d + [T_1 e^{-i\pi a/2} + T_2 e^{i\pi a/2}](z,0)_s$. Write this as $(0,z)_d + T(z,0)_s$ with T the Stokes constant for continuation through 2π. The fact that the solution can be written in this manner justifies the use of the Stokes plot obtained using only asymptotically large $|z|$, Fig. 5.3, i.e. the structure near the origin is not relevant to global continuation.

Note that the origin is a regular singular point of the differential equation, so the expression in domain 6 cannot be continued across the real axis and equated to the expression in domain 1, because the solution itself possesses a cut.

13.3 Integral Representations

13.3.1 *Fourier–Laplace representation*

The direct use of a Fourier–Laplace transformation produces no simplification since factors of z^2 require integration by parts twice and hence lead again to a second order differential equation. However, the leading order terms in this process are reminiscent of the indicial equation, so we try the representation

$$y(z) = z^\nu \int_C e^{isz} f(s) ds. \tag{13.19}$$

Substituting we find that all terms in $z^{\nu-2}$ cancel, leading to only a single integration by parts, and thus a first order differential equation for $f(s)$. We find

$$\begin{aligned} (2\nu+1)sf(s) - 2sf(s) - (s^2-1)f'(s) = 0, \\ f(s)(s^2-1)e^{isz}|_a^b = 0, \end{aligned} \tag{13.20}$$

giving

$$f(s) = (s^2-1)^{\nu-1/2}, \qquad (s^2-1)^{\nu+1/2} e^{isz}|_a^b = 0. \tag{13.21}$$

For $Re(z) > 0$ there is a solution for any ν given by $a, b = i\infty$. Two solutions can be realized by taking contours which circle the cuts originating at ± 1. These are known as the Hankel functions.

13.3.2 *The Hankel functions*

The Hankel functions are given by the two contours valid for arbitrary ν,

$$H_\nu^k(z) = \left(\frac{z}{2}\right)^\nu \frac{\Gamma(1/2-\nu)}{\pi i \Gamma(1/2)} \int_{C_k} e^{izs} (s^2-1)^{\nu-1/2} ds, \tag{13.22}$$

for $k = 1, 2$, with the contours C_1, C_2 shown in Fig. 13.2. These integrals are convergent for all $Re z > 0$ and are solutions to Bessel's equation by construction, and therefore must each be a linear combination of the solutions J_ν, N_ν. Rather than carry along undetermined normalization constants, we have chosen to use the standard normalization, and will verify that

$$\begin{aligned} H_\nu^1(z) &= J_\nu(z) + iN_\nu(z) \\ H_\nu^2(z) &= J_\nu(z) - iN_\nu(z). \end{aligned} \tag{13.23}$$

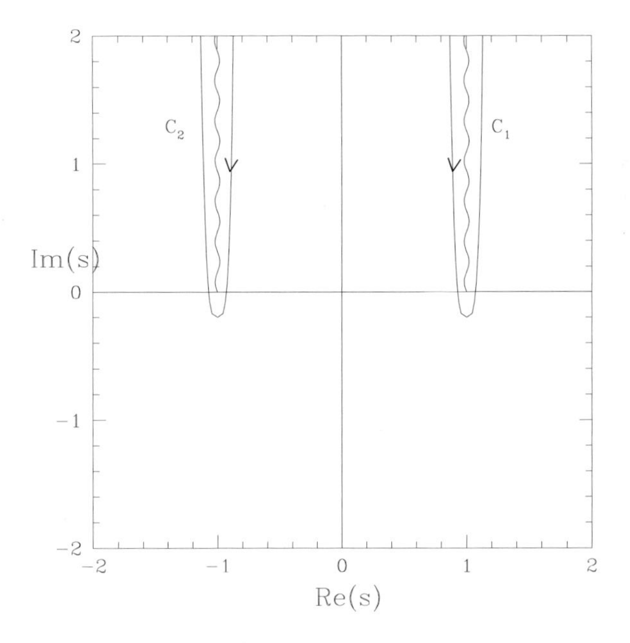

Fig. 13.2 Integration paths for the Hankel functions.

13.3.3 *Asymptotic limits*

Examine $H_\nu^1(z)$ for $z \to +\infty$. On the right side of the cut let $s = 1 + e^{\pi i/2}\rho$. Continuing to the left side we have $s = 1 + e^{-3\pi i/2}\rho$, giving

$$\begin{aligned} H_\nu^1(z) = \left(\frac{z}{2}\right)^\nu \frac{\Gamma(1/2-\nu)}{\pi i\Gamma(1/2)} e^{iz} \Bigg[&\int_0^\infty e^{-z\rho}(2e^{\pi i/2}\rho - \rho^2)^{\nu-1/2} i d\rho \\ &- \int_0^\infty e^{-z\rho}(2i\rho e^{-3\pi i/2} - \rho^2)^{\nu-1/2} i d\rho \Bigg]. \end{aligned} \tag{13.24}$$

For large z the dominant contribution to the integrals comes near $\rho = 0$ and we can drop the ρ^2 terms. Combining we find

$$H_\nu^1(z) \to -\left(\frac{z}{2}\right)^\nu \frac{\Gamma(\frac{1}{2}-\nu)}{\pi\Gamma(1/2)} e^{i(z-\frac{\pi\nu}{2}+\frac{\pi}{4})} 2sin(\pi\nu - \tfrac{\pi}{2}) \int_0^\infty e^{-z\rho}(2\rho)^{\nu-\frac{1}{2}} d\rho. \tag{13.25}$$

The integral reduces to a Gamma function, and making use of Eq. 8.51 we find

$$H_\nu^1(z) \to \sqrt{\frac{2}{\pi z}} e^{i(z-\pi\nu/2-\pi/4)}. \tag{13.26}$$

Then write $H_\nu^1(z) = \sqrt{\frac{2}{\pi z}} e^{i(z-\pi\nu/2-\pi/4)} W(z)$ and use Eq. 13.18, separating real and imaginary terms, giving the asymptotic series

$$H_\nu^1(z) \sim \sqrt{\frac{2}{\pi z}} e^{i(z-\pi\nu/2-\pi/4)} [\Sigma_1 - i\Sigma_2], \tag{13.27}$$

with

$$\Sigma_1 \sim \sum_n \frac{(-1)^n \Gamma(2n+\frac{1}{2}+\nu)\Gamma(2n+\frac{1}{2}-\nu)}{(2n)!(2z)^{2n}\Gamma(\frac{1}{2}+\nu)\Gamma(\frac{1}{2}-\nu)}$$
$$= 1 - \frac{(3^2-4\nu^2)(1-4\nu^2)}{2!2^6 z^2} + \frac{(7^2-\nu^2)(5^2-\nu^2)(3^2-4\nu^2)(1-4\nu^2)}{4!2^{12}z^4} \cdots, \tag{13.28}$$

and

$$\Sigma_2 \sim \sum_n \frac{(-1)^n \Gamma(2n+\frac{3}{2}+\nu)\Gamma(2n+\frac{3}{2}-\nu)}{(2n+1)!(2z)^{2n+1}\Gamma(\frac{1}{2}+\nu)\Gamma(\frac{1}{2}-\nu)}$$
$$= \frac{1-4\nu^2}{2^3 z} - \frac{(5^2-\nu^2)(3^2-\nu^2)(1-\nu^2)}{3!2^9 z^3} + \dots. \tag{13.29}$$

Similarly we find

$$H_\nu^2(z) \sim \sqrt{\frac{2}{\pi z}} e^{-i(z-\pi\nu/2-\pi/4)} [\Sigma_1 + i\Sigma_2]. \tag{13.30}$$

Examine $H_\nu^1(z)$ for small z. On the right side of the cut let $s = 1+e^{i\pi/2}w$ and continuing to the left side we have $s = 1 - e^{-i\pi/2}w$. Substituting into

Eq. 13.22 we find for the integral

$$\int_{C_1} = \int_0^\infty e^{iz(1+iw)}(-e^{-i\pi/2}w)^{\nu-1/2}(2 - e^{-i\pi/2}w)^{\nu-1/2}e^{-i\pi/2}dw$$
$$+ \int_0^\infty e^{iz(1+iw)}(e^{i\pi/2}w)^{\nu-1/2}(2 + e^{i\pi/2}w)^{\nu-1/2}e^{i\pi/2}dw. \quad (13.31)$$

Now let $u = zw$ and look only at the leading order for small z, giving

$$\int_{C_1} \to -\int_0^\infty e^{-u}e^{-i\pi(\nu-1/2)}(u/z)^{2\nu-1}idu/z$$
$$+ \int_0^\infty e^{-u}e^{i\pi(\nu-1/2)}(u/z)^{2\nu-1}idu/z, \quad (13.32)$$

or

$$\int_{C_1} \to -2sin[\pi(\nu - 1/2)]\Gamma(2\nu)/z^{2\nu}, \quad (13.33)$$

and using Eq. 8.55 and Eq. 8.51 we find

$$H_\nu^1(z) \to -\frac{\Gamma(\nu)}{\pi i}\left(\frac{z}{2}\right)^{-\nu}. \quad (13.34)$$

The full expansion near zero is then given using Eq. 13.5

$$H_\nu^1(z) \to -\frac{\Gamma(\nu)}{\pi i}\left(\frac{z}{2}\right)^{-\nu}\sum_{k=0}^\infty \frac{(-1)^k\Gamma(1-\nu)}{4^k k!\Gamma(k+1-\nu)}z^{2k} + cJ_\nu(z) \quad (13.35)$$

with the second J_ν term with unknown coefficient necessary because the leading asymptotic form does not determine this subdominant part. Similarly

$$H_\nu^2(z) \to \frac{\Gamma(\nu)}{\pi i}\left(\frac{z}{2}\right)^{-\nu}\sum_{k=0}^\infty \frac{(-1)^k\Gamma(1-\nu)}{4^k k!\Gamma(k+1-\nu)}z^{2k} + dJ_\nu(z). \quad (13.36)$$

Below we will find that $c = d = 1$.

13.3.4 *Relation to Bessel and Neumann functions*

To relate the Hankel and Bessel functions, first consider the sum of the two Hankel functions. To simplify the representation, deform the contour as shown in Fig. 13.3. For the vertical paths let $s = iu$. Then find for part a) and c) $s^2 - 1 = e^{-i\pi}(1 + u^2)$, part b) and d) $s^2 - 1 = e^{i\pi}(1 + u^2)$. The

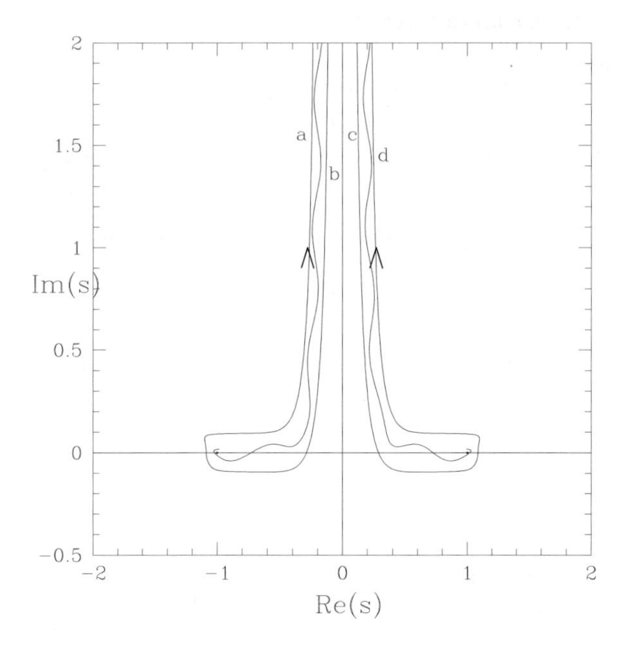

Fig. 13.3 Deformed integration paths for the Hankel functions.

contributions from the vertical paths are seen to cancel. Combining the horizontal path contributions we find for $\nu > -1/2$ they reduce to

$$\frac{H_\nu^1(z) + H_\nu^2(z)}{2} = \frac{(z/2)^\nu}{\Gamma(1/2)\Gamma(\nu+1/2)} \int_{-1}^{1} e^{izs}(1-s^2)^{\nu-1/2} ds. \quad (13.37)$$

To obtain the local expansion near $z = 0$, expand the exponential in Eq. 13.37, use the Euler beta function Eq. 8.63, and use Eq. 8.55, $(2k)! = 2^{2k}k!\Gamma(k+1/2)/\Gamma(1/2)$, giving the sum

$$\frac{H_\nu^1(z) + H_\nu^2(z)}{2} = \left(\frac{z}{2}\right)^\nu \sum_k \frac{(-1)^k z^{2k}}{2^{2k}k!\Gamma(\nu+k+1)}, \quad (13.38)$$

agreeing with Eq. 13.6, and thus the sum of the two Hankel functions is twice the Bessel function.

We also thus obtain, from Eqs. 13.27 and 13.30, for $z \to +\infty$,

$$J_\nu(z) \to \sqrt{\frac{2}{\pi z}}[cos(z - \tfrac{\pi\nu}{2} - \tfrac{\pi}{4})\Sigma_1 + sin(z - \tfrac{\pi\nu}{2} - \tfrac{\pi}{4})\Sigma_2]. \quad (13.39)$$

Now use Eq. 13.7 and substitute the leading asymptotic form of J_ν to find

$$N_\nu(z) \rightarrow \sqrt{\frac{2}{\pi z}} sin(z - \pi\nu/2 - \pi/4). \tag{13.40}$$

But $N_\nu(z) = aH^1_\nu(z) + bH^2_\nu(z)$ and using the asymptotic expressions for $H^1_\nu(z), H^2_\nu(z)$ we find

$$N_\nu(z) = \frac{H^1_\nu(z) - H^2_\nu(z)}{2i}, \tag{13.41}$$

completing the proof of Eq. 13.23, and giving the full asymptotic expansion

$$N_\nu(z) \rightarrow \sqrt{\frac{2}{\pi z}}[sin(z - \tfrac{\pi\nu}{2} - \tfrac{\pi}{4})\Sigma_1 - cos(z - \tfrac{\pi\nu}{2} - \tfrac{\pi}{4})\Sigma_2]. \tag{13.42}$$

Both J_ν and N_ν contain combinations of dominant and subdominant terms for z near the positive real axis. From the Stokes plot, Fig. 5.3, we know that continuation to $arg(z) = \pm\pi/2$ will contribute a new subdominant term, which will remain subdominant only up to $arg(z) = \pm\pi$, so these expansions for J_ν and N_ν are limited to $|arg(z)| < \pi$.

Finally, we obtain, from Eqs. 13.35 and 13.36, the local expansion for N_ν for small z and ν non integer,

$$N_\nu(z) \rightarrow \frac{\Gamma(\nu)}{\pi}\left(\frac{z}{2}\right)^{-\nu}\sum_{k=0}^{\infty}\frac{(-1)^k\Gamma(1-\nu)}{4^k k!\Gamma(k+1-\nu)}z^{2k}, \tag{13.43}$$

which can also be directly obtained from Eqs. 13.6 and 13.7.

Another useful integral representation for J_ν can be obtained by noting that in Eq. 13.37 $1 - s^2$ is even in s, and letting $s = cos\theta$ we find

$$J_\nu(z) = \left(\frac{z}{2}\right)^\nu \frac{1}{\Gamma(1/2)\Gamma(\nu+1/2)}\int_0^{\pi/2} cos(zcos\theta)sin^{2\nu}(\theta)d\theta, \tag{13.44}$$

valid for all $\nu > -1/2$. In particular

$$J_0(z) = \frac{1}{\pi}\int_0^{\pi/2} cos(zcos\theta)d\theta. \tag{13.45}$$

13.3.5 *The Wronskian*

Using Abel's formula, Eq. 2.10, we find that the Wronskian is proportional to $1/z$. Then using the expansions of J_ν and N_ν near $z = 0$ we find for the

Wronskian for the functions J_ν and N_ν

$$W_{JN}(z) = \frac{lim_{z\to 0}[z(J_\nu N_\nu' - J_\nu' N_\nu)]}{z} = -\frac{2}{\pi z}. \tag{13.46}$$

Similarly, use the relations between $J_\nu, N_\nu, H_\nu^1, H_\nu^2$ or the asymptotic expressions at large z to find the Wronskian for the functions $H_\nu^1(z), H_\nu^2(z)$

$$W_H(z) = \frac{lim_{z\to\infty}[z(H_\nu^1 H_\nu^{2'} - H_\nu^{1'} H_\nu^2)]}{z} = -\frac{4i}{\pi z}. \tag{13.47}$$

13.3.6 *Sommerfeld integral representation*

Any integral representation that gives a simple differential equation for $f(t)$ is useful. The Sommerfeld transformation is

$$y(z) = \int_C e^{zsinh(t)} f(t)dt. \tag{13.48}$$

Substitute and integrate by parts to eliminate the factor z using abbreviations of $s = sinh(t)$, $c = cosh(t)$, and using $\int ze^{zs} f dt = \int f \frac{1}{c}\frac{de^{zs}}{dt} dt$ we find

$$0 = \frac{zs^2 fe^{zs}}{c}|_a^b - \int_C z\frac{d}{dt}\left(\frac{s^2 f}{c}\right) e^{zs} dt - \int_C \frac{d}{dt}\left(\frac{sf}{c}\right) e^{zs} dt$$
$$+\frac{sfe^{zs}}{c}|_a^b + \frac{zfe^{zs}}{c}|_a^b - \int_C z\frac{d}{dt}\left(\frac{f}{c}\right) e^{zs} dt - \nu^2 \int_C fe^{zs} dt. \tag{13.49}$$

Combine using $s^2 + 1 = c^2$ giving

$$0 = \frac{zs^2 fe^{zs}}{c}|_a^b + \left(\frac{sf}{c}\right) e^{zs}|_a^b + \frac{zfe^{zs}}{c}|_a^b - \frac{1}{c}\frac{d}{dt}(cf)e^{zs}|_a^b, \tag{13.50}$$

$$0 = \frac{d}{dt}\frac{1}{c}\frac{d}{dt}(cf) - \frac{d}{dt}\left(\frac{sf}{c}\right) - \nu^2 f, \tag{13.51}$$

giving

$$f'' = \nu^2 f, \tag{13.52}$$

with solution $f = e^{\pm\nu t}$.

For the vanishing of the contribution from the end points it is required that $sinh(t) \to -\infty$ giving the solutions $t \to +\infty \pm i\pi$, or $t \to -\infty$. Three possible contours are shown in Fig. 13.4, giving solutions

$$y_k(z) = \frac{1}{\pi i}\int_{C_k} e^{zsinh(t)-\nu t} dt. \tag{13.53}$$

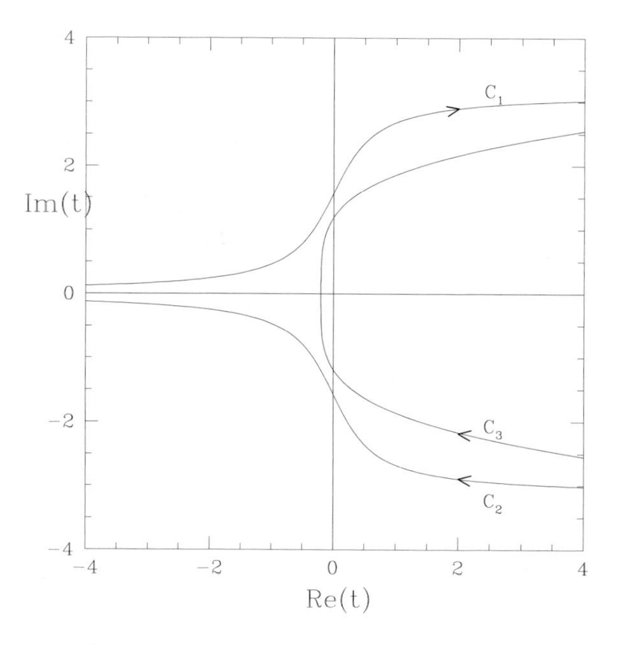

Fig. 13.4 Sommerfeld contours for the Bessel functions.

Of course $C_3 = C_1 + C_2$, so there are only two independent solutions. To identify these solutions evaluate the asymptotic limits for large z. Writing $\phi = z sinh(t) - \nu t$ we find saddle points located for large z at $z = \pm i\pi/2$. At the upper saddle $\phi'' = iz$ so the saddle is pointed towards $e^{i\pi/4}$ and $\phi_0 = iz - i\nu\pi/2$. The contour C_1 can be made to pass through this saddle giving

$$y_1(z) \simeq \sqrt{\frac{2\pi}{z}} e^{i(z-\pi\nu/2-\pi/4)}. \tag{13.54}$$

Similarly contour C_2 can be made to pass through the saddle at $-i\pi/2$ which points towards $e^{3i\pi/4}$, and we find

$$y_2(z) \simeq \sqrt{\frac{2\pi}{z}} e^{-i(z-\pi\nu/2-\pi/4)}, \tag{13.55}$$

and since the solutions are uniquely determined by their asymptotic behavior at large z we have for $k = 1, 2$

$$H_\nu^k(z) = \frac{1}{\pi i} \int_{C_k} e^{z sinh(t) - \nu t} dt. \tag{13.56}$$

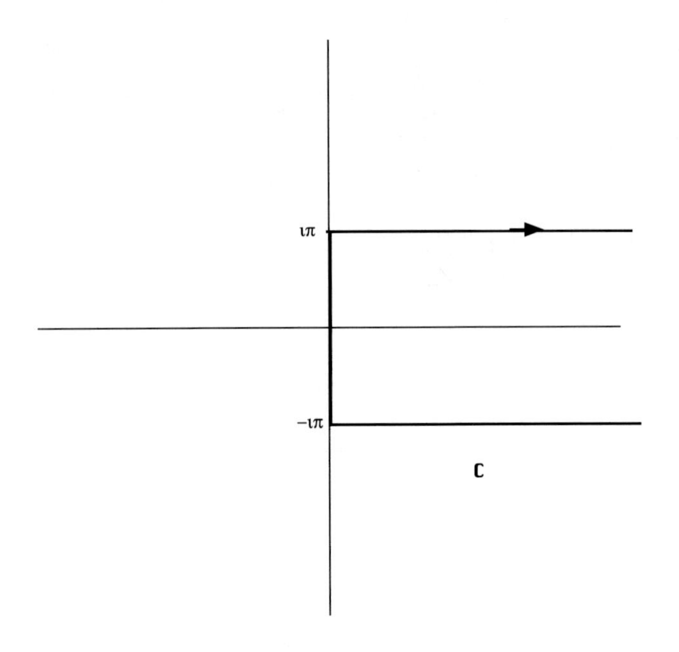

Fig. 13.5 Sommerfeld contour for the Bessel function for ν integer.

Now consider the representation of the Bessel function given by half the sum of the Hankel functions,

$$J_\nu(z) = \frac{1}{2\pi i} \int_{C_3} e^{z sinh(t) - \nu t} dt. \tag{13.57}$$

Take ν to be an integer. Deform the contour to horizontal and vertical paths passing through the points $-i\pi, i\pi$ as shown in Fig. 13.5. For ν integer the contributions along the two horizontal parts of the path exactly cancel. Changing variables using $t = is$ gives

$$J_n(z) = \frac{1}{2\pi} \int_{-\pi}^{\pi} e^{izsin(s) - ins} ds. \tag{13.58}$$

To find the asymptotic limit for small z from this representation, let $e^{is} = 2u/z$, giving $sin(s) = (2u/z - z/2u)/(2i)$, and thus

$$J_n(z) = \frac{1}{2\pi i} \left(\frac{z}{2}\right)^n \oint \frac{du}{u^{n+1}} e^{u - z^2/4u}, \tag{13.59}$$

and for $z \to 0$, using $e^u = \sum u^n/n!$, we find the usual limit for small z, $J_n(z) \to \left(\frac{z}{2}\right)^n \frac{1}{n!}$.

Now consider

$$\sum_{n=-\infty}^{\infty} J_n(z)e^{in\phi} = \frac{1}{2\pi}\int_{-\pi}^{\pi} e^{izsin(\theta)} \sum_{n} e^{in(\phi-\theta)} d\theta. \tag{13.60}$$

But

$$\frac{1}{2\pi}\sum_{n=-\infty}^{\infty} e^{in(\phi-\theta)} = \delta(\phi-\theta), \tag{13.61}$$

giving the important identity

$$e^{izsin\phi} = \sum_{n=-\infty}^{\infty} J_n(z)e^{in\phi}. \tag{13.62}$$

13.3.7 *Mellin integral representation*

The Mellin representation can be constructed either from the differential equation or directly from the local series expansion. Consider the convergent series representation for the Bessel function

$$J_\nu(z) = \left(\frac{z}{2}\right)^\nu \sum_0^\infty a_k \left(\frac{z}{2}\right)^{2k}, \tag{13.63}$$

with $a_k = (-1)^k/[k!\Gamma(\nu+k+1)]$, and reproduce this series as a contour integral of the form

$$J_\nu(z) = \frac{1}{2\pi i}\left(\frac{z}{2}\right)^\nu \int_{\gamma-i\infty}^{\gamma+i\infty} M(s)\left(\frac{z}{2}\right)^{-2s} ds, \tag{13.64}$$

with $0 < \gamma < 1$, so that the contour passes vertically upward to the right of $z = 0$ and encloses the entire left half plane. Then choose $M(s)$ to have first order poles at s equal to $0, -1, -2, -3\ldots$ with residue equal to a_s. From Eq. 8.5 the function $\Gamma(s)$ has first order poles at these points with residue $(-1)^s/\Gamma(1-s)$ and thus we take

$$M(s) = (-1)^s\Gamma(s)\Gamma(1-s)a_{-s} = \frac{\Gamma(s)}{\Gamma(1+\nu-s)}, \tag{13.65}$$

giving

$$J_\nu(z) = \frac{1}{2\pi i}\left(\frac{z}{2}\right)^\nu \int_{\gamma-i\infty}^{\gamma+i\infty} \frac{\Gamma(s)}{\Gamma(1+\nu-s)}\left(\frac{z}{2}\right)^{-2s} ds. \tag{13.66}$$

This integral representation also easily produces the asymptotic expression for large x. Write the integral as $\int e^{\phi} ds$ with

$$\phi = -2sln(z/2) + ln[\Gamma(s)] - ln[\Gamma(1+\nu-s)]. \tag{13.67}$$

Differentiating we find

$$\phi' = -2ln(z/2) + \frac{\Gamma'(s)}{\Gamma(s)} + \frac{\Gamma'(1+\nu-s)}{\Gamma(1+\nu-s)}. \tag{13.68}$$

Using the leading order Stirling approximation for the Gamma functions, Eq. 8.15, we find saddle points located at $s = 1/2 + \nu/2 \pm iz/2$, with the saddle in the lower half plane having a steepest descent direction of $\pm e^{i\pi/4}$ and that in the upper half plane of $\pm e^{-i\pi/4}$. Also $\phi \to -\infty$ for s tending to infinity along the real axis. Thus there is a contour connecting the end points and passing through the two saddle points. Adding the contributions from these two saddle points reproduces the leading order term of the asymptotic expression given by Eq. 13.39.

13.4 Generating Function

The integral representation given by Eq. 13.59 can be immediately used to find a generating function for the Bessel functions. To eliminate the factor of $(z/2)^n$ make the substitution $u = tz/2$ giving

$$J_n(z) = \frac{1}{2\pi i}\oint \frac{dt}{t^{n+1}} e^{\frac{z}{2}(t-1/t)}. \tag{13.69}$$

It then follows that

$$e^{\frac{z}{2}(t-1/t)} = \sum_n J_n(z) t^n. \tag{13.70}$$

Note that substituting $t = e^{i\phi}$ we recover Eq. 13.62.

13.5 Matching Local Solutions

Figure 13.6 shows a plot of the local expansions about $x = 0$ of $J_0(x)$, including terms up to $a_5 x^{10}$, $a_{10} x^{20}$, and $a_{15} x^{30}$, along with the large x asymptotic expression, including only the leading order term $\sqrt{\frac{2}{\pi x}} cos(x - \pi/4)$. From this it is obvious that the large x asymptotic expansion is much

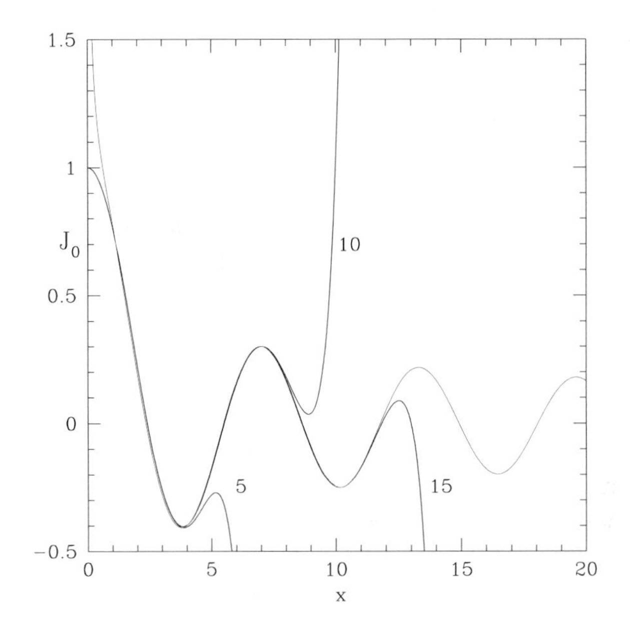

Fig. 13.6 Local approximations to $J_0(x)$.

better than the expansion about $x = 0$ for finding even the smallest zeros of the Bessel functions.

13.6 Imaginary Argument

The Bessel functions of imaginary argument are used sufficiently often that they have their own designation

$$I_\nu(z) = e^{-i\pi\nu/2} J_\nu(e^{i\pi/2} z), \qquad -\pi < arg(z) < \pi/2, \tag{13.71}$$

$$K_\nu(z) = \frac{\pi i}{2} e^{i\pi\nu/2} H^1_\nu(iz). \tag{13.72}$$

They satisfy the modified Bessel equation

$$y'' + \frac{y'}{z} - \left(1 + \frac{\nu^2}{z^2}\right) y = 0. \tag{13.73}$$

13.7 Gaussian–Bessel integrals

Sometimes using the series expansion of a function allows the exchange of integration and summation, providing a simple way to prove certain identities. Let

$$I = \int_0^\infty e^{-x^2} J_\nu(bx) x^{\nu+1} dx. \tag{13.74}$$

But substituting the series expansion gives

$$I = \int_0^\infty e^{-x^2} \sum \frac{(bx)^{2k+\nu}(-1)^k}{2^{2k+n} k! \Gamma(\nu+k+1)!} x^{\nu+1} dx. \tag{13.75}$$

Substitute $v = x^2$ and integrate, giving

$$I = (1/2) \sum \frac{(b/2)^{2k+\nu}(-1)^k}{k!} = \frac{b^\nu}{2^{\nu+1}} e^{-b^2/4}. \tag{13.76}$$

An important identity that appears in the derivation of a plasma dispersion relation[2] is

$$\int_0^\infty e^{-ax^2} J_n^2(bx) x dx = \frac{1}{2a} e^{-b^2/(2a)} I_n\left(\frac{b^2}{2a}\right), \tag{13.77}$$

where I_n is the modified Bessel function, described in the previous section.

Consider first the left-hand side. Expand the Bessel functions in power series giving

$$J_n^2(bx) = \left(\frac{bx}{2}\right)^{2n} \sum_{k=0}^\infty \sum_{s=0}^\infty \frac{(-1)^{k+s}(bx)^{2k+2s}}{2^{2k+2s} k! s! \Gamma(n+k+1)\Gamma(n+s+1)}. \tag{13.78}$$

Now let $m = k + s$ giving

$$J_n^2(bx) = \left(\frac{bx}{2}\right)^{2n} \sum_{m=0}^\infty \frac{(-1)^m (bx/2)^{2m} \Gamma(2n+2m+1)}{m! \Gamma(n+m+1)^2 \Gamma(2n+m+1)}, \tag{13.79}$$

where one sum has been carried out by the use of an identity given first in 1303 by the Chinese mathematician Zhu Shijie but known as the Vander-

[2]M.N. Rosenbluth, "Topics in Microinstabilities", Eq. 2.12, in Proceedings of the International School of Physics Enrico Fermi, Course XXV, *Theory of Plasma,* Academic Press NY, 1962.

monde identity[3]

$$\sum_{k=0}^{m}\frac{(m+n)!(m+n)!}{k!(m+n-k)!(n+k)!(m-k)!}=\frac{(2m+2n)!}{m!(m+2n)!}. \tag{13.80}$$

Now the integration over x can be carried out, being

$$\int_0^\infty e^{-ax^2}x^{2m+2n+1}dx=\frac{1}{2a^{m+n+1}}\Gamma(m+n+1),$$

so

$$\int_0^\infty e^{-ax^2}J_n^2(bx)xdx=\frac{1}{2a}\sum_{m=0}^{\infty}\frac{(-1)^m w^{m+n}\Gamma(2n+2m+1)}{2^{m+n}m!\Gamma(n+m+1)\Gamma(2n+m+1)} \tag{13.81}$$

with $w=b^2/2a$, and using Eq 8.55

$$\int_0^\infty e^{-ax^2}J_n^2(bx)xdx=\frac{1}{2a}\sum_{m=0}^{\infty}\frac{(-1)^m w^{m+n}2^{n+m}\Gamma(n+m+1/2)}{\sqrt{\pi}m!\Gamma(2n+m+1)}. \tag{13.82}$$

Now find a series for the right-hand side of Eq. 13.77. Write $I_n(w)=e^w f(w)$ with $w=b^2/2a$ and substitute into Eq. 13.73 to find the differential equation for $f(w)$

$$f''+(2+1/w)f'+(1/w-n^2/w^2)f=0, \tag{13.83}$$

so the series expansion of $f(w)=\sum_m a_m w^{n+m}$ gives

$$a_m=-\frac{2(n+m-1/2)}{m(m+2n)}a_{m-1} \tag{13.84}$$

with solution

$$a_m=\frac{(-1)^m 2^m\Gamma(n+m+1/2)(2n)!}{m!(m+2n)!\Gamma(n+1/2)}a_0. \tag{13.85}$$

Now use the fact that for small w we have $f(w)\simeq(w/2)^n/\Gamma(n+1)$ so

$$a_0=\frac{1}{2^n\Gamma(n+1)}=\frac{2^n\Gamma(n+1/2)}{\sqrt{\pi}(2n)!} \tag{13.86}$$

---

[3]In general we have $(1+x)^{a+b}=\sum_{m=0}^{a+b}\binom{a+b}{m}x^m$. But also write this as $(1+x)^a(1+x)^b=\sum_{k=0}^{a}\binom{a}{k}x^k\sum_{i=0}^{b}\binom{b}{i}x^i$ and let $i+k=m$ giving $\sum_{m=0}^{a+b}\sum_{k=0}^{m}\binom{a}{k}\binom{b}{m-k}x^m$. Thus $\binom{a+b}{m}=\sum_{k=0}^{m}\binom{a}{k}\binom{b}{m-k}$. To obtain the desired result let $a=b=m+n$.

giving

$$f(w) = \sum_{m=0}^{\infty} \frac{(-1)^m 2^{m+n} w^{m+n} \Gamma(n+m+1/2)}{\sqrt{\pi} m! \Gamma(2n+m+1)} \tag{13.87}$$

and the identity is proven. Note that in addition we have found that $2a$ times the left-hand side of Eq. 13.77 satisfies the differential equation Eq. 13.83.

13.8 Problems

1. To use dominant balance to find roots of a polynomial of the form $\sum c_{p,q}\epsilon^q x^p$ one must be careful that the coefficients $c_{p,q}$ are of order 1 with respect to ϵ. Show that the expansion for $J_0(x)$ can be written as

$$J_0(x) \simeq 1 - \epsilon^2 x^2 + \epsilon^6 x^4 - \frac{8}{9}\epsilon^{11}x^6 + \frac{8}{9}\epsilon^{17}x^8 - \ldots,$$

with $\epsilon = 1/2$. Use a Kruskal–Newton graph to find dominant balance and iterate to find the first zero. Compare with numerical tables. Now use the asymptotic expression for large x, writing

$$J_0(x) \simeq \sqrt{\frac{2}{\pi x}}[cos(z - \tfrac{\pi}{4})\Sigma_1 + sin(z - \tfrac{\pi}{4})\Sigma_2],$$

using the lowest order expressions for Σ_1, Σ_2 and compare the accuracy of a determination of the first zero with that using the local expansion about $x = 0$.

2. Many identities can be directly obtained from an integral representation. To prove Weber's integral

$$\int_0^\infty \frac{J_\nu(t)dt}{t^{\nu-\mu+1}} = \frac{\Gamma(\mu/2)}{2^{\nu-\mu+1}\Gamma(\nu - \mu/2 + 1)},$$

for $0 < Re(\mu) < Re(\nu) + 1/2$, substitute the representation Eq. 13.37, first integrate over t, and then over s. (This method produces restrictions on μ, ν which can be avoided by using the contours for the Hankel functions.)

3. Other integral representations for the Bessel functions can be obtained by using different kernels. Try

$$y(z) = z^\nu \int e^{-\frac{z^2}{4t}} f(t)dt.$$

Find $f(t)$ and a contour giving a solution to Bessel's equation.

4. Using the representation of problem 3, write

$$0 = \int \frac{d}{dt}[te^{-\frac{z^2}{4t}} f(t)]dt.$$

Substitute the function $f(t)$ and carry out the differentiation, giving a sum of three terms. Express each term as a Bessel function, giving an identity relating three Bessel functions.

5. We derived the integral representation for the Bessel function J_ν

$$J_\nu(x) = \frac{1}{2\pi i}\int_C e^{x sinh(t) - \nu t},$$

with the contour $C = C_3$ as shown in Fig. 13.4. Find the asymptotic behavior of $J_x(x)$ for large x. This is not a typo! The subscript is x, and it matters.

6. Use the integral representation

$$J_\nu(x) = \frac{(x/2)^\nu}{\sqrt{\pi}\Gamma(\nu + 1/2)}\int_0^\pi cos(x cos\theta) sin^{2\nu}\theta d\theta,$$

valid for $\nu > -1/2$ to show that the Bessel function $J_\nu(x)$ satisfies $J_\nu(x) \sim (x/2)^\nu/\Gamma(\nu+1)$ for $\nu \to \infty$.

7. Construct the Mellin representation of the Bessel function, Eq. 13.66, beginning with the differential equation.

8. We derived the integral representation for the Bessel function J_n

$$J_n(z) = \frac{1}{2\pi i}\left(\frac{z}{2}\right)^n \oint \frac{du}{u^{n+1}} e^{u - z^2/4u}.$$

Divide this equation by z^n and differentiate with respect to z^2, obtaining the recursion relation

$$\frac{dJ_n(z)}{dz} = \frac{n}{z}J_n(z) - J_{n+1}(z),$$

and in particular

$$\frac{dJ_0(z)}{dz} = -J_1(z).$$

9. Use Eqs. 13.27 and 13.30 to find

$$H^1_{1/2} = -i\sqrt{\frac{2}{\pi z}}e^{iz}$$

$$H^2_{1/2} = i\sqrt{\frac{2}{\pi z}}e^{-iz}.$$

10. Use Eq. 13.62 and the identity $e^{i(a+b)sin\phi} = e^{iasin\phi}e^{ibsin\phi}$ to find Neumann's addition theorem:

$$J_n(a+b) = \sum_{m=-\infty}^{\infty} J_m(a)J_{n-m}(b).$$

11. Show that

$$J_{1/2}(z) = \left(\frac{2}{\pi z}\right)^{1/2} sin(z),$$

$$J_{3/2}(z) = \left(\frac{2}{\pi z}\right)^{1/2} \left(\frac{sin(z)}{z} - cos(z)\right).$$

12. Verify directly that $2a$ times the left-hand side of Eq. 13.77 satisfies Eq. 13.83.

Chapter 14

Weber and Hermite

Wilhelm Weber was born in 1804 in Wittenberg, Saxony, now Germany. He studied and taught at the University of Halle. In 1831 he was appointed chair at Göttingen where he collaborated with Gauss. While there he developed sensitive magnetometers and other magnetic instruments. One of his important works was the *Atlas des Erdmagnetismus*, a series of magnetic maps, which was responsible for the institution of magnetic observatories. Gauss and Weber constructed the first electromagnetic telegraph in 1833, connecting the observatory with the institute for physics in Göttingen. In 1837, when Victoria became Queen of Britain, Weber was one of seven professors to protest the revocation of the liberal constitution in Hanover by her uncle, and was dismissed. He remained at Göttingen without a position until he became professor at Leipzig in 1843. In 1855 he and Dirichlet became directors of the observatory in Göttingen. He worked on electrodynamics and the electrical structure of matter, and his work on the electrostatic and electrodynamic units of charge was very important for Maxwell's formulation of electromagnetic theory.

Charles Hermite was born in Dieuze in Lorraine, France in 1822. He studied at the École Polytechnique but after one year was refused the right to continue because of a physical disability. Nevertheless he continued his studies and research and eventually was a professor there. He made contributions to number theory, algebra, the theory of orthogonal polynomials, and of elliptic functions. He found the roots to a fifth order polynomial in terms of elliptic integrals. In 1873 he proved that e is transcendental, and the same method was used by Lindemann in 1882 to prove that π is transcendental. He is well known to physicists for Hermitian matrices and the Hermite polynomials. Poincaré was a student of Hermite, who remarked that he did not proceed in a logical fashion, but that methods were born in his mind in some mysterious way.

Solutions to the Weber–Hermite equation

$$y'' + \left(p + \frac{1}{2} - \frac{z^2}{4}\right) y = 0 \tag{14.1}$$

are generally known as the parabolic cylinder functions. First note that if $y(z)$ is a solution, then so is $y(-z)$, and if $y_p(z)$ is a solution, then so is $y_{-p-1}(iz)$. These relations can be used to construct additional solutions.

14.1 Local Analysis at Infinity

Infinity is an irregular singular point. Writing $y \sim e^S$ we find

$$S'' + (S')^2 + \left(p + \frac{1}{2} - \frac{z^2}{4}\right) = 0, \tag{14.2}$$

with the solution $S' = \pm z/2$. Going one step further by writing $S' = \pm z/2 + h'$ we find two possible asymptotic forms, leading to the two functions

$$D_p(z) \sim z^p e^{-z^2/4}, \qquad D_{-p-1}(z) \sim (iz)^{-p-1} e^{z^2/4}, \tag{14.3}$$

related by the substitution $p \to -p-1$, $z \to iz$. From this analysis we do not yet know how solutions at $+\infty$ relate to solutions at $-\infty$, in general a solution which tends to zero in one direction will become infinite in the other direction. Take the first solution

$$D_p(z) = z^p e^{-z^2/4} W(z), \tag{14.4}$$

assumed to have this subdominant behavior at plus infinity, and find the differential equation for $W(z)$

$$z^2 W'' + (2pz - z^3) W' + p(p-1) W = 0. \tag{14.5}$$

We are interested only in solutions which begin with $W(0) = 1$ so we assume a series of the form $W = \sum a_n z^{-n}$ and find the conditions

$$a_1 = 0, \qquad (n+2) a_{n+2} + (n-p)(n-p+1) a_n = 0, \tag{14.6}$$

with solution the asymptotic series

$$D_p(z) \sim z^p e^{-z^2/4} \sum_0^\infty \frac{(-1)^k \Gamma(p+1) z^{-2k}}{2^k k! \Gamma(p-2k+1)}. \tag{14.7}$$

This series truncates, giving an exact solution, if p is a non-negative integer. In this case the expression is valid for all z and also remains subdominant for both positive and negative z, otherwise it is an asymptotic series valid only for positive z.

The second solution, given by the substitution $p \to -p-1$, $z \to iz$, and using Eq. 8.51, has the asymptotic form (we choose normalization to make the function real) for large positive z of

$$D_{-p-1}(z) \sim (z)^{-p-1} e^{z^2/4} \sum_0^\infty \frac{(-1)^k \Gamma(p+2k+1) z^{-2k}}{2^k k! \Gamma(p+1)}. \tag{14.8}$$

Unlike the series for $D_p(z)$, this series does not truncate for integer p. Also note that it is identically zero for p a negative integer.

14.2 Local Analysis at Zero

Zero is an analytic point. Instead of finding a series for $y(z)$, which gives a three term recursion relation, it is convenient to extract the behavior at large z, choosing $y(z) = e^{-z^2/4} f(z)$. This gives the simpler differential equation for $f(z)$

$$f'' - zf' + pf = 0. \tag{14.9}$$

Writing $f = \sum_k a_k z^k$ we find

$$a_{k+2} = -\frac{p-k}{(k+2)(k+1)} a_k. \tag{14.10}$$

We then have two Taylor series

$$f_0(z) = \sum_n \frac{2^n \Gamma(n - p/2)}{(2n)! \Gamma(-p/2)} z^{2n},$$
$$f_1(z) = \sum_n \frac{2^n \Gamma(n + 1/2 - p/2)}{(2n+1)! \Gamma(1/2 - p/2)} z^{2n+1}. \tag{14.11}$$

Both series truncate if p is a non-negative integer, and in fact one of them is identically zero. Note that $f \sim e^{z^2/2}, const$ for $z \to \infty$ so that extracting the asymptotic behavior $e^{-z^2/4}$ is of no help in relating behavior at infinity to that near zero. Both the normalization and the combination of these functions for the standard solution $D_p(z)$ is chosen with regard to the behavior at infinity, so at this point we cannot decide on the correct expansion near zero.

The WKB analysis of this equation was carried out in Chapter 12.

14.3 Euler Integral Representation

Extract the asymptotic form $y = e^{-z^2/4}f$, giving $f'' - zf' + pf = 0$, and attempt an Euler representation,

$$f(z) = \int_C (z-t)^\mu h(t)dt. \tag{14.12}$$

Substitute and use $d^2/dz^2(z-t)^\mu = -\mu d/dt(z-t)^{\mu-1}$ and integrate by parts, giving

$$(z-t)^{\mu-1}h|_a^b + \int_C dt(z-t)^{\mu-1}[\mu h'(t) - z\mu h + zph - pth] = 0, \tag{14.13}$$

and note that terms in z in the brackets in the integrand cancel if $\mu = p$, so that the equation for $h(t)$ reduces to

$$h' = th, \qquad h(z-t)^{p-1}|_a^b = 0, \tag{14.14}$$

giving $h = e^{t^2/2}$. Change integration variable $t \to -it$ giving

$$y(z) = e^{-z^2/4}\int_C (z+it)^p e^{-t^2/2}dt, \tag{14.15}$$

so that $a = -\infty, b = \infty$ gives a solution. To identify it, look at the limit $z \to \infty$. For large positive z there is a horizontally oriented saddle point at $t = ip/z$, a vertically oriented saddle at $t = iz - ip/z$, and a cut originating at $t = iz$ if p is non integer, as shown in Fig. 14.1.

For large positive z evaluation of the contribution from the saddle point gives

$$y(z) = e^{-z^2/4}z^p\sqrt{2\pi}, \tag{14.16}$$

and thus using the standard normalization $D_p(z) \to z^p e^{-z^2/4}$,

$$D_p(z) = \frac{e^{-z^2/4}}{\sqrt{2\pi}}\int_{-\infty}^{\infty}(z+it)^p e^{-t^2/2}dt. \tag{14.17}$$

Notice that we immediately find an integral representation for the Hermite polynomials defined through $D_n(z) = 2^{-n/2}e^{-z^2/4}H_n(z/\sqrt{2})$,

$$H_n(z) = \frac{\sqrt{2^n}}{\sqrt{\pi}}\int_{-\infty}^{\infty}(z+it)^n e^{-t^2}dt. \tag{14.18}$$

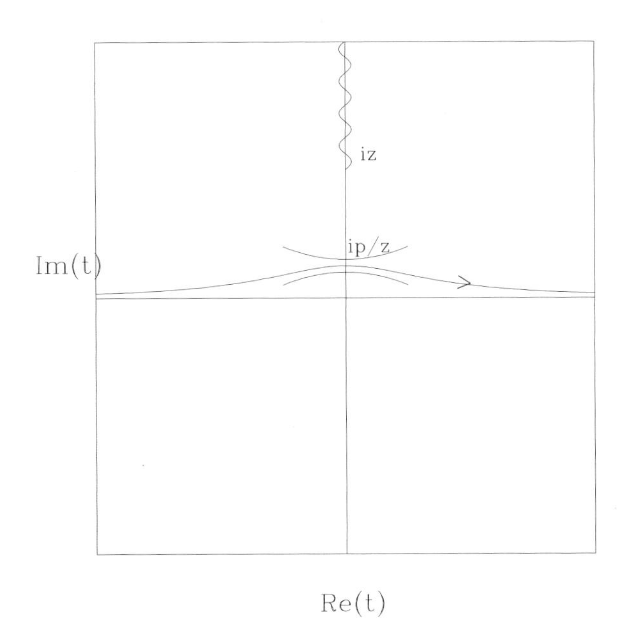

Fig. 14.1 Saddle point for large positive z evaluation using the Euler integral representation.

For large negative z, if p is not an integer, the integration path is as shown in Fig. 14.2. Now there are contributions from both saddle points, and the lower saddle gives an exponentially increasing solution. If p is an integer there is no cut, and the integration contour is not deformed so as to pass through the lower saddle, giving only the exponentially decreasing solution.

14.4 Fourier–Laplace Integral Representations

The direct use of a Fourier–Laplace transformation produces no simplification since factors of z^2 require integration by parts twice and hence lead again to a second order differential equation. Factor out the leading asymptotic behavior at $z \to \infty$, given by $y \to e^{\pm z^2/4}$. Write $y = e^{-z^2/4} f$, and substitute, giving $f'' - zf' + pf = 0$. Now use a Fourier–Laplace representation for f

$$f(z) = \int_C e^{-zt} g(t) dt. \tag{14.19}$$

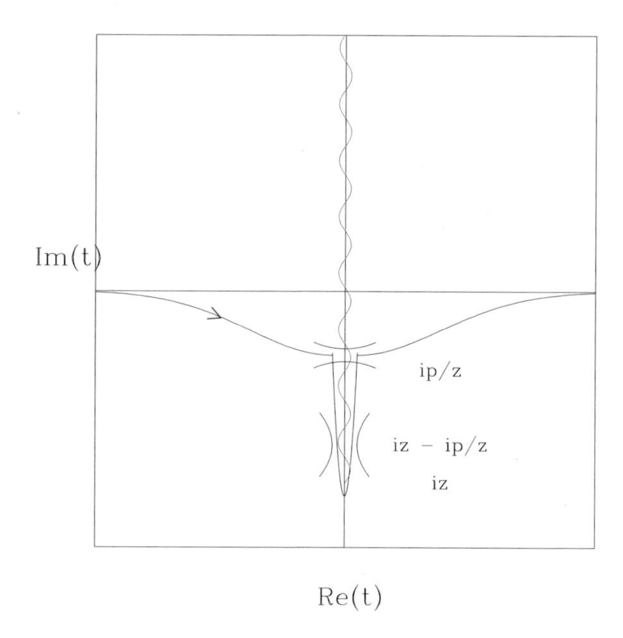

Fig. 14.2 Saddle point for large negative z evaluation using the Euler integral representation.

Substitute and integrate by parts, giving

$$t^2 g + (tg)' + pg = 0, \qquad tge^{-tz}|_a^b = 0, \tag{14.20}$$

with solution $g = e^{-t^2/2}t^{-p-1}$, with the integration path shown in Fig. 14.3. Thus a solution is

$$y(z) = e^{-z^2/4} \int_C e^{-zt-t^2/2}(-t)^{-p-1}dt. \tag{14.21}$$

This function is an analytic function of p. To evaluate it, first take $Re(p) < 0$. Then the integration path can be moved to the real axis and the integral becomes

$$y(z) = 2isin(p\pi)e^{-z^2/4} \int_0^{\infty} e^{-zt-t^2/2}(-t)^{-p-1}dt. \tag{14.22}$$

The asymptotic behavior for large positive z comes from the end point $t = 0$. Setting $u = zt$ we find

$$y(z) \to 2isin(p\pi)\Gamma(-p)z^p e^{-z^2/4}, \tag{14.23}$$

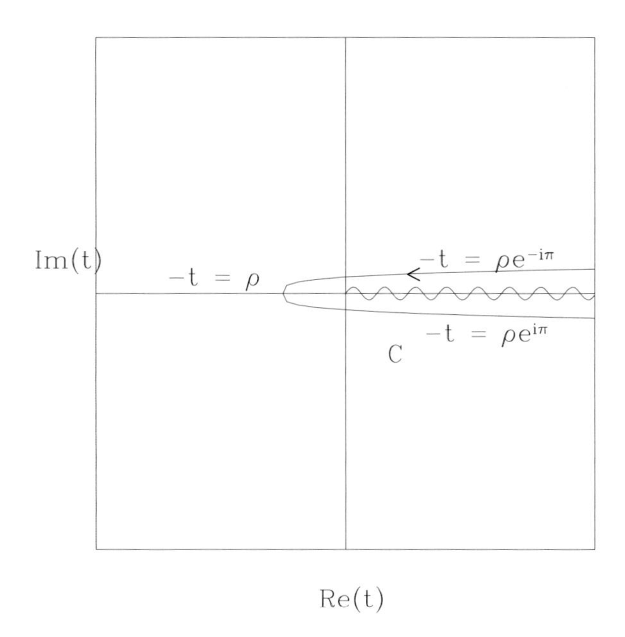

Fig. 14.3 Integration path for a Fourier–Laplace representation of the parabolic cylinder function.

and thus this contour gives a solution well-behaved at $+\infty$, denoted by $D_p(z)$. The standard normalization for the parabolic cylinder function is that $D_p(z) \to z^p e^{-z^2/4}$ for $z \to +\infty$, giving the integral representation, using the identity Eq. 8.51

$$D_p(z) = -\frac{\Gamma(1+p)}{2\pi i} e^{-z^2/4} \int_C e^{-zt-t^2/2}(-t)^{-p-1}dt. \tag{14.24}$$

Now we can find the behavior near $z = 0$ by expanding e^{-zt} in the integrand and integrating term by term, giving for $p < 0$

$$\begin{aligned} D_p(z) = &-\frac{e^{-z^2/4}}{\Gamma(-p)}[2^{-p/2-1}\Gamma(-p/2) \\ &+z2^{-p/2-1/2}\Gamma(1/2-p/2)+\ldots] \end{aligned} \tag{14.25}$$

and using Eq. 8.55 we identify the expansion near $z = 0$ to be $D_p(z) = e^{-z^2/4}[af_0 + bf_1]$, with $a = 2^{-p/2-1}\Gamma(-p/2)\Gamma(-p)$, $b = 2^{(p+1)/2}\Gamma(1/2 -$

$p/2)/\Gamma(-p)$, and f_0, f_1 the Taylor series given in 14.11, giving

$$D_p(z) = e^{-z^2/4} \sum_n \frac{2^{n-1-p/2}\Gamma(n-p/2)}{(2n)!\Gamma(-p)} z^{2n}$$

$$+e^{-z^2/4} \sum_n \frac{2^{n-(p+1)/2}\Gamma(n+1/2-p/2)}{(2n+1)!\Gamma(-p)} z^{2n+1}. \qquad (14.26)$$

Note that this expression is analytic in p, and thus valid for all p, requiring special care for p a positive integer. Using Eq. 8.55 to eliminate $\Gamma(-p)$ we find

$$D_p(z) = \sqrt{\pi} e^{-z^2/4} \sum_n \frac{2^{n+p/2}\Gamma(n-p/2)}{(2n)!\Gamma(-p/2)\Gamma(-p/2+1/2)} z^{2n}$$

$$+\sqrt{\pi} e^{-z^2/4} \sum_n \frac{2^{n+(p+1)/2}\Gamma(n+1/2-p/2)}{(2n+1)!\Gamma(-p/2)\Gamma(-p/2+1/2)} z^{2n+1}. \qquad (14.27)$$

For $p = n$, integer the integral has the form

$$D_n(z) = -\frac{n! e^{-z^2/4}}{2\pi i} \int_C \frac{e^{-zt-t^2/2}}{t^{n+1}} dt, \qquad (14.28)$$

and writing $t = v - z$ we have

$$D_n(z) = (-1)^n \frac{n!}{2\pi i} e^{-z^2/4} \int_C \frac{e^{-v^2/2}}{(v-z)^{n+1}} dv, \qquad (14.29)$$

where the contour now circles z in a positive sense, giving

$$D_n(z) = (-1)^n e^{-z^2/4} \frac{d^n}{dz^n} e^{-z^2/2}. \qquad (14.30)$$

Defining the Hermite polynomials

$$H_n(z) = (-1)^n e^{z^2} \frac{d^n}{dz^n} e^{-z^2}, \qquad (14.31)$$

we have

$$D_n(z) = (2)^{-n/2} e^{-z^2/4} H_n(z/\sqrt{2}), \qquad (14.32)$$

giving immediately also an integral representation for the Hermite polynomials

$$H_n(z) = (-1)^n \frac{n!}{2\pi i} \int_C \frac{e^{-v^2}}{(v-z)^{n+1}} dv. \qquad (14.33)$$

The second solution is then obtained by the substitution $p \to -p-1$ and $z \to iz$ giving

$$D_{-p-1}(z) = -\frac{\Gamma(-p)}{2\pi i} e^{z^2/4} \int_C e^{-izt-t^2/2} (-t)^p dt. \qquad (14.34)$$

For large z, writing $(-t)^p = \rho^p e^{i\pi p}$ this becomes

$$D_{-p-1}(z) \simeq -\frac{\Gamma(-p)}{2\pi i} e^{z^2/4} 2i sin(p\pi) \int_0^\infty e^{-iz\rho-\rho^2/2} (\rho)^p d\rho. \quad (14.35)$$

The dominant contribution comes from the end point giving for $p > -1$

$$D_{-p-1}(z) \simeq -\frac{\Gamma(-p)}{\pi} e^{z^2/4} sin(p\pi) \int_0^\infty e^{-u} \frac{(u)^p}{(iz)^{p+1}} du, \qquad (14.36)$$

and finally

$$D_{-p-1}(z) \simeq e^{z^2/4} (iz)^{-p-1}. \qquad (14.37)$$

A second Fourier–Laplace integral representation can be found by extracting the growing asymptotic form, $y = e^{z^2/4} f$. This analysis was carried out in Chapter 12 giving the integral representation

$$D_p(z) = \frac{2^{p+1/2}}{\sqrt{\pi}} e^{-i\pi p/2} e^{z^2/4} \int_{-\infty}^{\infty} e^{2izt-2t^2} t^p dt, \qquad (14.38)$$

with asymptotic form for $z \to -\infty$

$$D_p(z) \simeq z^p e^{-z^2/4} - \frac{\sqrt{2\pi} e^{z^2/4} e^{i\pi p} z^{-p-1}}{\Gamma(-p)}. \qquad (14.39)$$

Note that the second term vanishes if p is a positive integer.

Again the second solution is obtained by the substitution $p \to -p-1$ and $z \to iz$ giving

$$D_{-p-1}(z) = \frac{2^{-p-1/2}}{\sqrt{\pi}} e^{-i\pi p/2} e^{-z^2/4} \int_{-\infty}^{\infty} e^{-2zt-2t^2} t^{-p-1} dt. \quad (14.40)$$

14.5 Orthogonality

The $D_m(x)$ form an orthogonal set. To prove this use the differential equation to obtain

$$D_n(z) D_m''(z) - D_m(z) D_n''(z) + (m-n) D_n(z) D_m(z) = 0. \quad (14.41)$$

Integrating we have

$$(m-n)\int_{-\infty}^{\infty} D_m(z)D_n(z)dz = [D_n(z)D'_m(z) - D_m(z)D'_n(z)]_{-\infty}^{\infty} = 0, \tag{14.42}$$

by the large z expansion for D_n valid for all integer n. Thus when $m \neq n$

$$\int_{-\infty}^{\infty} D_m(z)D_n(z)dz = 0. \tag{14.43}$$

But when $n = m$

$$\begin{aligned}(n+1)\int_{-\infty}^{\infty} D_n^2(z)dz &= \int_{-\infty}^{\infty} D_n(z)[D'_{n+1}(z) + \frac{z}{2}D_{n+1}(z)]dz = \\ D_n(z)D_{n+1}(z)|_{-\infty}^{\infty} &+ \int_{-\infty}^{\infty} [\frac{z}{2}D_n(z)D_{n+1}(z) - D_{n+1}(z)D'_n]dz = \\ &\int_{-\infty}^{\infty} D_{n+1}^2(z)dz,\end{aligned} \tag{14.44}$$

by using the recurrence relation, integrating and using the recurrence relation again. Then from induction

$$\int_{-\infty}^{\infty} D_n^2(z)dz = n!\int_{-\infty}^{\infty} D_0^2(z)dz = \sqrt{2\pi}n!, \tag{14.45}$$

and so if a function $f(z)$ is expandable in terms of parabolic cylinder functions,

$$f(z) = \sum a_n D_n(z), \tag{14.46}$$

then

$$a_n = \frac{1}{\sqrt{2\pi}n!}\int_{-\infty}^{\infty} D_n(z)f(z)dz. \tag{14.47}$$

14.6 The Wronskian

We evaluate the Wronskian for the first Fourier–Laplace integral representation, Eq. 14.24. The Wronskian is independent of z, so evaluate it at $z = 0$,

$$D_p(0) = \frac{1}{\Gamma(-p)}\int_0^{\infty} e^{-t^2/2}t^{-p-1}dt. \tag{14.48}$$

Change variables to $w = t^2/2$ giving

$$D_p(0) = \frac{2^{-p/2-1}\Gamma(-p/2)}{\Gamma(-p)}. \tag{14.49}$$

Similarly we find

$$D_p'(0) = -i\frac{2^{-p/2-1/2}\Gamma(1/2 - p/2)}{\Gamma(-p)}, \tag{14.50}$$

and

$$D_{-p-1}(0) = \frac{2^{p/2-1/2}\Gamma(1/2 + p/2)}{\Gamma(1+p)}. \tag{14.51}$$

In like manner

$$D_{-p-1}'(0) = -i\frac{2^{p/2}\Gamma(1 + p/2)}{\Gamma(1+p)}, \tag{14.52}$$

giving

$$W = \frac{i}{2}\frac{\Gamma(1+p/2)\Gamma(-p/2) - \Gamma(1/2-p/2)\Gamma(1/2+p/2)}{\Gamma(1+p)\Gamma(-p)}, \tag{14.53}$$

which simplifies to

$$W = i[cos(\pi p/2) - sin(\pi p/2)]. \tag{14.54}$$

14.7 Mellin Integral Representation

We look for a Mellin representation for the function $D_p(z)$, which has the asymptotic form $D_p \simeq z^p e^{-z^2/4}$ for large positive z. Begin with the differential equation

$$y'' + \left(p + \frac{1}{2} - \frac{z^2}{4}\right) y = 0, \tag{14.55}$$

and extract the exponential behavior at infinity, writing $y = e^{-z^2/4} f$, giving

$$f'' - zf' + pf = 0. \tag{14.56}$$

Now assume a Mellin integral representation, writing

$$f = \frac{1}{2\pi i}\int_{\gamma-i\infty}^{\gamma+i\infty} M(s) z^{-s} ds, \tag{14.57}$$

with $0 < \gamma < 1$. Substituting we find a two term relation

$$s(s+1)M(s) + (p+s+2)M(s+2) = 0. \tag{14.58}$$

Choose the solution

$$M(s) = N(p)\Gamma(s)2^{-s/2}\Gamma(-s/2 - p/2), \tag{14.59}$$

with $N(p)$ a normalization constant. This gives for large u with $s = t + iu$ and $z = re^{i\phi}$ the integrand

$$\Gamma(s)2^{-s/2}\Gamma(-s/2 - p/2)z^{-s} \sim e^{\pm\phi u - 3\pi|u|/4}, \tag{14.60}$$

giving a convergent integral for $|\phi| < 3\pi/4$. This is consistent with the Stokes analysis. The function D_p is subdominant for z positive real. In Fig. 12.19, continuing to domain 2 the solution becomes dominant. Passing into domain 3 a new subdominant term is acquired, and this term then becomes dominant at the boundary between domain 4 and 5. The same sequence occurs in the lower half plane, giving the limit $|\phi| < 3\pi/4$.

The integrand has a singularity at $s = -p$, giving $f \sim z^p$ to leading order. Writing $-s/2 - p/2 = -k + \epsilon$, using $\Gamma(-k+\epsilon) = (-1)^k/(\epsilon k!)$ and changing variables to $t = s/2$ we determine the normalization constant and find the Mellin representation

$$D_p(z) \simeq \frac{e^{-z^2/4}}{2\pi i\Gamma(-p)2^{p/2}} \int_{\gamma-i\infty}^{\gamma+i\infty} \Gamma(2t)\Gamma(-t-p/2)(2z^2)^{-t}dt. \tag{14.61}$$

From this representation it is straightforward to write the contributions from all the poles at negative t arising from $\Gamma(2t)$ and obtain again the expansion for small z given by Eq. 14.27. But the integrand also has poles at $t = n - p/2$, with $n = 0, 1, 2, \ldots$, so moving the integration contour to the right produces also the asymptotic expansion for $D_p(z)$ for large z, Eq. 14.7.

14.8 Problems

1. Show that

$$\frac{\Gamma(1+p/2)\Gamma(-p/2) - \Gamma(1/2-p/2)\Gamma(1/2+p/2)}{2\Gamma(1+p)\Gamma(-p)} = cos(\pi p/2) - sin(\pi p/2).$$

2. From Eq. 14.61 find the convergent expansion of $D_p(z)$ for small z and the asymptotic expansion for large z.

3. Choose a kernel of the form $e^{z^2 t/4}$ and derive an integral representation

$$D_\nu(z) \sim \int e^{z^2 t/4} f(t) dt.$$

Find the function $f(t)$, the integration contour, and the normalization. This representation is due to Whittaker.

4. Show that for n a positive integer

$$D_n(x) = \frac{(-1)^\mu 2^{n+2} e^{x^2/4}}{\sqrt{2\pi}} \int_0^\infty t^n e^{-2t^2} Cs(2xt) dt,$$

with $Cs(z) = cos(z)$ if n is even and $Cs(z) = sin(z)$ if n is odd. Also $\mu = n/2$ or $\mu = (n-1)/2$, whichever is integer.

5. Using Eq. 14.24 and

$$0 = \int_C \frac{d}{dt}\left[e^{-zt-t^2/2}(-t)^{-n-1}\right] dt,$$

show that

$$D_{n+1}(z) - zD_n(z) + nD_{n-1}(z) = 0.$$

6. Differentiate Eq. 14.24 to find that

$$D_n'(z) + \frac{1}{2} z D_n(z) - nD_{n-1}(z) = 0.$$

7. Obtain the results of problem 5 and problem 6 using the local expansions about $z = 0$.

8. Since there are only two independent solutions to the differential equation, we must have

$$D_n(z) = aD_{-n-1}(iz) + bD_{-n-1}(-iz).$$

Use the local expansions about $z = 0$ to find that

$$D_n(z) = \frac{\Gamma(n+1)}{\sqrt{2\pi}} \left[e^{n\pi i/2} D_{-n-1}(iz) + e^{-n\pi i/2} D_{-n-1}(-iz) \right].$$

9. Use the Euler representation Eq. 14.17

$$D_p(z) = \frac{e^{-z^2/4}}{\sqrt{2\pi}} \int_{-\infty}^{\infty} (z+it)^p e^{-t^2/2} dt$$

to find the asympotic form of $D_p(z)$ for $z \to -\infty$.

Chapter 15

Whittaker and Watson

Edmund Taylor Whittaker was born in 1873 in Lancashire, England. He went to Cambridge in 1892 with a scholarship. He became a fellow of Trinity College and made important changes in the topics taught at Cambridge, writing *A Course of Modern Analysis* in 1902, joined for a later edition in 1915 by a former student of his, Neville Watson. This is a marvelous reference text, laying out in logical sequence essentially all of what was known in mathematical analysis at the time. Whittaker became Astronomer Royal of Ireland in 1906 and moved to Dunsink Observatory. In 1918 Watson found solutions to the problem of electromagnetic waves propagating in a dielectric around a conducting earth, which were found not to explain the observed long-distance propagation of radio waves. In 1902 Heaviside predicted that a conducting layer existed in the upper atmosphere (celebrated in verse by T.S. Eliot [1939] in *Old Possum's book of Practical Cats*), and Watson showed that this model gave results which agreed with experiment. Watson went on to publish his exhaustive *The Theory of Bessel Functions* in 1922.

Whittaker worked primarily in analysis but also on celestial mechanics and the history of applied mathematics and physics. He found a general solution to the Laplace equation in three dimensions. He was knighted in 1945.

The Whittaker function is a solution to the equation

$$\frac{d^2y}{dz^2} + \left[\frac{\frac{1}{4} - \mu^2}{z^2} + \frac{\lambda}{z} - \frac{1}{4}\right] y = 0.$$

Many of the functions of mathematical physics are special cases of the Whittaker functions, also called confluent hypergeometric functions. The error function, the incomplete Gamma function, the logarithmic integral function, parabolic cylinder functions, and Bessel functions are all special cases. Note that if $f_{\lambda,\mu}(z)$ is a solution, then so is $f_{-\lambda,\mu}(-z)$. We can take $Re\mu \geq 0$ without loss of generality.

15.1 Local Analysis at Infinity

The point $z = \infty$ is an irregular singular point. Thus we attempt a solution of the form

$$y(z) = e^{S(z)}. \tag{15.1}$$

Substituting into the differential equation we find

$$z^2[S'' + (S')^2] - \mu^2 + \lambda z + \frac{1-z^2}{4} = 0. \tag{15.2}$$

Dominant balance gives $S = \pm(z/2 - \lambda ln(z))$ and thus $y \simeq z^{-\lambda}e^{z/2}$ or $y \simeq z^{\lambda}e^{-z/2}$. To complete the analysis at large z we require an asymptotic series. Using

$$y(z) = e^{-z/2}g(z), \tag{15.3}$$

gives

$$z^2g'' - z^2g' + (1/4 - \mu^2 + \lambda z)g = 0. \tag{15.4}$$

Then write $g = \sum a_n z^{\alpha-n}$, giving $\alpha = \lambda$ and the recursion relation

$$a_{n+1} = -\frac{(n-\lambda+1/2-\mu)(n-\lambda+1/2+\mu)}{n+1}a_n. \tag{15.5}$$

Similarly using $y = e^{z/2}g(z)$ gives $\alpha = -\lambda$ and

$$a_{n+1} = \frac{(n+\lambda+1/2-\mu)(n+\lambda+1/2+\mu)}{n+1}a_n. \tag{15.6}$$

The normalization of the subdominant solution is chosen to be

$$y \simeq e^{-z/2}z^{\lambda}, \tag{15.7}$$

for large z with $|argz| < \pi$ so as to remain in the domain of subdominancy, and thus $a_0 = 1$, giving for the subdominant solution the asymptotic series

$$W_{\lambda,\mu}(z) = e^{-z/2}\sum(-1)^n\frac{\Gamma(n-\lambda+1/2-\mu)\Gamma(n-\lambda+1/2+\mu)}{n!\Gamma(-\lambda+1/2-\mu)\Gamma(-\lambda+1/2+\mu)}z^{\lambda-n}. \tag{15.8}$$

From this equation we note that

$$W_{\lambda,\mu}(z) = W_{\lambda,-\mu}(z). \tag{15.9}$$

The second series, which aside from normalization is seen simply to be $W_{-\lambda,\mu}(-z)$, is

$$y_{-\lambda,\mu}(-z) = e^{z/2} \sum \frac{\Gamma(n+\lambda+1/2-\mu)\Gamma(n+\lambda+1/2+\mu)}{n!\Gamma(\lambda+1/2-\mu)\Gamma(\lambda+1/2+\mu)} z^{-\lambda-n}, \tag{15.10}$$

and these two solutions are clearly linearly independent unless $\lambda = 0$.

15.2 Local Analysis at Zero

Note $z = 0$ is a regular singular point. A solution of the form

$$y_\alpha(z) = \sum_0^\infty a_n z^{n+\alpha}, \tag{15.11}$$

leads to a three term recursion relation, which is not convenient. Again extract the leading order behavior at infinity using Eq. 15.3. A solution of the form $g = \sum b_n z^{n+\alpha}$ gives $\alpha = 1/2 \pm \mu$, and if 2μ is not an integer provides two linearly independent solutions denoted by $M_{\lambda,\pm\mu}(z)$, with

$$M_{\lambda,\mu}(z) = \frac{\Gamma(1+2\mu)e^{-z/2}z^{1/2+\mu}}{\Gamma(1/2+\mu-\lambda)} \sum \frac{\Gamma(n+1/2+\mu-\lambda)}{n!\Gamma(n+1+2\mu)} z^n. \tag{15.12}$$

However if 2μ is integer only the solution $M_{\lambda,\mu}$ is given by this series. Either $M_{\lambda,-\mu}$ is not defined, or in the case $\mu = 0$ does not give a second solution.

Note that $M_{-\lambda,\pm\mu}(-z)$ are also solutions, and must be expressible in terms of $M_{\lambda,\pm\mu}(z)$. But looking at the behavior near $z = 0$ we readily find that

$$M_{-\lambda,\mu}(-z) = e^{i\pi(1/2+\mu)} M_{\lambda,\mu}(z). \tag{15.13}$$

15.3 Euler Integral Representation

The $M_{\lambda,\mu}$ do not give two solutions if 2μ is an integer, so we wish to construct a representation without this restriction. Attempt an Euler representation, first extracting the behavior near zero

$$y(z) = z^\alpha e^{-z/2} g(z) \tag{15.14}$$

with $\alpha = 1/2 \pm \mu$. Substituting into the differential equation we find

$$g'' + \left(\frac{2\alpha}{z} - 1\right) g' + \left[\frac{1/4 - \mu^2 + \alpha^2 - \alpha}{z^2} + \frac{\lambda - \alpha}{z}\right] g = 0. \tag{15.15}$$

Now attempt a solution using the Euler integral representation

$$g(z) = \int_C (z+t)^p f(t) dt. \tag{15.16}$$

But note that this representation also determines the behavior at $z = \infty$, and thus the power p is fixed to be $p = \lambda - \alpha$. Further note that the terms in Eq. 15.15 involve powers of z, not different powers of $z + t$, so it is more convenient to use $\alpha = \lambda$ giving

$$g(z) = \int_C \left(1 + \frac{t}{z}\right)^p f(t) dt \tag{15.17}$$

so that dg/dz produces factors of t/z^2.

Substituting this expression into Eq. 15.15, performing integration by parts, and writing some factors of $(1 + t/z)$ as two terms, all terms can be made to be proportional to $(1+t/z)^{p-2}$. There are terms in $1/z^4$, $1/z^3$, and $1/z^2$ which must each be set to zero. The terms in $1/z^4$ do not involve any derivatives of f, giving simply an algebraic equation involving the power p,

$$p(p+1) - 2\lambda p + 1/4 - \mu^2 + \alpha^2 - \alpha = 0 \tag{15.18}$$

giving $p = \lambda - \alpha$. The equations arising from the $1/z^3$ and $1/z^2$ terms each give first order differential equations for the function f. To make them consistent with one another it is necessary to add a perfect derivative which has contributions to each. Note that choosing $G = tf(1 + t/z)^{p-1}$ we have

$$\frac{dG}{dt} = (tf)'(1 + t/z)^{p-2}(1 + t/z) + (p-1)(1 + t/z)^{p-2}\frac{tf}{z}, \tag{15.19}$$

and thus adding the expression $(A/z^2)dG/dt$, which gives zero upon integration, contributes $A(tf)'$ to the terms in $(1 + t/z)^{p-2}/z^2$ and adds $A[(tf)'t + (p-1)tf]$ to the terms in $(1 + t/z)^{p-2}/z^3$. Setting $A = \lambda - \alpha$ these equations become equivalent to one another and give

$$f(t) = (-t)^{-\lambda - 1/2 + \mu} e^{-t}. \tag{15.20}$$

The integration by parts requires also that

$$(-t)^{-\lambda-1/2+\mu}e^{-t}\left(1+\frac{t}{z}\right)^{\lambda-3/2+m}|_a^b = 0. \tag{15.21}$$

Thus the contour can be taken counterclockwise around the cut along the real axis with both end points at $+\infty$ as shown in Fig. 15.1, with

$$y(z) = z^{\lambda}e^{-z/2}\int_C \left(1+\frac{t}{z}\right)^{\lambda-1/2+\mu}(-t)^{-\lambda-1/2+\mu}e^{-t}dt. \tag{15.22}$$

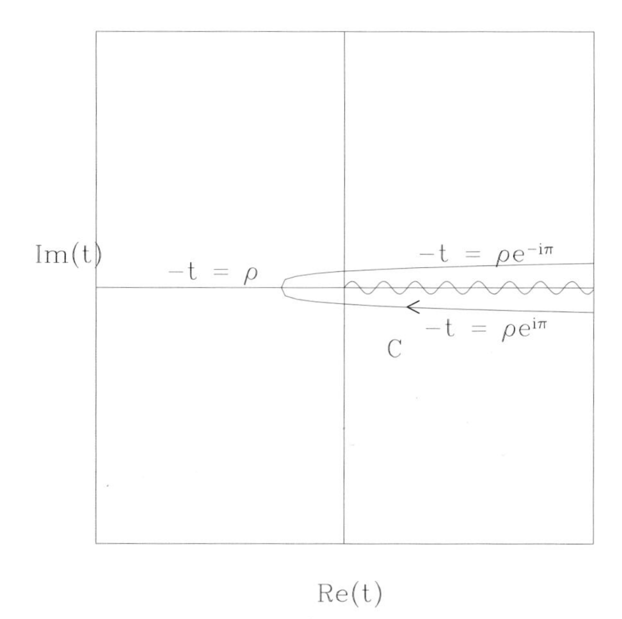

Fig. 15.1 Integration path for the Euler representation of the Whittaker function.

Evaluating the asymptotic behavior for large z and requiring the solution to asymptotically equal $z^{\lambda}e^{-z/2}$ gives finally the normalization. This solution is subdominant at large z so the asymptotic behavior uniquely determines it, it must be identical to that given by the series in section 15.1. Since this representation is valid for all values of λ, μ we use it to define the functions

$$W_{\lambda,\mu}(z) = \frac{-\Gamma(\frac{1}{2}+\lambda-\mu)z^{\lambda}}{2\pi i e^{z/2}}\int_C \left(1+\frac{t}{z}\right)^{\lambda-1/2+\mu}(-t)^{-\lambda-1/2+\mu}e^{-t}dt. \tag{15.23}$$

This equation defines $W_{\lambda,\mu}(z)$ for all z with the exception of the negative real axis.

If $Re(\lambda - 1/2 - \mu) \leq 0$ the integral can be moved to the real axis, giving

$$W_{\lambda,\mu}(z) = \frac{z^{\lambda}e^{-z/2}}{\Gamma(\frac{1}{2} - \lambda + \mu)} \int_0^{\infty} \left(1 + \frac{t}{z}\right)^{\lambda - 1/2 + \mu} t^{-\lambda - 1/2 + \mu} e^{-t} dt. \tag{15.24}$$

For small z we find

$$W_{\lambda,\mu}(z) \simeq \frac{z^{1/2-\mu}e^{-z/2}\Gamma(2\mu)}{\Gamma(1/2 + \mu - \lambda)}. \tag{15.25}$$

15.4 Relation Between $M_{\lambda,\mu}(z)$ and $W_{\lambda,\mu}(z)$

There must be a relation[1] between $M_{\lambda,\mu}(z)$ and $W_{\lambda,\mu}(z)$ of the form

$$W_{\lambda,\mu}(z) = AM_{\lambda,\mu}(z) + BM_{\lambda,-\mu}(z) \tag{15.26}$$

which we can discover using the asymptotic expression for small z, Eq. 15.25, and since for small z

$$M_{\lambda,\mu} \simeq z^{1/2+\mu}, \tag{15.27}$$

the $M_{\lambda,-\mu}$ dominates over $M_{\lambda,\mu}$ in Eq. 15.26 giving

$$B = \frac{\Gamma(2\mu)}{\Gamma(1/2 + \mu - \lambda)}. \tag{15.28}$$

Now write Eq. 15.26 as

$$W_{\lambda,-\mu}(z) = AM_{\lambda,-\mu}(z) + BM_{\lambda,\mu}(z). \tag{15.29}$$

Now the dominant term on the right-hand side for small z is the $AM_{\lambda,-\mu}$ term, and substituting the small z expression for $W_{\lambda,-\mu}(z)$ we find

$$A = \frac{\Gamma(-2\mu)}{\Gamma(1/2 - \mu - \lambda)}, \tag{15.30}$$

[1]Compare the simplicity of this derivation, which makes use of the asymptotic expressions, to that given by Whittaker and Watson.

giving

$$W_{\lambda,\mu}(z) = \frac{\Gamma(-2\mu)}{\Gamma(1/2-\mu-\lambda)}M_{\lambda,\mu}(z) + \frac{\Gamma(2\mu)}{\Gamma(1/2+\mu-\lambda)}M_{\lambda,-\mu}(z). \tag{15.31}$$

15.5 Fourier–Laplace Integral Representation

A direct Fourier–Laplace representation leads to a second order differential equation for $f(t)$, so we extract the indicial behavior. Attempt a Fourier–Laplace representation of the form

$$y(z) = z^{\alpha}\int_C e^{tz}f(t)dt, \tag{15.32}$$

with $\alpha = 1/2 \pm \mu$. Substituting and integrating by parts we find a first order differential equation for $f(t)$,

$$(2\alpha t - 2t + \lambda)f(t) = (t-1/2)(t+1/2)f'(t), \tag{15.33}$$

giving

$$f(t) = (1/2-t)^{\alpha+\lambda-1}(t+1/2)^{\alpha-\lambda-1}, \tag{15.34}$$

and thus

$$y(z) = z^{\alpha}\int_C e^{tz}(1/2-t)^{\alpha+\lambda-1}(t+1/2)^{\alpha-\lambda-1}dt, \tag{15.35}$$

with the requirement that $(1/2-t)^{\alpha+\lambda}(t+1/2)^{\alpha-\lambda}e^{tx}|_a^b = 0$ with a, b the ends of the contour. For $\alpha > |\lambda|$ the contour can be chosen between $t = \pm 1/2$.

To identify the solution write this as

$$y(z) = z^{\alpha}e^{-z/2}\int_{-1/2}^{1/2} e^{(t+1/2)z}(1/2-t)^{\alpha+\lambda-1}(t+1/2)^{\alpha-\lambda-1}dt \tag{15.36}$$

which we convert to a series in z by expanding the exponential,

$$y(z) = z^{\alpha}e^{-z/2}\int_{-1/2}^{1/2}\sum \frac{z^n(t+1/2)^n}{n!}(1/2-t)^{\alpha+\lambda-1}(t+1/2)^{\alpha-\lambda-1}dt, \tag{15.37}$$

or $y(z) = z^\alpha e^{-z/2} \sum a_n z^n$ with

$$a_n = \frac{1}{n!} \int_{-1/2}^{1/2} (1/2 - t)^{\alpha+\lambda-1} (t + 1/2)^{\alpha-\lambda+n-1} dt. \tag{15.38}$$

Upon substituting $u = 1/2 - t$ we find

$$a_n = \frac{\Gamma(\alpha+\lambda)\Gamma(\alpha-\lambda+n)}{\Gamma(2\alpha+n)n!}, \tag{15.39}$$

and comparing with Eq. 15.12 we see that

$$y(z) = \frac{\Gamma(\mu+\lambda+1/2)\Gamma(1/2+\mu-\lambda)}{\Gamma(1+2\mu)} M_{\lambda,\mu}(z), \tag{15.40}$$

and thus

$$M_{\lambda,\mu}(z) = \frac{\Gamma(1+2\mu) z^{\mu+1/2}}{\Gamma(\mu+\lambda+\frac{1}{2})\Gamma(\frac{1}{2}+\mu-\lambda)} \int_{-1/2}^{1/2} e^{tz} (\tfrac{1}{2} - t)^{\mu+\lambda-\frac{1}{2}} (t + \tfrac{1}{2})^{\mu-\lambda-\frac{1}{2}} dt. \tag{15.41}$$

15.6 Mellin Integral Representation

The series Eq. 15.12 behaves for large n as $z^n/n!$, i.e. it is exponentially large, which means that a Mellin representation is not useful. Extract this exponential behavior by writing

$$y_{\lambda,\mu}(z) = e^{z/2} g(z), \tag{15.42}$$

and substitute into the differential equation, giving

$$z^2 g'' + z^2 g' + (1/4 - \mu^2) g + \lambda z g = 0. \tag{15.43}$$

The substitution $g(z) = \sum b_n z^{n+\alpha}$ then gives the series (with $\alpha = 1/2 + \mu$)

$$M_{\lambda,\mu}(z) = \frac{\Gamma(1+2\mu) e^{z/2} z^{1/2+\mu}}{\Gamma(1/2+\mu+\lambda)} \sum \frac{(-1)^n \Gamma(n+1/2+\mu+\lambda)}{n!\Gamma(n+1+2\mu)} z^n. \tag{15.44}$$

Within a constant, this is simply $M_{-\lambda,\mu}(-z)$, see Eq. 15.13. The alternating signs now make this series behave asymptotically as $(-1)^n z^n/n!$ or e^{-z}.

Using this series we find the Mellin representation

$$M_{\lambda,\mu}(z) = \frac{e^{z/2}}{2\pi i} \int_{\gamma-i\infty}^{\gamma+i\infty} M(s) z^{-s+\alpha} ds, \tag{15.45}$$

with $M(s) = b_{-s}(-1)^s\Gamma(s)\Gamma(1-s)$, $\alpha = 1/2+\mu$, and $0 < \gamma < 1$. This gives

$$M_{\lambda,\mu}(z) = \frac{\Gamma(1+2\mu)e^{z/2}}{2\pi i\Gamma(1/2+\mu+\lambda)} \int_{\gamma-i\infty}^{\gamma+i\infty} \frac{\Gamma(-s+1/2+\mu+\lambda)\Gamma(s)}{\Gamma(-s+1+2\mu)} z^{-s+\alpha} ds. \tag{15.46}$$

Convergence is given by the behavior at large u with $s = t + iu$, and $z = re^{i\phi}$, and the integrand behaves as

$$\frac{\Gamma(-s+\alpha+\lambda)\Gamma(s)}{\Gamma(-s+1+2\mu)} z^{-s+\alpha} \sim e^{(\pm\phi-\pi)|u|}, \tag{15.47}$$

with the $\pm$ coming from the two integration limits, restricting the phase of z to $|\phi| < \pi$.

To find the large z behavior look for saddle points. Write the integrand as e^{ϕ} with

$$\phi = -sln(z) + ln[\Gamma(-s+\alpha+\lambda)] + ln[\Gamma(s)] - ln[\Gamma(-s+2\alpha)], \tag{15.48}$$

giving

$$\phi' \simeq ln\left(\frac{s(s-2\alpha)}{z(s-\alpha-\lambda)}\right), \tag{15.49}$$

and we see that for large z there is a vertically oriented saddle point[2] at $s \simeq z + \alpha - \lambda$. Evaluating the contribution from the saddle point we find

$$e^{\phi_0} \simeq z^{\lambda-\alpha} e^{-z} e^{i\pi(\lambda-\alpha)} \sqrt{\frac{2\pi}{z}}, \tag{15.50}$$

giving the leading contribution from the saddle point

$$S(z) \simeq \frac{\Gamma(2\mu+1)}{\Gamma(\mu+1/2+\lambda)} e^{-z/2} z^{\lambda} e^{i\pi(\lambda-\alpha)}. \tag{15.51}$$

In addition the integrand has a sequence of poles, with the dominant leftmost pole located at $s = \alpha - \lambda$. The integration contour, the location of the saddle at $s \simeq z$, and the poles along the real axis are shown schematically in Fig. 15.2. Using the fact that the residue of $\Gamma(-t)$ at integer t is

[2] To avoid the complication of the singularities in $\Gamma(-s+\alpha+\lambda)$ and $\Gamma(-s+2\alpha)$ we avoid values of z near $\alpha+\lambda+k$ and $2\lambda+k$ with k any non-negative integer. We already know the behavior of the solutions at large z, which are smooth, so there can be no problem arising from this avoidance of small z domains.

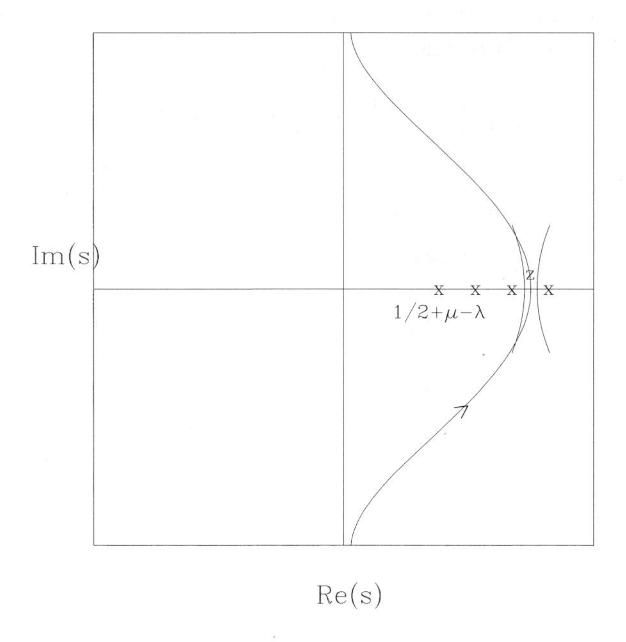

Fig. 15.2 Integration path for the Mellin representation of the Whittaker function.

$(-1)^t/\Gamma(1+t)$ we find the leading behavior from the first pole due to $\Gamma(0)$ at $s = 1/2 + \mu - \lambda$, with the leading contribution

$$P(z) \simeq \frac{\Gamma(1+2\mu)e^{z/2}z^{-\lambda}}{\Gamma(1/2+\mu-\lambda)}. \tag{15.52}$$

Adding the contribution from the pole and the saddle point we find the leading behavior

$$M_{\lambda,\mu}(z) \simeq \frac{\Gamma(2\mu+1)}{\Gamma(1/2+\mu-\lambda)}e^{z/2}z^{-\lambda} + \frac{\Gamma(2\mu+1)}{\Gamma(1/2+\mu+\lambda)}e^{i\pi(1/2+\mu+\lambda)}e^{-z/2}z^{\lambda}. \tag{15.53}$$

15.7 Special Cases

15.7.1 *The error function*

The error function is defined as

$$Erfc(z) = \int_z^\infty e^{-s^2}ds. \tag{15.54}$$

Use Eq. 15.24 and substitute $s = z^2(w^2 - 1)$ and then $w = t/x$ in the integral for $W_{\frac{1}{4},-\frac{1}{4}}(z^2)$ to find

$$Erfc(z) = \frac{1}{2\sqrt{z}} e^{-z^2/2} W_{\frac{1}{4},-\frac{1}{4}}(z^2). \tag{15.55}$$

15.7.2 *The logarithmic integral function*

The logarithmic integral function is defined as

$$li(z) = \int_0^z \frac{ds}{ln(s)}. \tag{15.56}$$

Write $t - ln(z) = -ln(s)$ in Eq. 15.24 for $W_{-\frac{1}{2},0}(-ln(z))$ to find

$$li(z) = -\frac{z^{1/2}}{(-ln(z)^{1/2}} z^{1/2} W_{-\frac{1}{2},0}(-ln(z)). \tag{15.57}$$

15.8 Problems

1. Find the Mellin representation for $W_{\lambda,\mu}(z)$,

$$W_{\lambda,\mu}(z) = \frac{e^{-z/2}}{2\pi i}\int_{\gamma-i\infty}^{\gamma+i\infty} \frac{\Gamma(s-\lambda)\Gamma(-s-\mu+1/2)\Gamma(-s+\mu+1/2)}{\Gamma(-\lambda-\mu+1/2)\Gamma(-\lambda+\mu+1/2)} z^s ds,$$

and use it to evaluate the behavior at large z.

2. Show that

$$W_{\lambda,\mu}(z) = \sqrt{z}W_{\lambda-1/2,\mu-1/2}(z) + (1/2+\lambda-\mu)W_{\lambda,\mu-1}(z).$$

3. Show that

$$z^{-1/2-\mu}M_{\lambda,\mu}(z) = (-z)^{-1/2-\mu}M_{-\lambda,\mu}(-z),$$
$$for \qquad 2\mu \neq -1, -2, -3\ldots.$$

4. Show for large λ

$$W_{-\lambda,\mu}(z) \simeq \left(\frac{z}{4\lambda}\right)^{1/4} e^{\lambda-\lambda ln(\lambda)-2\sqrt{\lambda z}}.$$

5. Show for large λ

$$W_{\lambda,\mu}(z) \simeq -\left(\frac{4z}{\lambda}\right)^{1/4} e^{-\lambda+\lambda ln(\lambda)} sin(2\sqrt{\lambda z}-\pi\lambda-\pi/4).$$

6. Show for large λ

$$M_{\lambda,\mu}(z) \simeq \frac{1}{\sqrt{\pi}}\Gamma(2\mu+1)\lambda^{-\mu-1/4}z^{1/4}cos(2\sqrt{\lambda z}-\pi\mu-\pi/4).$$

7. Show that (see section 13.6)

$$M_{0,\mu}(z) = 2^{2\mu}\Gamma(\mu+1)\sqrt{z}I_{\mu}(z/2).$$

8. Show that (see section 13.6)

$$W_{0,\mu}(z) = \sqrt{\left(\frac{z}{\pi}\right)} K_\mu(z/2).$$

9. Use Eq. 15.13 to prove the identity

$$\sum_0^k \frac{(-1)^n \Gamma(n+1/2+\mu-\lambda)}{(k-n)!n!\Gamma(n+1+2\mu)} = \frac{\Gamma(k+1/2+\mu+\lambda)\Gamma(1/2+\mu-\lambda)}{k!\Gamma(k+1+2\mu)\Gamma(1/2+\mu+\lambda)}.$$

Chapter 16

Inhomogeneous Differential Equations

Born in Albany in 1927, Marshall Nicholas Rosenbluth enlisted in the navy during World War II, graduated from Harvard in 1946 and received his PhD at age 22 at the University of Chicago, where he was a student of Enrico Fermi. Eleven students earned PhDs under Fermi's guidance while at Chicago, including T.D. Lee, M.L. Goldberger, S. Trieman, O. Chamberlain, J. Steinberger, J. Rainwater, J. Friedman, and G. Chew. Of these eleven students, five went on to receive Nobel Prizes. They also formed most of the first generation of top level theoretical physicists educated in the US. As a young man Rosenbluth helped invent the hydrogen bomb, was exposed to radioactive fallout in a nuclear test, and soon thereafter devoted himself to the problem of turning the hot plasmas of nuclear fusion into a source of nearly limitless electrical power.

He also developed the Monte Carlo simulation technique, now a standard research tool in statistical mechanics, chemistry, biochemistry and other fields. He was responsible for the basic theory of the free electron laser.

During the Cold War, Dr. Rosenbluth promoted science exchanges with the Soviet Union, and in 1965 organized a year long meeting between Soviet, European, and American physicists to work on fusion-related physics problems in Trieste, Italy. This was the first extensive collaboration of Soviet and western physicists on fusion problems, leading to lifelong friendships and working relations between Soviet and American fusion physicists.

Marshall was incredibly ingenious in inventing physical models which captured the essence of a problem, and in finding analytic solutions to them. He was very much at home with all forms of analysis and gave many elegant solutions to major problems in the physics of controlled nuclear fusion.

To find an integral solution to an inhomogeneous differential equation one can either use the Green's function technique, or require that the terms left over from the usual integration by parts equal the inhomogeneous term. On occasion, as shown in the next section, Fourier representation can lead directly to a closed form analytic solution.

The Stokes phenomena result from the nature of the WKB solutions, and the analytic continuation of the function $Q^{1/2}(z)$ through the complex plane. Inhomogeneous solutions do not share these properties as they result from an entirely different dominant balance than that giving the WKB approximation. Thus in performing an analytic continuation through the complex z plane, the inhomogeneous solution does not participate in the Stokes phenomena. However, even if the asymptotic behavior is given by the inhomogeneous solution, there may be hidden in this behavior some homogeneous subdominant solution, which can become dominant upon analytic continuation.

16.1 The Driven Oscillator

Consider an oscillator driven at two driving frequencies with one drive term resonant at the natural frequency and the second one non-resonant. The differential equation takes the form

$$y''(t) + y(t) = ae^{it} + be^{i\omega t}, \tag{16.1}$$

where the real part is understood, and a, b can be complex. Use the Fourier–Laplace kernel, i.e. write

$$y(t) = \int_C e^{izs} f(s) ds, \tag{16.2}$$

giving upon substitution

$$f(s) = a\frac{\delta(s-1)}{1-s^2} + b\frac{\delta(s-\omega)}{1-s^2}. \tag{16.3}$$

Substituting into Eq. 16.2 the second term is immediately integrated. For the first term use $\delta(s-1)/(s-1) = -\delta'(s-1)$, integrate by parts and take the real part, to find

$$y(t) = Re(a)\left[\frac{tsin(t)}{2} + \frac{cos(t)}{4}\right] + Im(a)\left[\frac{tcos(t)}{2} + \frac{sin(t)}{4}\right]$$
$$+Re(b)\left[\frac{cos(\omega t)}{1-\omega^2}\right] + Im(b)\left[\frac{sin(\omega t)}{1-\omega^2}\right]. \quad (16.4)$$

To this particular solution it is, of course, possible to add the general solution of the homogeneous equation.

16.2 The Driven Weber Equation

Consider the inhomogeneous Weber equation

$$y'' + (\alpha - 4z^2)y = 1 \quad (16.5)$$

where we wish to find α such that the solution tends to zero at $\pm\infty$. Dominant balance for large z for the homogeneous equation gives

$$y \sim e^{\pm z^2} z^{-(1/4\pm\alpha/4)}, \quad (16.6)$$

and balance for the inhomogenous equation gives $y \sim 1/(4z^2)$. Obtain an integral solution of the form

$$y = \int e^{z^2 t} f(t)dt \quad (16.7)$$

giving

$$f(t) = A\frac{(1-t)^{(\alpha-6)/8}}{(1+t)^{(\alpha+6)/8}}, \qquad 4A(t^2-1)e^{z^2 t}\frac{(1-t)^{(\alpha-6)/8}}{(1+t)^{(\alpha+6)/8}}|_a^b = 1 \quad (16.8)$$

with a, b the integration end points, giving

$$y(z) = \frac{1}{4}\int_0^{-\infty} e^{z^2 t}\frac{(1-t)^{(\alpha-6)/8}}{(1+t)^{(\alpha+6)/8}}dt \quad (16.9)$$

as the solution that tends to zero for large positive z. Note that this integral is not real. For $t < -1$ the denominator is complex. But note that the real part of $y(z)$ satisfies the inhomogeneous Weber equation and the imaginary part the homogeneous equation. Thus we continue to work with the complex solution, and will absorb the imaginary part into the

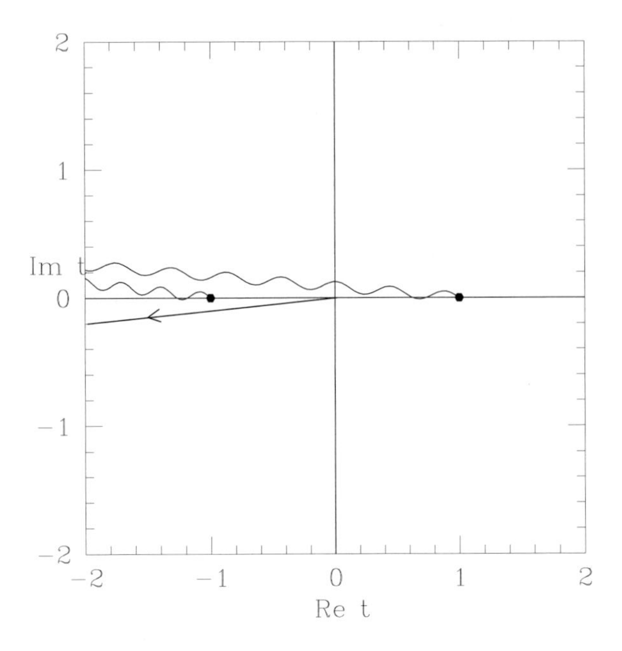

Fig. 16.1 Integration path for the inhomogeneous Weber equation for $z > 0$.

homogeneous solution. The dominant contribution for large z comes from $t \simeq 0$ giving

$$y(z) \simeq -\frac{1}{4z^2}. \tag{16.10}$$

Shown in Fig 16.1 is the integration path for positive real z. Now analytically continue to negative z through the lower half plane. Then keeping $z^2 t < 0$ we find the contour wraps around the cut starting at $t = 1$ as shown in Fig. 16.2. To evaluate the additional contribution from the cut write $1 - t = \rho e^{i\pi}$ along the real axis to the right of $t = 1$, $1 - t = \rho e^{i2\pi}$ above the cut and $1 - t = \rho$ below the cut. The contribution for $z \to -\infty$ comes from the end point $\rho = 0$ giving

$$y_{cut} = \frac{2ie^{z^2}}{z^{(1/4+\alpha/4)}} \frac{e^{i\pi(\alpha-6)/8}}{2^{(\alpha+10)/2}} \Gamma(\alpha/8 + 1/4) sin(\pi(\alpha - 6)/8), \tag{16.11}$$

which is a part of the exponentially large homogeneous solution of the differential equation.

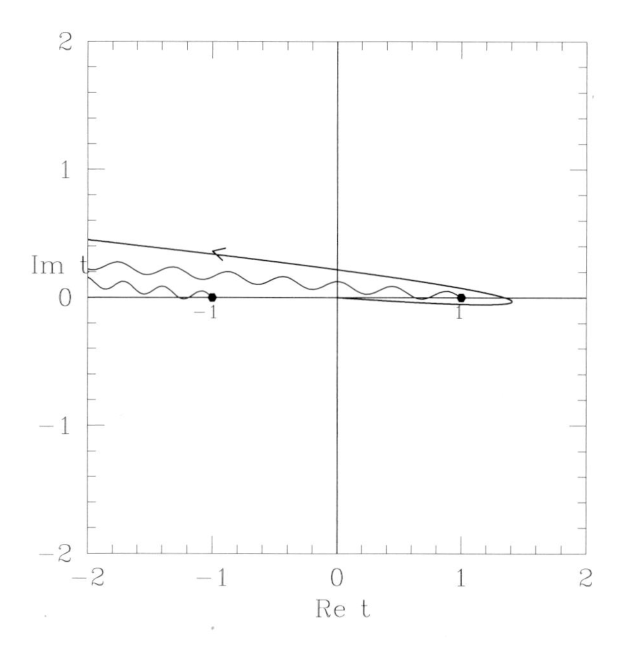

Fig. 16.2 Integration path for the inhomogeneous Weber equation for $z < 0$.

Requiring that $y \to 0$ as $z \to -\infty$ we find the condition

$$\Gamma(\alpha/8 + 1/4)sin(\pi(\alpha - 6)/8) = 0 \tag{16.12}$$

satisfied for $(\alpha-6)/8 = n$ but also $(\alpha+2)/8 = n+1 > 0$ giving $n = 0, 1, 2...$ with $\alpha = 8n + 6$.

16.3 The Struve Equation

The Struve equation is an inhomogeneous Bessel equation

$$y'' + \frac{y'}{z} + \left(1 - \frac{\nu^2}{z^2}\right) y = \frac{(z/2)^{\nu-1}}{\sqrt{\pi}\Gamma(\nu + 1/2)}, \tag{16.13}$$

and the desired solution is the particular solution.

16.3.1 *Local analysis at zero*

The local solution to the homogeneous equation is just that for the Bessel equation with two independent solutions for ν non integer, of the form

$\sum a_k z^{\alpha+k}$ with $\alpha = \pm\nu$, giving a function of the form z^α times an even function. The inhomogeneous term instead gives a function of the form z^α times an odd function. Writing

$$y_\alpha(z) = \sum_{k=0}^{\infty} a_k z^{k+\alpha}, \tag{16.14}$$

we find $\alpha = \nu$, $a_0 = 0$, and $a_1 = [2^{\nu-1}\sqrt{\pi}\Gamma(\nu+1/2)(1+2\nu)]^{-1}$ and $n(n+2\nu)a_n = -a_{n-2}$ generates the odd series. The Struve function is the particular solution

$$H_\nu(z) = \sum_{k=0}^{\infty} \frac{(-1)^k(\frac{z}{2})^{2k+\nu+1}}{\Gamma(k+3/2)\Gamma(\nu+k+3/2)}. \tag{16.15}$$

16.3.2 *Local analysis at infinity*

Write

$$y = \frac{2(z/2)^\nu}{\sqrt{\pi}\Gamma(\nu+1/2)} W(z), \tag{16.16}$$

giving

$$zW'' + (2\nu+1)W' + zW = 1, \tag{16.17}$$

and the series expansion $W = \sum a_n z^{-n}$ gives $a_{n+2} = -a_n n(n-2\nu)$, and $a_0 = 0$, $a_1 = 1$. We then find

$$y(z) \simeq \frac{(z/2)^\nu}{\pi} \sum_{k=0}^{\infty} \frac{\Gamma(\frac{1}{2}+k)}{\Gamma(\nu+\frac{1}{2}-k)} \left(\frac{z}{2}\right)^{-2k-1} \tag{16.18}$$

plus any constant times a solution of the homogeneous equation. To find the homogeneous part, use the equation for W to construct an integral solution to Eq. 16.13 valid for any ν

$$y(z) = \frac{2}{\sqrt{\pi}\Gamma(\nu+1/2)} \left(\frac{z}{2}\right)^\nu \int_{-\infty}^{0} e^{zt}(1+t^2)^{\nu-1/2} dt. \tag{16.19}$$

The behavior for large z is easily seen to be

$$y(z) \sim \frac{1}{\sqrt{\pi}\Gamma(\nu+1/2)} \left(\frac{z}{2}\right)^{\nu-1}. \tag{16.20}$$

Then look at the behavior at $z = 0$. We find $y(z) \sim (z/2)^{-\nu}\Gamma(\nu)/\pi$ and thus identify the homogeneous part, giving

$$H_\nu(z) \simeq \frac{(z/2)^\nu}{\pi} \sum_{k=0}^{\infty} \frac{\Gamma(\frac{1}{2}+k)}{\Gamma(\nu+\frac{1}{2}-k)} \left(\frac{z}{2}\right)^{-2k-1} + N_\nu(z). \qquad (16.21)$$

16.4 Resistive Reconnection

Consider the equation

$$y'' - z^2 y = z, \qquad (16.22)$$

with boundary condition $y \to -1/z$ for $z \to \infty$, which occurs in the theory of resistive tearing modes in a plasma (see, for example, White [2001], Eq. 5.34). As written the equation involves z^2, which would introduce second order derivatives under Fourier–Laplace transformation, not leading to simplification. A simple change of variables can remedy this. Let $y = zg(x)$ with $x = z^2/2$. Substituting we find

$$2xg'' + 3g' - 2xg = 1, \qquad (16.23)$$

an equation containing only x and thus leading to a first order differential equation under transformation. Now make a Fourier–Laplace transformation of the function g

$$g(x) = \int_C e^{-tx} f(t)dt, \qquad (16.24)$$

and substitute this into Eq. 16.23. Integrating by parts we find

$$\int_C dte^{tx} \left[tf(t) + 2(t^2-1)f'(t)\right] - 2f(t)(t^2-1)e^{tx}|_a^b = 1, \qquad (16.25)$$

with the ends of the integration contour C at a, b. We then have a solution provided

$$f(t) = c(1-t^2)^{-1/4}, \qquad 2f(t)(1-t^2)e^{tx}|_a^b = 1. \qquad (16.26)$$

There is only one end point making $f(t)(1-t^2)e^{tx}$ independent of x, namely $t = 0$. We thus find the two contours $a = 0$, $b = \pm 1$, giving $c = -1/2$. The two solutions have different asymptotic behavior for $z \to \infty$. Choosing the solution which tends to $-1/z$ (given by $b = 1$) we find

$$y(z) = -\frac{z}{2}\int_0^1 e^{-tz^2/2}(1-t^2)^{-1/4}dt. \tag{16.27}$$

16.5 Resistive Internal Kink

Another example arising from the theory of resistive modes in a plasma (see, for example, White [2001], Eq. 5.52) is

$$y'' - \frac{2y'}{z} - (z^2 + A)y = -z^2 y_\infty, \tag{16.28}$$

with A a constant and $y \to y_\infty$ for $z \to \pm\infty$. This equation again involves z^2 and thus would lead to a second order differential equation using the Fourier–Laplace transform. Change variables using $y = g(x)$ with $x = z^2/2$, giving

$$2xg''(x) - g'(x) - (2x + A)g(x) = -2xg_\infty, \tag{16.29}$$

an equation containing only x and thus leading to a first order differential equation under transformation. The x dependence of the inhomogeneous term can be removed by the Fourier–Laplace transformation

$$g(x) = g_\infty + \int_C e^{tx} f(t)dt. \tag{16.30}$$

Integrating by parts we find

$$\int_C dte^{tx}[(A + 5t)f(t) + 2(1-t^2)f'] - 2f(t)e^{tx}|_a^b = Ag_\infty, \tag{16.31}$$

with the ends of the integration contour C at a, b. We then have a solution provided

$$f(t) = c(1-t)^{-A/4-5/4}(1+t)^{A/4-5/4}, \qquad 2f(t)e^{tx}|_a^b = Ag_\infty, \tag{16.32}$$

giving $a = 0$, $b = \pm 1$, $c = A/2$. We must choose $b = -1$ to prevent y from diverging at $z \to \pm\infty$. Changing variable from t to $-t$ we then find

$$\frac{y(z)}{y_\infty} = 1 - \frac{A}{2}\int_0^1 dte^{-tz^2/2}(1+t)^{-A/4-5/4}(1-t)^{A/4-5/4}. \tag{16.33}$$

16.6 A Causal Inhomogeneous Problem

The evolution in space (x) and time (t) of a parametric instability[1] is described by the equations

$$\begin{aligned}\partial_t a_1 + i(x/2)a_1 + \partial_x a_1 &= \gamma a_2, \\ \partial_t a_2 + i(x/2)a_2 - \partial_x a_2 &= \gamma a_1,\end{aligned} \tag{16.34}$$

with uncoupled solutions ($\gamma = 0$) of $a_1 = e^{-ix^2/4}$, $a_2 = e^{ix^2/4}$, describing right moving and left moving waves for $x > 0$, respectively (the convention is $e^{i\omega t}$). A causal solution is desired, i.e. the response of the system to an initial perturbation at a point x_0 in space at time $t = 0$. Perform a Fourier–Laplace transformation in the time variable, $a_j(x,t) = \int exp(pt)a_j(x,p)dp$. For any differential equation of the form $\partial_t a + L_x a = 0$ with L_x an operator in x, a causal solution (one which is zero for $t < 0$ and equal to $a(x,0)$ for $t = 0$) is obtained by the Laplace inversion of $(p + L_x)a(x,p) = a(x,0)$, given by (written symbolically)

$$a(x,t) = \int dp e^{pt} \frac{a(x,0)}{p + L_x}, \tag{16.35}$$

with the contour taken vertically to the right of any zeros of the denominator, so that for $t < 0$ the integration contour can be closed in the right half plane, giving zero. Thus we have the two equations

$$\begin{aligned}pa_1 + i(x/2)a_1 + \partial_x a_1 &= \gamma a_2 + a_1(x,0), \\ pa_2 + i(x/2)a_2 - \partial_x a_2 &= \gamma a_1 + a_2(x,0),\end{aligned} \tag{16.36}$$

along with the prescription for the inversion of the Laplace transform. Taking $a_1(x,0) = 0$ and $a_2(x,0) = -\delta(x - x_0)$ and eliminating a_2 we find for $a_1(x,p)$ (dropping the subscript)

$$a'' + \left[\frac{x^2}{4} - ipx - p^2 + \frac{i}{2} + \gamma^2\right] a = \gamma\delta(x - x_0). \tag{16.37}$$

Make the substitution $x = y + 2ip$, giving

$$a'' + \frac{y^2}{4} + \left(\lambda + \frac{i}{2}\right) a = \gamma\delta(y - y_0) \tag{16.38}$$

with $\lambda = \gamma^2$, taken to be large.

[1] This is a simplified version of Rosenbluth, White and Liu, *PRL* 31, 1190 [1973].

To construct the Green's function we seek solutions of the homogeneous equation well-behaved at $\pm\infty$. For $+\infty$ extract the asymptotic behavior of a right moving wave, $a = e^{-iy^2/4}f(y)$ giving $f'' - iyf' + \lambda f = 0$. Then seek an integral representation of the form $f = \int exp(-\alpha yu)g(u)du$, giving $g = u^{i\lambda-1}e^{i\alpha^2u^2/2}$. To ensure convergence for any y take $\alpha = e^{i\pi/4}$, giving the integral solution

$$a_+(y) = e^{-iy^2/4}\int_0^\infty e^{-\alpha yu}u^{i\lambda-1}e^{-u^2/2}du, \tag{16.39}$$

which is well-behaved for $y \to \infty$, convergent, and the end point term resulting from integration by parts $e^{-\alpha yu}u^{i\lambda}e^{-u^2/2}|_0^\infty = 0$. This solution is in fact a parabolic cylinder function of complex argument, but since what is desired is the integral representation this fact is not particularly useful. The normalization is not relevant, it will drop out of our final expressions.

For $-\infty$ extract the asymptotic behavior of a left moving wave, $a = e^{iy^2/4}f(y)$ giving $f'' + iyf' + (\lambda + i)f = 0$. We then find the integral solution of the form $f = \int exp(\beta ys)g(s)ds$ with $\beta = e^{-i\pi/4}$

$$a_-(y) = e^{iy^2/4}\int_0^\infty e^{\beta ys}s^{-i\lambda}e^{-s^2/2}ds, \tag{16.40}$$

well behaved for $y \to -\infty$, and convergent with the end point term resulting from integration by parts $e^{\beta ys}s^{-i\lambda+1}e^{-s^2/2}|_0^\infty = 0$.

The Green's function solution for the inhomogeneous equation Eq. 16.38 is then

$$A(x,p) = \frac{a_+(y)a_-(y_0)\Theta(y-y_0) + a_-(y)a_+(y_0)\Theta(y_0-y)}{W(y_0)}, \tag{16.41}$$

with $W(y)$ the Wronskian, and for $f'' + p_1f' + p_0f = 0$ we have $W = exp(-\int p_1dx)$ and in this case $p_1 = 0$ so $W = const$. Evaluate the Wronskian for $y = 0$. Note that $A(x,p)$ is independent of the normalizations of a_+, a_-.

We have

$$a_+(0) = \int_0^\infty u^{i\lambda-1}e^{-u^2/2}du = 2^{i\lambda/2}\Gamma(i\lambda/2)/2. \tag{16.42}$$

Similarly $a'_+(0) = -\alpha 2^{i\lambda/2}\Gamma(i\lambda/2 + 1/2)/\sqrt{2}$. For a_- we have

$$a_-(0) = \int_0^\infty s^{-i\lambda}e^{-s^2/2}ds = 2^{-i\lambda/2}\Gamma(-i\lambda/2 + 1/2)/\sqrt{2}. \tag{16.43}$$

Similarly $a'_-(0) = \beta 2^{-i\lambda/2}\Gamma(1 - i\lambda/2)$.

Using Eq. 8.51 we find

$$\Gamma(1/2 - i\lambda/2)\Gamma(1/2 + i\lambda/2) = \frac{\pi}{cosh(\pi\lambda/2)}, \tag{16.44}$$

and finally

$$W = \frac{\pi\alpha e^{-\pi\lambda/2}}{sinh(\pi\lambda)}. \tag{16.45}$$

Now find the solution of the original equation, Eq. 16.34, given by the Laplace transform $A(x,t) = \int dpe^{pt}A(x,p)$. Substitute the integral representations for $x > x_0$, reinsert $x = y + 2ip$, and deform the integration contours through $s \to se^{-i\pi/4}$, $u \to ue^{i\pi/4}$. This transformation contributes a factor of $e^{-\pi\lambda/2}$ to $A(x,t)$, cancelling this factor in the Wronskian.

Then upon inverting the Laplace transform, and integrating over p we find a factor

$$2\pi\delta[t - x + x_0 - 2(s + u)], \tag{16.46}$$

allowing an immediate integration over s. Then substitute $u = z(t - x + x_0)/2$ giving $s = (1 - z)(t - x + x_0)/2$, there remaining a single integration over z. Note that in the causal domain $t - (x - x_0) > 0$, and the range of z vanishes if this is negative, giving $A = 0$. We have taken $x > x_0$ so $A = 0$ for $t < (x - x_0)$, as it should. The range of z is restricted to $0, 1$ if $t - (x - x_0) > 0$. Note also that for fixed x, t the range of integration in $s(u)$ is finite.

We then find for $x > x_0$, upon reinserting the $exp(ix^2/4 - ix_0^2/4)$, and using Eq. 8.51, the amplitude in response to a delta function initial condition at $x = x_0, t = 0$ is given for large λ by

$$A(x,t) = \gamma e^{\pi\lambda}e^{-i\Phi}\int_0^1 \frac{dz}{z}\left(\frac{z}{1-z}\right)^{i\lambda} e^{-i\lambda(z-1/2)/Q}, \tag{16.47}$$

with $\Phi = (t - x + x_0)(x + x_0)/4$, and $Q = 4\lambda/[(t - x + x_0)(t + x - x_0)]$ and the causal domain in which this expression is defined is for $t > 0$, $-t < x - x_0 < t$, where $Q > 0$, the amplitude being zero outside this domain. It may seem artificial to include the factor of λ in the definition of Q, but as we will see, changes in the nature of the solution occur for Q (including the λ factor) of order unity.

Write the integrand as e^{ϕ}/z with

$$\phi(z) = i\lambda[ln(z) - ln(1 - z) - (z - 1/2)/Q], \tag{16.48}$$

from which we find that there are two saddle points at

$$z = 1/2 \pm \sqrt{1/4 - Q}, \tag{16.49}$$

with real or complex locations depending on whether $Q > 1/4$. First consider $Q > 1/4$, corresponding for a given x to small t, a location near the causal front of the developing wave. In this case deform the integration contour to pass through the saddle at $z = 1/2 + i\sqrt{Q - 1/4}$, as shown in Fig. 16.3. The saddle in the lower half plane has a vertical orientation and cannot be used to connect the end points of the integration. At the upper

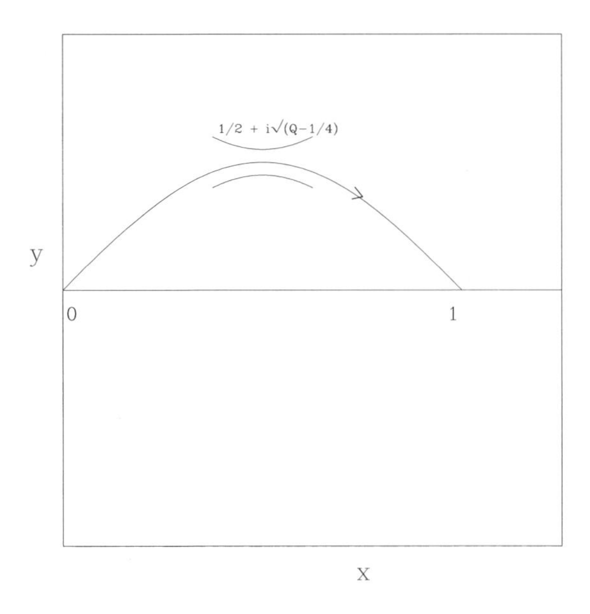

Fig. 16.3 Integration path for $Q > 1/4$.

saddle point we have

$$e^{\phi} = \left(\frac{1 + i\sqrt{4Q - 1}}{1 - i\sqrt{4Q - 1}}\right)^{i\lambda} e^{\lambda\sqrt{Q-1/4}/Q}, \tag{16.50}$$

and $\phi'' = -\lambda\sqrt{4Q - 1}/Q^2$. Note that for large Q, z is large and at the saddle $z/(1 - z) \sim e^{i\pi}$. Substituting $sin\theta = 1/(2\sqrt{Q})$ we then find

$$\left(\frac{1 + i\sqrt{4Q - 1}}{1 - i\sqrt{4Q - 1}}\right)^{i\lambda} = e^{2\lambda\theta} e^{-\pi\lambda}, \tag{16.51}$$

giving, except for a complex phase not changing the amplitude,

$$A(x,t) \simeq \sqrt{2\pi}\left(\frac{1}{2\sqrt{Q}}(1-\frac{1}{4Q})\right)^{-1/2} e^{W}. \tag{16.52}$$

Here

$$W = \frac{\lambda}{\sqrt{Q}}\left(1-\frac{1}{4Q}\right)^{1/2} + 2\lambda sin^{-1}\left(\frac{1}{2\sqrt{Q}}\right), \tag{16.53}$$

and for $Q \gg 1$ and $x = x_0$ this reduces to

$$W \simeq \frac{2\lambda}{\sqrt{Q}} \simeq \gamma t, \tag{16.54}$$

and thus

$$A(x,t) \sim \frac{e^{\gamma t}}{\sqrt{t}}, \tag{16.55}$$

showing exponential growth for small time, after the initial decay of the delta function singularity.

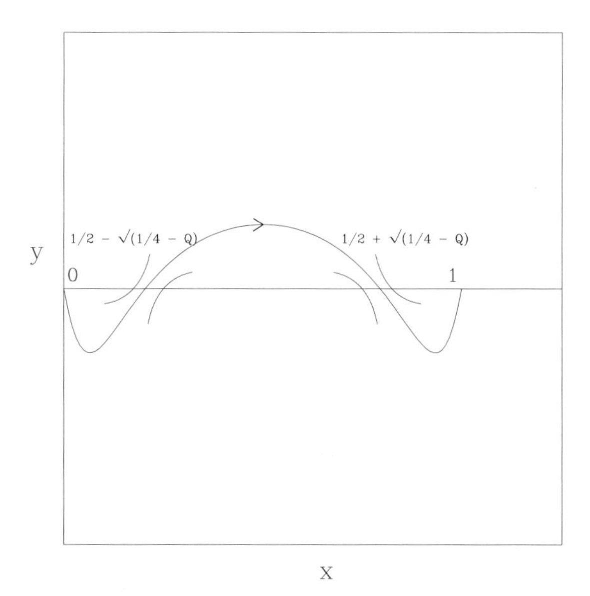

Fig. 16.4 Integration path for $Q < 1/4$.

For later time, when $Q < 1/4$, the saddle points are on the real axis with $0 < z < 1$ as shown in Fig. 16.4, where $z/(1-z)$ is real, and thus at the left saddle point we have

$$e^{\phi} = \left(\frac{1-\sqrt{1-4Q}}{1+\sqrt{1-4Q}}\right)^{i\lambda} e^{-i\lambda\sqrt{1/4-Q}/Q}, \tag{16.56}$$

and $\phi'' = i\lambda\sqrt{1-4Q}/Q^2$. The factor $1/z$ makes this left saddle point dominant, giving

$$A(x,t) \simeq e^{\pi\lambda} e^{i\lambda/2Q}, \tag{16.57}$$

showing saturation at a large maximum amplitude, this convective saturation being due to the dominance of wave lengths longer than the interaction length.

16.7 Driven Oscillator

Consider the simple equation

$$y'' + z^2 y = z^2. \tag{16.58}$$

By dominant balance, the asymptotic form of the general solution is

$$y \sim a\frac{e^{iz^2/2}}{\sqrt{z}} + b\frac{e^{-iz^2/2}}{\sqrt{z}} + 1 \tag{16.59}$$

and every solution satisfies $y \to 1$ for $z \to \pm\infty$.

If, on the other hand,

$$y'' - z^2 y = z^2, \tag{16.60}$$

dominant balance gives the asymptotic form of the solution

$$y \sim a\frac{e^{z^2/2}}{\sqrt{z}} + b\frac{e^{-z^2/2}}{\sqrt{z}} + 1. \tag{16.61}$$

If the boundary conditions are $y \to 1$ for $z \to \infty$ then we obtain $a = 0$, but no condition can be put on b.

In these simple cases we can make the substitution

$$y = f + 1 \tag{16.62}$$

giving

$$f'' \pm z^2 f = 0. \tag{16.63}$$

Thus it is clear that the exact inhomogeneous part, $y = 1$ does not participate in any Stokes phenomena, it is independent of any continuation through the complex z plane.

16.8 A Driven Overdamped Oscillator

Consider the differential equation

$$y'' - z^2 y = z \tag{16.64}$$

with boundary conditions given for $z \to \pm\infty$. By dominant balance the general solution must at large z have the form

$$y \sim a\frac{e^{-z^2/2}}{\sqrt{z}} + b\frac{e^{z^2/2}}{\sqrt{z}} - \frac{1}{z}. \tag{16.65}$$

Make the substitution $y = zf(w)$ with $w = z^2/2$. Boundary conditions are given at $w \to +\infty$ only and

$$2wf'' + 3f' - 2wf = 1. \tag{16.66}$$

Look for a Fourier–Laplace solution

$$f = \int_a^b e^{wt} g(t)dt. \tag{16.67}$$

Integrating by parts we find a solution provided

$$g = A(t^2 - 1)^{-1/4}, \qquad 2A(t^2 - 1)^{3/4} e^{wt}|_a^b = 1. \tag{16.68}$$

For the homogeneous solutions the integration limits are restricted to $a, b = \pm 1$ or $b = -\infty$.. All asymptotic limits for large w are given by end point contributions, there are no saddle points. Choose

$$f_{h1}(w) = \frac{1}{\Gamma(3/4)\sqrt{2}} \int_{-\infty}^{-1} e^{wt}(t^2 - 1)^{-1/4} dt. \tag{16.69}$$

Let $t = -1 - s$,

$$f_{h1}(w) = \frac{e^{-w}}{\Gamma(3/4)\sqrt{2}} \int_0^{\infty} e^{-sw}(2s + s^2)^{-1/4} ds \tag{16.70}$$

giving the asymptotic form

$$f_{h1}(w) \simeq \frac{e^{-w}}{(2w)^{3/4}}, \qquad y_{h1}(z) \simeq \frac{e^{-z^2/2}}{\sqrt{z}}. \tag{16.71}$$

The second solution is given by

$$f_{h2}(w) = \frac{1}{\Gamma(3/4)\sqrt{2}} \int_0^1 e^{wt}(1-t^2)^{-1/4} dt. \tag{16.72}$$

Let $t = 1 - s$,

$$f_{h2}(w) = \frac{e^w}{\Gamma(3/4)\sqrt{2}} \int_0^1 e^{-sw}(2s+s^2)^{-1/4} ds \tag{16.73}$$

giving the asymptotic form

$$f_{h2}(w) \simeq \frac{e^{w}}{(2w)^{3/4}}, \qquad y_{h2}(z) \simeq \frac{e^{z^2/2}}{\sqrt{z}}. \tag{16.74}$$

For the inhomogeneous solution we must take $a = 0$, giving $2A(-1)^{3/4} = -1$. Solutions exist for $b = \pm 1$ or $b = -\infty$. But $b = -\infty$ reproduces f_{h1} from the $(-\infty, 0)$ part, and an integral in the domain $(0, 1)$ reproduces f_{h2}, so we must take for the inhomogeneous solution

$$f_i(w) = \frac{1}{2} \int_0^{-1} e^{wt}(1-t^2)^{-1/4} dt. \tag{16.75}$$

The asymptotic limit comes from the end point $t = 0$ and gives $y_i(z) \simeq -1/z$.

Now we can continue these solutions through the complex z plane. The inhomogeneous solution is immediately seen to be asymptotically equal to $-1/z$ for $z \to \pm\infty$; it is unchanged by the continuation because the integration contour converges for all finite z and need not be deformed at all. In fact the integral defining the inhomogeneous solution is exactly the same for $z = \pm r$, not only asymptotically for large z.

Now continue the subdominant solution in the upper half plane from $z = +\infty$ to $z = -\infty$. First consider the Stokes analysis. Begin in domain 1 of Fig. 16.5 with a subdominant solution Eq. 16.69. The continuation is

$$\begin{array}{ll} 1) & (0,z)_s \\ 2) & (0,z)_d \\ 3) & (0,z)_d + i\sqrt{2}(z,0)_s \\ 4) & (0,z)_s + i\sqrt{2}(z,0)_d \end{array} \tag{16.76}$$

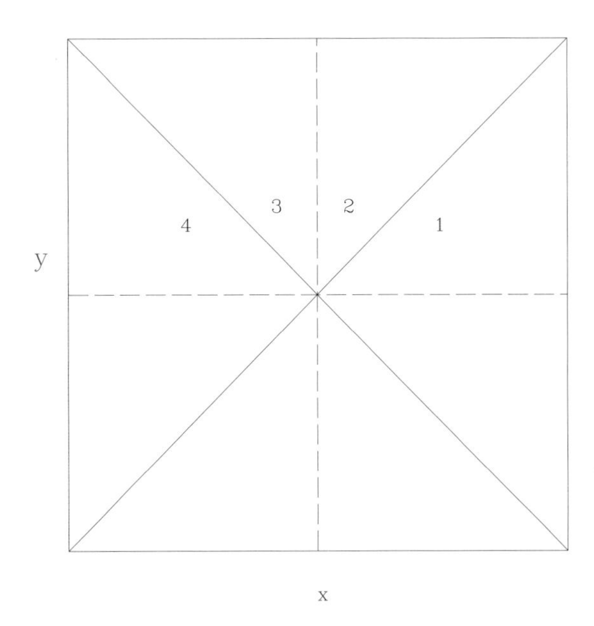

Fig. 16.5 Stokes diagram for Eq. 16.64.

Analytically continuing the integral Eq. 16.69 from $z = +\infty$ to $z = -\infty$ keeping $wt < 0$ the integration contour in t wraps clockwise around the cuts originating at -1 and 1 once, as shown in Fig. 16.6, giving as a contribution from near $t = 1$

$$(e^{-i\pi/4} - e^{-i3\pi/4})f_{h2} = \sqrt{2}f_{h2} \tag{16.77}$$

reproducing the Stokes analysis. (Note that f_{h1} and f_{h2} are defined with a sign difference inside the fourth root.)

If boundary conditions are taken to be $f \sim -1/z$ for $z \to +\infty$ the solution must have the form

$$y = y_i(z) + cy_{h1}(z) \tag{16.78}$$

with c an arbitarary constant. Continuing to $z \to -\infty$ we then find

$$y = y_i(z) + cy_{h1}(z) + ci\sqrt{2}y_{h2}(z), \tag{16.79}$$

and thus only $c = 0$ satisfies the boundary conditions.

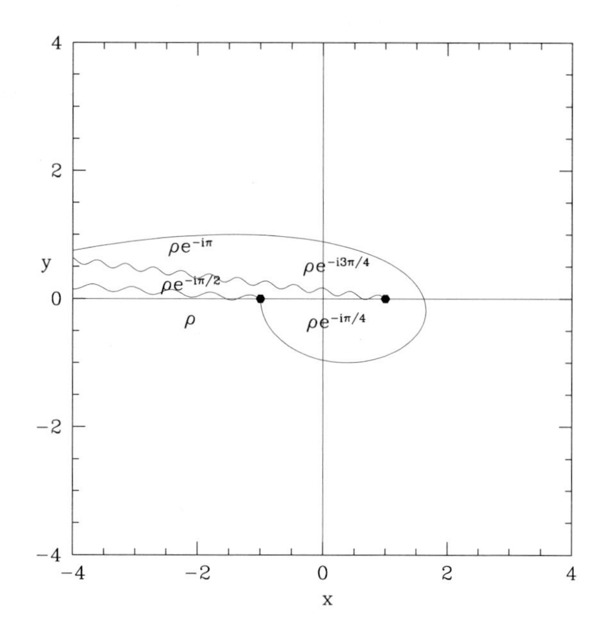

Fig. 16.6 Analytic continuation of Eq. 16.69, showing the integration contour for negative real z, and the phase of $(t^2-1)^{-1/4}$.

16.9 Stokes Modification of Boundary Conditions

It can happen that boundary conditions obtained from dominant balance including the inhomogeneous term are impossible to impose at both plus and minus infinity, because of Stokes phenomena. Consider the equation

$$a\psi''(z) + \psi(z) = \frac{1}{1+z^2} \tag{16.80}$$

with boundary condition $\psi(z) \to 1/(1+z^2)$ for $z \to +\infty$. Perform a Fourier transformation

$$\int_{-\infty}^{\infty} dze^{-kz}[a\psi''(z) + \psi(z)] = \int_{-\infty}^{\infty} dze^{-kz}\frac{1}{1+z^2} \tag{16.81}$$

and write

$$\psi(z) = \int_{-\infty}^{\infty} dze^{ikz}\phi(k). \tag{16.82}$$

Then using

$$\int_{-\infty}^{\infty} e^{-ikz}\frac{dz}{1+z^2} = \pi e^{-|k|}, \qquad \int_{-\infty}^{\infty} e^{ikz}e^{-|k|}dk = \frac{2}{1+z^2} \quad (16.83)$$

we find

$$\phi(k) = \frac{e^{-|k|}}{2(1-ak^2)} \quad (16.84)$$

and an exact solution to the inhomogeneous differential equation

$$\psi(z) = \frac{1}{2}\int_{-\infty}^{\infty} e^{ikz}\frac{e^{-|k|}}{1-ak^2} = Re\int_0^{\infty} e^{ikz}\frac{e^{-k}}{1-ak^2}. \quad (16.85)$$

Now look at $z \to +\infty$. Writing the integrand as $exp(ikz-k-ln(1-ak^2))$ we find a saddle point for large $|z|$ at

$$k = \frac{1}{\sqrt{a}} - \frac{i}{z} \quad (16.86)$$

giving a contribution of

$$\psi_s(z) = \frac{cos(z/\sqrt{a})}{\sqrt{2a}}. \quad (16.87)$$

In addition there is an end point contribution from $k \simeq 0$, obtained to lowest order by dropping the ak^2 term in the denominator and using Eq. 16.83, giving

$$\psi_e(z) = \frac{1}{1+z^2} \quad (16.88)$$

so for large positive z the integral solution is asymptotic to

$$\psi(z) \simeq \frac{1}{1+z^2} + \frac{cos(z/\sqrt{a})}{\sqrt{2a}}. \quad (16.89)$$

We recognize the second term as a solution to the homogeneous equation, and thus the solution

$$\psi(z) = Re\int_0^{\infty} e^{ikz}\frac{e^{-k}}{1-ak^2} - \frac{cos(z/\sqrt{a})}{\sqrt{2a}} \quad (16.90)$$

is asymptotic to $1/(1+z^2)$ for large positive z, and thus satisfies the requisite boundary condition.

Now continue to negative z. The integration path must be moved to the new saddle point above the real axis, as shown in Fig. 16.7, producing an

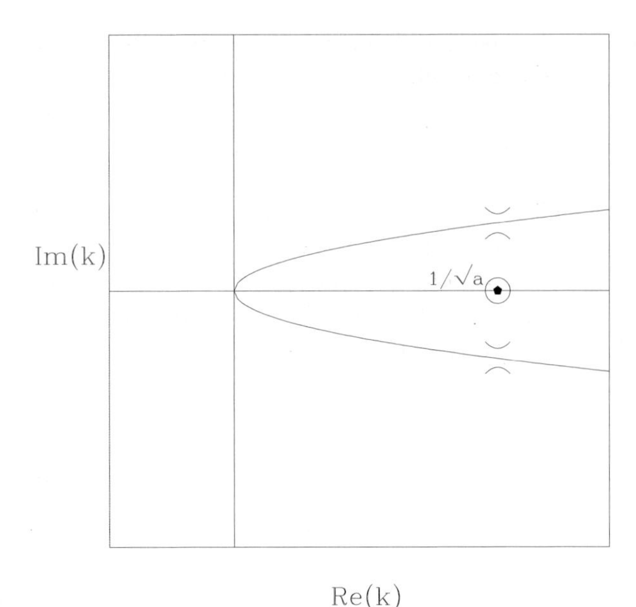

Fig. 16.7 Integration paths for $z > 0$ (lower) and $z < 0$ (upper) showing saddle points.

additional circular path around the pole at $k = 1/\sqrt{a}$. Thus for negative z we have

$$\psi(z) = Re \int_0^\infty \frac{e^{ikz-k}}{1-ak^2} - \frac{cos(z/\sqrt{a})}{\sqrt{2a}} + \frac{\pi e^{-1/\sqrt{a}}}{\sqrt{a}} sin(z/\sqrt{a}), \quad (16.91)$$

with the integration path now passing through the upper saddle point. Evaluating again the end point and saddle point contributions we have for large negative z

$$\psi(z) \simeq \frac{1}{1+z^2} + \frac{\pi e^{-1/\sqrt{a}}}{\sqrt{a}} sin(z/\sqrt{a}) \quad (16.92)$$

so it is impossible to require $\psi(z) \to 0$ for $z \to \pm\infty$.

16.10 Inhomogeneous Weber equation

Consider the equation

$$y'' + (\alpha - 4z^2)y = 1. \quad (16.93)$$

Find values of α so that $y \to 0$ for $z \to \pm\infty$. From dominant balance we find that there are two possible behaviors at large z,

$$y \sim -\frac{1}{4z^2}, \qquad y \sim \frac{e^{z^2}}{z^{1/2+\alpha/4}}, \tag{16.94}$$

where the second form is one of the solutions to the homogeneous equation. The second homogeneous solution, behaving as $y \sim e^{-z^2}/z^{1/2-\alpha/4}$, is not a possible asymptotic form of the inhomogeneous equation, although it can of course be present along with one of the other solutions. There is a Taylor expansion around $z = 0$. Further note that if y is real, it must be an even function of z.

An integral solution valid for arbitrary α is obtained using the variable $x = z^2$, with

$$y = \int_a^b e^{xt} f(t) dt \tag{16.95}$$

giving

$$f(t) = A\frac{(1-t)^{(\alpha-6)/8}}{(1+t)^{(\alpha+6)/8}}, \qquad 4(t^2-1)e^{xt} f(t)|_a^b = 1, \tag{16.96}$$

with

$$y(z) = \frac{1}{4}\int_0^{-\infty} e^{z^2 t} \frac{(1-t)^{(\alpha-6)/8}}{(1+t)^{(\alpha+6)/8}} dt. \tag{16.97}$$

This integral is clearly even in z. The asymptotic form for large z is given by the end point contribution near $t = 0$, easily seen to be $y \sim -1/(4z^2)$ so this solution satisfies the boundary conditions at ∞. However it is readily seen that while this integral gives solutions for $z > 0$ and for $z < 0$, they are not smoothly connected at $z = 0$. Consider

$$y'(z) = \frac{z}{2}\int_0^{-\infty} e^{z^2 t} t \frac{(1-t)^{(\alpha-6)/8}}{(1+t)^{(\alpha+6)/8}} dt. \tag{16.98}$$

Evaluating $y'(0)$ requires a limiting procedure, since the integral diverges upon setting $z = 0$. Introducing the variable $u = -z^2 t$ and using the fact that the dominant contribution to the integral for small z comes from large t we find $y'(0) = \sqrt{\pi} e^{i\pi(\alpha+6)/8}/2$, where we have taken the integration contour to pass below the cut originating at $t = -1$. Thus we conclude that Eq. 16.97 has a cusp at $z = 0$ and thus does not represent the analytic continuation of the same solution throughout the z plane.

Take the cut originating at $t = 1$ to the right, so that integral Eq. 16.97 from 0 to -1 is real. The contribution from $t < -1$ gives a complex subdominant contribution coming from the vicinity of $t = -1$, giving for large positive z

$$y_{sub}(z) \sim \frac{e^{-z^2}}{4z^{1/2-\alpha/4}} 2^{(\alpha-6)/8} e^{i\pi(\alpha+6)/8} \Gamma(1/4 - \alpha/8). \qquad (16.99)$$

Demanding that the real part of $y(z)$ be even gives $y'(0)$ pure imaginary, or $\alpha = 8n - 2$ with n integer. The subdominant term is then proportional to $\Gamma(1/2 - n)$ and is pure imaginary. Note that the subdominant term can be eliminated entirely by taking $b = -1$, provided that $\alpha > 2$. But this then gives $\alpha = 8n + 6$ for $n = 0, 1, 2....$

Alternatively, consider Eq. 16.97 to give the solution for positive z and carry out an analytic continuation to negative z. As z rotates clockwise through the lower half plane, t must rotate counterclockwise through 2π, to keep $z^2 t$ negative throughout the continuation. The final contour for negative z is shown in Fig. 16.8, giving for $z < 0$ the representation

$$y(z) = \frac{1}{4} \int_0^{-\infty} e^{z^2 t} \frac{(1-t)^{(\alpha-6)/8}}{(1+t)^{(\alpha+6)/8}} dt$$
$$+ \frac{(1 - e^{2\pi i(\alpha-6)/8})}{4} \int_0^1 e^{z^2 t} \frac{(1-t)^{(\alpha-6)/8}}{(1+t)^{(\alpha+6)/8}} dt \qquad (16.100)$$

where in the first integral the contour passes above the cuts originating at $t = -1$ and $t = 1$.

For $z \to -\infty$ the dominant contribution to the first integral comes from $t \sim 0$. For $z > 0$ the cut originating at $t = 1$ can be taken out to $t = +\infty$ so $1 - t$ is real near $t = 0$. But the continuation to $z < 0$ wraps the contour around the cut as shown in Fig. 16.8 so $1 - t \sim e^{2\pi i}$ for $z < 0$. This integral gives a contribution of $-1/(4z^2) e^{2\pi i(\alpha-6)/8}$. The second integral is dominated by the end point $t = 1$, giving

$$y(z) \sim -\frac{e^{2\pi i(\alpha-6)/8}}{4z^2} + \frac{ie^{z^2} e^{\pi i(\alpha-2)/8}}{z^{1/2+\alpha/4} 2^{(\alpha+8)/2}} \Gamma(\frac{\alpha+2}{8}) sin(\frac{\pi(\alpha-6)}{8}), \qquad (16.101)$$

and the second term is recognized as the dominant homogeneous solution, resulting from Stokes phenomenon.

For $\alpha = 8n + 6$ with $n = 0, 1, 2...$ the exponentially large term is zero and the boundary conditions can be met. Furthermore for these values of α the asymptotic form for large negative z is $-1/(4z^2)$ and $y'(0)$ is pure imaginary, so that $Re y(z)$ is an even solution of the inhomogeneous

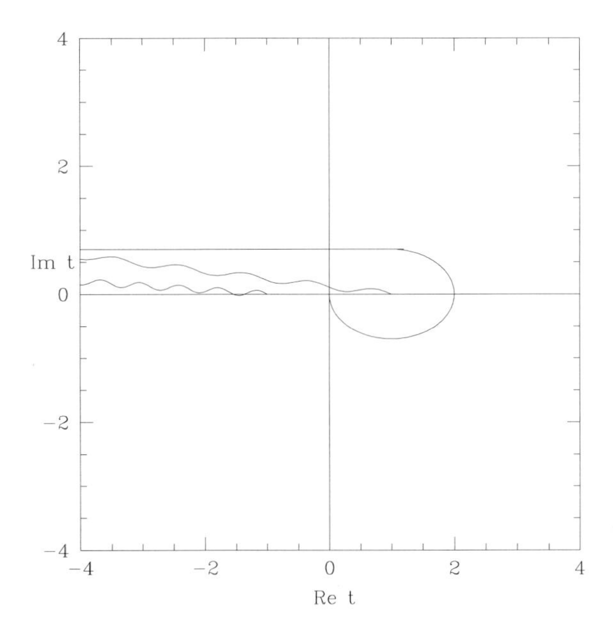

Fig. 16.8 Integration path for $z < 0$.

equation. For these values of α the second limit of the contour can be taken as $b = -1$, giving the real solution

$$y(z) = \frac{1}{4}\int_0^{-1} e^{z^2 t}\frac{(1-t)^n}{(1+t)^{n+3/2}}dt. \tag{16.102}$$

In retrospect, WKB analysis gives some insight into this problem. The real part of $y(z)$ must satisfy the inhomogeneous equation, is even in z, and has asymptotic behavior of $-1/(4z^2)$. For general α the solution is complex. But the imaginary part of $y(z)$ satisfies the homogeneous equation, and thus for $y(z)$ to tend to zero for $z \to \pm\infty$ the imaginary part must be subdominant in both directions. But we also know that the imaginary part of $y(z)$ is an odd function. For $\alpha > 0$ the Stokes diagram for this equation is that of the bound state problem, Fig. 5.4. The WKB condition gives

$$\int \sqrt{\alpha - 4z^2}dz = (m + 1/2)\pi \tag{16.103}$$

and evaluating the integral and taking $m = 2n + 1$ odd we find again $\alpha = 8n + 6$.

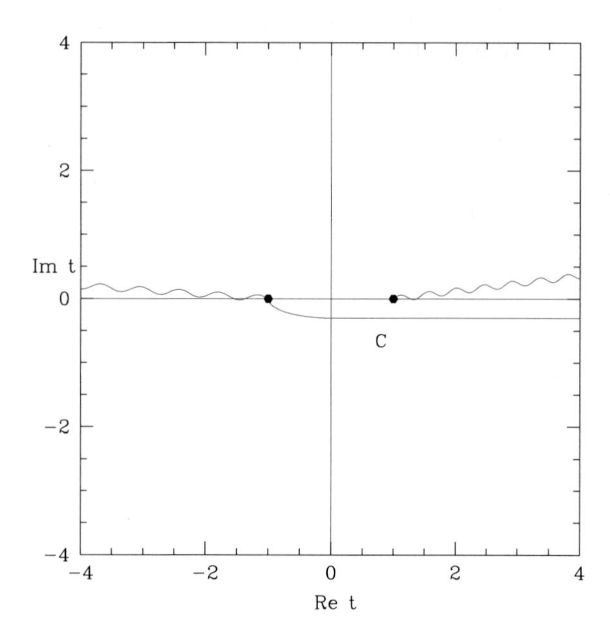

Fig. 16.9 Integration path for the continuation of the subdominant term, for $z \sim e^{-i\pi/2}$.

The analytic continuation through the complex z plane produces Stokes phenomenon regarding the homogeneous solution, but not with the usual Stokes constant. Consider the analytic continuation of the subdominant solution, with the integral beginning at $t = -1$. The Stokes constant is easily evaluated by looking at the contribution to the integral when z has rotated through $-\pi/2$, shown in Fig. 16.9. Break the integral into three parts. The integral from -1 to zero gives the dominant contribution that is the analytic continuation of the original subdominant term Eq. 16.99, with asymptotic form

$$y_{dom}(z) \sim -\frac{e^{r^2}}{4r^{1/2-\alpha/4}} 2^{(\alpha-6)/8} \Gamma(1/4 - \alpha/8) \tag{16.104}$$

with $z = re^{-i\pi/2}$.

The integral from zero to one then gives a real part asymptotically giving the inhomogeneous solution. This part of the solution does not participate in Stokes phenomenon, since $1/z^2$ has the same behavior for large z of any phase.

The complex contribution arising from the integral from one to infinity gives the new subdominant term asymptotic to

$$y_{sub}(z) \sim \frac{ie^{-r^2}}{4r^{1/2+\alpha/4}} 2^{-(\alpha+6)/8} e^{i\pi(\alpha-6)/8} \Gamma(1/4 + \alpha/8). \qquad (16.105)$$

Comparing with the analysis in section 12.6, we see that if the solution for positive z is $(z, b)_s$ then the solution for $z \sim i$ is $(z, b)_d + S(b, z)_s$, and thus we find that

$$S = \frac{i2^{-(\alpha+6)/4} e^{i\pi(\alpha-6)/8} \Gamma(1/4 + \alpha/8)}{\Gamma(1/4 - \alpha/8)}, \qquad (16.106)$$

which for $\alpha = 8n + 6$ becomes upon using Eq. 8.55

$$S = \frac{i2^{2n+3}(-1)^n \Gamma(n+1)}{\Gamma(-n-1/2)} = i8\sqrt{\pi}\Gamma(2n+1) \qquad (16.107)$$

to be compared with Eq. 12.51. Thus the Stokes constant is modified by the presence of the inhomogeneous term. As noted previously, the Bohr–Sommerfeld condition is independent of the value of the Stokes constant.

16.11 Problems

1. Solve

$$y'' + xy' - 3y = sin(x).$$

2. a) Find two integral solutions to

$$xy'' + xy = 1$$

and the behavior of each as $x \to +\infty$.
b) Does there exist a solution with $y \to 0$ for $x \to \pm\infty$?

3. Find the small z expansion of the Struve function using Eq. 16.19 to show that this representation is equivalant to Eq. 16.15.

4. Find the large z expansion of the Struve function using Eq. 16.19 to show that this representation is equivalant to Eq. 16.21.

5. Use the expansion for small z Eq. 16.15 to find a Mellin integral representation for the Struve function.

6. Find an integral solution to

$$y'' - 4zy' + z^2y = z.$$

7. Show that

$$H_{1/2}(z) = \frac{\sqrt{2}}{\sqrt{\pi z}}(1 - cos(z)).$$

8. Show that

$$\frac{d}{dz}[z^\nu H_\nu(z)] = z^\nu H_{\nu-1}(z).$$

9. Show that

$$H_{-(n+1/2)}(z) = (-1)^n J_{n+1/2}(z). \qquad [n = 0, 1, 2 \ldots]$$

10. Use the Green's function to find two independent solutions to the inhomogeneous Airy equation

$$w''(z) - zw(z) = \frac{1}{\pi}.$$

11. Use a Fourier–Laplace integral representation to find two independent solutions to the inhomogeneous Airy equation

$$w''(z) - zw(z) = \frac{1}{\pi}.$$

Chapter 17

The Riemann Zeta Function

Georg Friedrich Bernhard Riemann was born in 1826 in Breselenz, Hanover. He was shy, and sufffered from numerous nervous breakdowns. In high school he studied the Bible extensively, and tried to mathematically prove the correctness of the book of Genesis. He enrolled at the University of Göttingen to study theology, but soon transferred to mathematics, and then moved to Berlin University in 1847, studying with Dirichlet. In 1849 he returned to Göttingen where he completed his PhD with Gauss, but was also for 18 months the assistant of Weber, from whom he learned theoretical physics. He was also strongly influenced by Listing, from whom he learned topology. His thesis was on the theory of complex variables, and introduced topology to analytic function theory through the construction of Riemann surfaces. On Gauss's recommendation Riemann was appointed to a position in Göttingen. He made important contributions to the fields of calculus, geometry, and number theory. He formulated the Cauchy–Riemann conditions (see Chapter 3) for analyticity of a complex function, and extended geometry to include non-Euclidean metrics, introducing the curvature tensor and showing finally that the parallel postulate of Euclidean geometry could not be proven from the other axioms. He introduced a collection of numbers at each point in a space to describe how it was bent or curved. He found that in four dimensions ten numbers were necessary to describe the properties of a manifold. The Riemann hypothesis concerning the zeta function, to be discussed in this chapter, is still unproven.

17.1 Introduction

The Riemann zeta function is defined for $Res > 1$ as

$$\zeta(s) = \sum_{n=1}^{\infty} \frac{1}{n^s}, \tag{17.1}$$

and this representation converges absolutely and is analytic in this half plane. This function is also transcendental, it satisfies no differential equation, and it is of great importance in number theory.

Consider the product of this function with the Gamma function, Eq. 8.1

$$\Gamma(s)\zeta(s) = \sum_{n=1}^{\infty} \int_0^{\infty} dte^{-t}t^{s-1}\frac{1}{n^s}. \tag{17.2}$$

Change the order of integration and summation, and let $t = n\rho$, giving

$$\Gamma(s)\zeta(s) = \sum_{n=1}^{\infty} \int_0^{\infty} d\rho e^{-n\rho}\rho^{s-1} = \int_0^{\infty} d\rho e^{-\rho}\frac{\rho^{s-1}}{1-e^{-\rho}}. \tag{17.3}$$

Now consider the integral

$$I = \int_C dw \frac{e^{-w}(-w)^{s-1}}{1-e^{-w}}, \tag{17.4}$$

with the integration contour as shown in Fig. 17.1. For $Re(s) > 0$ the contour can be moved to the real axis, giving

$$I = \int_0^{\infty} d\rho[-e^{-i\pi(s-1)} + e^{i\pi(s-1)}]\frac{e^{-\rho}(\rho)^{s-1}}{1-e^{-\rho}} = \tag{17.5}$$

$$2isin(\pi(s-1))\Gamma(s)\zeta(s). \tag{17.6}$$

But using Eq. 8.51 we have

$$\zeta(s) = -\frac{\Gamma(1-s)}{2\pi i}\int_C dw \frac{e^{-w}(-w)^{s-1}}{1-e^{-w}}, \tag{17.7}$$

and the integral is analytic for all s. Thus the only possible singularities of $\zeta(s)$ are the poles given by $\Gamma(1-s)$ at $s = 1, 2, 3\ldots$. But we already know that $\zeta(s)$ is analytic for $Res > 1$, so we conclude that $\zeta(s)$ is singular only at $s = 1$, the other poles must be cancelled by zeros of the integral. To

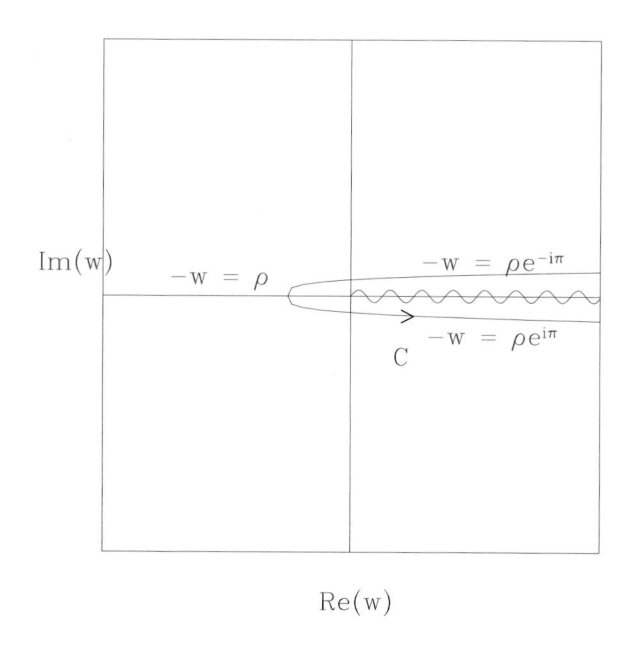

Fig. 17.1 Integration contour for $\zeta(s)$.

find the residue at this point set $s = 1$ inside the integral in Eq. 17.7. The contour collapses to a single circular path around $w = 0$, giving

$$lim_{s\to 1}\frac{\zeta(s)}{\Gamma(1-s)} = -1, \tag{17.8}$$

and thus $\zeta(s)$ has a simple pole at $s = 1$ with residue 1, and no other singularities.

17.2 $\zeta(s)$ and $\zeta(1-s)$

In this section we prove a remarkable identity due to Riemann relating the values of the zeta function at s and $1-s$. Consider the integral

$$I = \frac{1}{2\pi i}\int_C dt \frac{e^{-t}(-t)^{s-1}}{1-e^{-t}}, \tag{17.9}$$

over the contour shown in Fig. 17.2, consisting of a square passing through $x = \pm(2n+1)\pi$, and $y = \pm(2n+1)\pi$, with an excursion along the real axis to avoid the cut. The integrand is analytic inside this contour except for

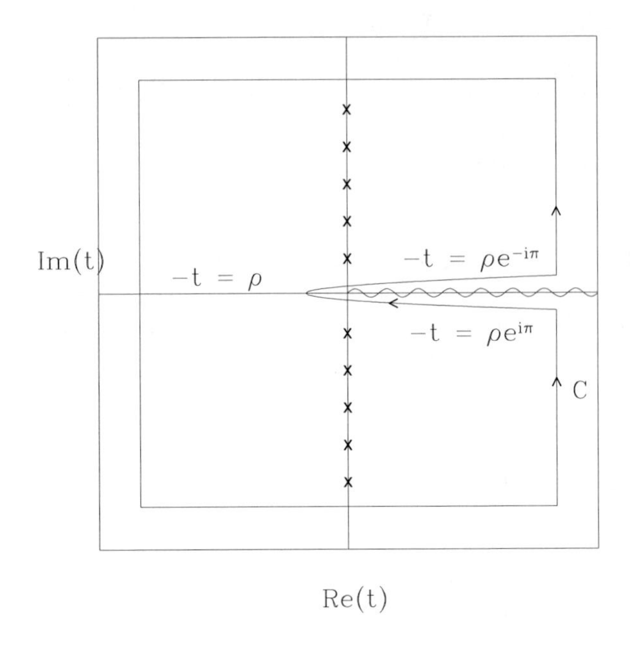

Fig. 17.2 Contour for integral representation of the zeta function.

the poles at $t = 2\pi ik$ with k a nonzero integer. Thus we immediately find that

$$I = \sum_{k=1}^{n}[(2\pi ik)^{s-1} + (-2\pi ik)^{s-1}] = \frac{2sin(\pi s/2)}{(2\pi)^{1-s}} \sum_{1}^{n} \frac{1}{k^{1-s}}. \quad (17.10)$$

Now split the integral into the contribution from along the real axis I_R, and that from the edges of the square I_S. Along the square at the right side $|e^{-t}| \ll 1$ and thus $|e^{-t}/(1-e^{-t})| \ll 1$. On the left side $|e^{-t}| \gg 1$ and thus $|e^{-t}/(1-e^{-t})| \simeq 1$. Along the top and bottom $y = \pm(2n+1)\pi$ and so $e^{-t} = -e^{-x}$ and $e^{-t}/(1-e^{-t}) < 1$. Thus the contribution from the sides of the square is bounded,

$$I_S < \frac{1}{2\pi}\int_S dt|(-t)^{s-1}| < \frac{4}{2\pi}((2n+1)\pi)^s, \quad (17.11)$$

which tends to zero for large n provided $Res < 0$. Now consider the contribution from along the real axis. Substituting $-t = \rho e^{\pm i\pi}$ as shown in

Fig. 17.2 and as in Eq. 17.6 we find

$$I_R = -\frac{sin(\pi(s-1))}{\pi}\int_0^\infty d\rho \frac{e^{-\rho}(\rho)^{s-1}}{1-e^{-\rho}} = \frac{sin(\pi s)}{\pi}\Gamma(s)\zeta(s). \quad (17.12)$$

Combining these expressions and taking the limit $n \to \infty$ we find Riemann's relation

$$2^{1-s}\Gamma(s)\zeta(s)cos(s\pi/2) = \pi^s\zeta(1-s). \quad (17.13)$$

This expression is analytic in s and thus can be continued to all s except where there are poles. Using Eqs. 8.55 and 8.51 this equation can also be put in the form

$$\pi^{-s/2}\Gamma\left(\frac{s}{2}\right)\zeta(s) = \pi^{-(1-s)/2}\Gamma\left(\frac{1-s}{2}\right)\zeta(1-s). \quad (17.14)$$

This is known as the functional equation for zeta, and it has a very important consequence. Let $s = 1/2 + iy$, so that $s^* = 1 - s$. Then this equation takes the form

$$\pi^{-s/2}\Gamma\left(\frac{s}{2}\right)\zeta(s) = \pi^{-s^*/2}\Gamma\left(\frac{s^*}{2}\right)\zeta(s^*), \quad (17.15)$$

showing that $\pi^{-s/2}\Gamma(s/2)\zeta(s)$ is real, and this simplifies the problem of searching for zeros along this line.

17.3 The Euler Product for $\zeta(s)$

Consider the definition

$$\zeta(s) = \sum_{n=1}^\infty \frac{1}{n^s} = 1 + \frac{1}{2^s} + \frac{1}{3^s} + \frac{1}{4^s} + \ldots, \quad (17.16)$$

and note that

$$\frac{1}{2^s}\zeta(s) = \frac{1}{2^s} + \frac{1}{4^s} + \frac{1}{6^s} + \ldots. \quad (17.17)$$

Subtract these two equations giving

$$\zeta(s)\left[1 - \frac{1}{2^s}\right] = 1 + \frac{1}{3^s} + \frac{1}{5^s} + \ldots, \quad (17.18)$$

consisting only of terms prime to 2. Similarly, multiplying by successive prime factors and subtracting find

$$\zeta(s)\left[1-\frac{1}{2^s}\right]\left[1-\frac{1}{3^s}\right]\dots\left[1-\frac{1}{p^s}\right]=1+\sum{}'\frac{1}{n^s}, \tag{17.19}$$

where $\sum'$ means only values of n prime to $2, 3, \dots p$ appear in the sum, i.e. the sum begins with the first prime after p.

But

$$\left|\sum{}'\frac{1}{n^s}\right|<\sum{}'\frac{1}{n^{1+\delta}}<\sum_{n=p+1}^{\infty}\frac{1}{n^{1+\delta}}\to 0, \tag{17.20}$$

for $p\to\infty$ for any $\delta>0$. Thus for $Res>1$

$$\zeta(s)\prod_p\left(1-\frac{1}{p^s}\right)\to 1, \tag{17.21}$$

with $\prod_p$ denoting the product over all primes p. This product converges for $Res>1$ because it consists of some of the factors of $\prod_2^{\infty}(1-1/n^s)$, which converges. Thus for $Res>1$

$$\frac{1}{\zeta(s)}=\prod_p\left(1-\frac{1}{p^s}\right), \tag{17.22}$$

and this expression intimately relates the Riemann zeta function to the distribution of prime numbers. This expression appears for the first time in Euler's book *Introductio in Analysin Infinitorum* (Introduction to the Analysis of the Infinite), published in 1748.

17.4 Distribution of Prime Numbers

Let $\pi(r)$ be the number of prime numbers less than r. We show that the asymptotic distribution of prime numbers is related to the singularity of the zeta function. Consider the representation Eq. 17.22 and take $s=1+\epsilon$ for $\epsilon\ll 1$. Then

$$\epsilon\simeq\prod_p\left(1-\frac{1}{p^{1+\epsilon}}\right). \tag{17.23}$$

Taking the logarithm of this equation we have

$$ln(\epsilon) \simeq \sum_p ln\left(1 - \frac{1}{p^{1+\epsilon}}\right) = \int_2^\infty d\pi(r) ln\left(1 - \frac{1}{r^{1+\epsilon}}\right). \tag{17.24}$$

For small ϵ this expression tends to infinity, and the right-hand side must do so through the divergence of the integral at $r \to \infty$. Use the asymptotic expression for the integrand to find this divergent behavior,

$$ln(\epsilon) \simeq -\int d\pi(r)\frac{1}{r^{1+\epsilon}} = -\int \frac{d\pi(r)}{r} e^{-\epsilon lnr}. \tag{17.25}$$

Now use l'Hôpital[1] to differentiate, giving

$$\frac{1}{\epsilon} \simeq \int \frac{d\pi(r) ln(r)}{r} e^{-\epsilon ln(r)}. \tag{17.26}$$

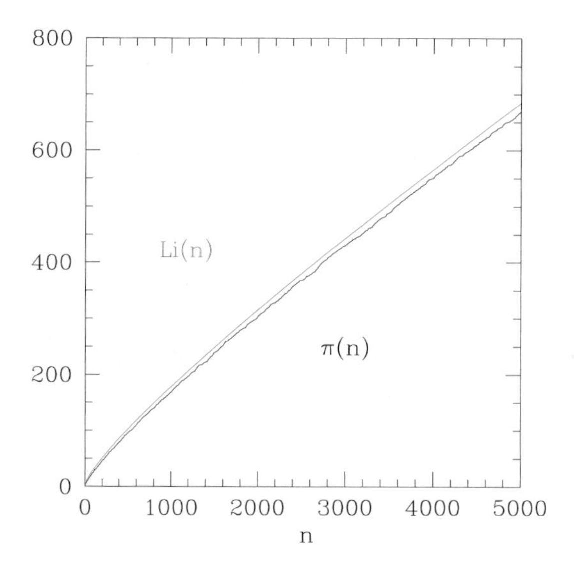

Fig. 17.3 Distribution of prime numbers.

Now let $w = \epsilon ln(r)$ giving

$$\frac{1}{\epsilon} \simeq \frac{1}{\epsilon} \int^\infty \left(\frac{d\pi(r)}{dr} ln(r)\right) e^{-w} dw. \tag{17.27}$$

[1]Actually this method of finding an indeterminate limit was stolen from Johan I Bernoulli.

But $r = e^{w/\epsilon}$ and $lnr \to 1/\epsilon$ for $\epsilon \to 0$ so we must have $d\pi/dr \simeq 1/ln(r)$ for small ϵ. Thus for large r

$$\pi(r) \simeq \int^{r} \frac{dr}{ln(r)} = Li(r), \tag{17.28}$$

which gives the asymptotic distribution of the prime numbers, first conjectured in 1791 by Gauss and finally proven in 1896 by Hadamard and Poussin. Primes up to $n = 5000$ are shown in Fig. 17.3. A plot of the difference between $Li(n)$ and $\pi(x)$ is shown up to $n = 3 \times 10^6$ in Fig. 17.4. Note from the plots that $Li(x) > \pi(x)$ up to the values shown. In fact this is true for x up to one trillion, an observation which led to the overestimated prime conjecture, namely that this inequality is true for all x. But in 1914 Littlewood showed that $Li(x) - \pi(x)$ assumes positive and negative values infinitely often, and in 1986 Te Riele showed that there were 10^{180} successive integers x with $6.62 \times 10^{370} < x < 6.69 \times 10^{370}$ for which $\pi(x) > Li(x)$, showing that reliance on numerical calculations can be misleading, even if a conjecture is verified for a very large number of cases.

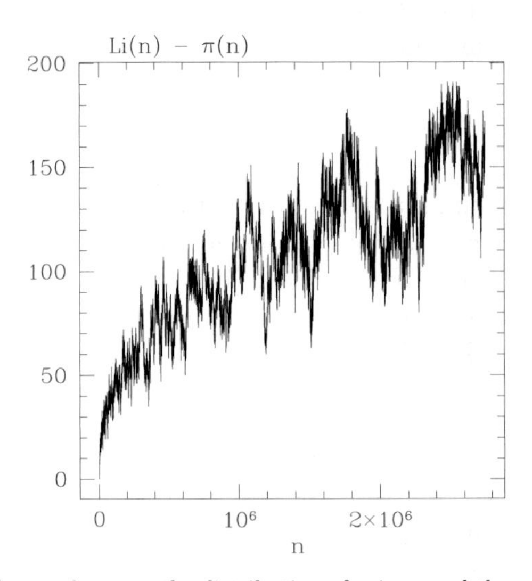

Fig. 17.4 Difference between the distribution of primes and the analytic expression.

Riemann did much more than simply find the asymptotic distribution. Take the log of Eq. 17.22, and expand the log, giving

$$log\zeta(s) = \sum p^{-s} + \frac{1}{2}\sum p^{-2s} + \frac{1}{3}\sum p^{-3s} + \ldots. \tag{17.29}$$

Then replace

$$p^{-s} \to s\int_p^\infty x^{-s-1}ds, \qquad p^{-2s} \to s\int_{p^2}^\infty x^{-s-1}ds, \ldots \tag{17.30}$$

giving

$$\frac{1}{s}log\zeta(s) = \int_1^\infty dx\Pi(x)x^{-s-1}, \tag{17.31}$$

with

$$\Pi(x) = \pi(x) + \frac{1}{2}\pi(x^{1/2}) + \frac{1}{3}\pi(x^{1/3}) + \ldots. \tag{17.32}$$

By Fourier inversion he expresses $\Pi(x)$ as a complex integral

$$\Pi(x) = \int_{a-i\infty}^{a+i\infty} ds\frac{log\zeta(s)}{s}x^s ds, \tag{17.33}$$

for $a > 1$, and computes it using the calculus of residues. The residues occur at the singularities of $log\zeta(s)$ at $s = 1$ and at the zeros of $\zeta(s)$. Finally inverting Eq. 17.32 through

$$\pi(x) = \sum_m \frac{(-1)^\mu}{m}\Pi(x^{1/m}), \tag{17.34}$$

where m consists of all natural numbers not divisible by any square other than 1, and in which μ denotes the number of prime factors of m, the distribution of the prime numbers is related to the zeros of $\zeta(s)$.

For this reason the zeros of the zeta function are of great interest in number theory. In 1859 Riemann made the hypothesis (Bombieri [2000]) that all complex zeros of $\zeta(s)$ lie on the line $s = 1/2 + iy$. The first few zeros are at $y = 14, 21, 25, 30.5, 33, \ldots$. This conjecture has been checked numerically up to the first 1,500,000,001 zeros above the real axis, and so far all of them lie on this line. But the hypothesis, which now has a prize of one million dollars, is still unproven. The validity of the Riemann hypothesis implies that the deviation of the prime numbers from the asymptotic limit $Li(x)$ is

$$\pi(x) = Li(x) + O(\sqrt{x}logx), \tag{17.35}$$

and a single zero straying from the line $s = 1/2 + iy$ would modify the distribution of primes in a significant manner.

A non-rigorous calculation found that the correlation function for two zeros of the zeta function was a simple expression (Montgomery [1973]), and then Freeman Dyson recognized that this function was the two point correlation function for the distribution of eigenvalues of an $N \times N$ random Hermitian matrix, given by Eq. 6.16. Extensive calculations by Andrew Odlyzko then showed agreement between the distribution of zeros of the zeta function and matrix eigenvalues, also for higher order correlation functions. See "Stalking the Riemann Hypothesis" by Dan Rockmore, Pantheon (2005). Thus it appears that the zeros of the zeta function are given by the eigenvalues of some particular Hermitian matrix.

17.5 Public Key Codes

Prime numbers provide a means of encoding messages whereby the means of encoding can be public knowledge, but knowing how to encode a message does not allow one to decode it. This is of great use because with previous codes the instructions allowed both processes, which meant that every time a code was changed and the new encoding procedure sent to the users there was a risk of the instructions being intercepted and the code being broken. The method was invented in 1977 by Ronald Rivest, Adi Shamir, and Leonard Adelman at MIT. Any integer can be uniquely factored into a product of prime factors. Two integers are said to be relatively prime if they share no common prime factors. To make the code, choose two large prime numbers p and q and publish their product $N = pq$ along with a number k which is relatively prime to the product $(p-1)(q-1)$. The combination (N, k) is the public key, which can be distributed without fear of allowing anyone to decode messages. To code a message it must first be broken into a sequence of numbers each being less than N in a standard way. The message then has the form

$$a, b, c, \dots \tag{17.36}$$

with $a, b, c\dots$ each a number less than N. Then the message is encripted by raising each number to the power k mod N. Thus

$$\begin{aligned} a \rightarrow A &= a^k \; mod \; N \\ b \rightarrow B &= b^k \; mod \; N \\ &\dots \end{aligned} \tag{17.37}$$

and the message $A, B\ldots$ is transmitted. To decode this message it is necessary to know both the numbers p and q. The decoding is done with the use of Euler's $\phi(N)$ function, defined to be the number of integers less than N which are relatively prime to N. For example,

$$\begin{aligned} \phi(11) &= 10 \\ \phi(12) &= 4 \\ \phi(p) &= p - 1 \quad \mathit{for} \quad p \quad \mathit{prime}. \end{aligned} \tag{17.38}$$

The Euler theorem states that if p and q are primes, z an integer less than pq, and k, s, t integers such that

$$ks = 1 + t(p-1)(q-1) \tag{17.39}$$

then

$$z^{ks} = z \; mod \; pq. \tag{17.40}$$

Furthermore, if p and q are primes with $N = pq$, then we have

$$\phi(N) = (p-1)(q-1), \tag{17.41}$$

which we obtain by direct counting.

Knowing both p and q it is possible to invert the encoding, i.e. to produce the map $A \to a$, $B \to b, \ldots$. To do this, note that if two integers m and n are relatively prime then by the Euclidean algorithm, which is essentially repeated long division, there exist unique integers s and t such that

$$m \cdot s - n \cdot t = 1. \tag{17.42}$$

Setting $m = k$ and $n = (p-1)(q-1)$ gives

$$k \cdot s = (p-1)(q-1) \cdot t + 1, \tag{17.43}$$

and thus given p and q it is possible to find s. But then note by the Euler theorem that $k \cdot s = 1 \; mod \; N$, i.e. the number s is the $mod \; N$ multiplicative inverse of k. Thus we have that

$$A^s = (a^k)^s = a^{ks} = a^{t\phi(N)}a = a \; mod \; N, \tag{17.44}$$

which is unique because a is less than N. Thus to decode the message each number is simply raised to the power s. Given only N and k it is a very time-consuming task to find s, because N must be factored into its prime factors. If the primes p and q are chosen to be very large, this process can take years of computing even for the fastest computers, and in the meantime a new public key will have been issued, making the calculation irrelevant.

17.6 Stirling Revisited

A derivation of the Stirling approximation using a Mellin transformation not only gives the full expansion to all orders, but also provides three constants regarding $\zeta(s)$, namely the next order term at the pole at $s = 1$ as well as $\zeta(0)$ and $\zeta'(0)$. These values previously were obtained in a much more arduous fashion using Hermite's formula.[2] Begin with the Euler product representation for the Gamma function, Eq. 8.48

$$\Gamma(x+1) = \prod_{1}^{\infty} \frac{(1+\frac{1}{n})^x}{\left(1+\frac{x}{n}\right)}, \tag{17.45}$$

and use the definition of the Euler–Mascheroni constant $\gamma = \sum_1^m (1/n) - ln(m)$ (with the limit of large m understood) and the fact that $\Pi_1^{m-1}(1 + 1/n) = m$ to rewrite this as

$$\Gamma(x+1) = e^{-\gamma x} \prod_{1}^{\infty} \frac{e^{x/n}}{\left(1+\frac{x}{n}\right)}. \tag{17.46}$$

Taking the log of each side we have

$$ln[\Gamma(x+1)] = -\gamma x + \sum_{n=1}^{\infty} \left[\frac{x}{n} - ln\left(1+\frac{x}{n}\right)\right]. \tag{17.47}$$

Now expand the logarithm in powers of x, giving

$$ln[\Gamma(x+1)] = -\gamma x + \sum_{n=1}^{\infty} \left[\frac{x}{n} + \sum_{s=1}^{\infty} \left(\frac{-x}{n}\right)^s \frac{1}{s}\right]. \tag{17.48}$$

Changing order of summation we have

$$ln[\Gamma(x+1)] = -\gamma x + \sum_{s=2}^{\infty} (-x)^s \frac{\zeta(s)}{s}. \tag{17.49}$$

Now represent this sum using a Mellin transformation. Write

$$ln[\Gamma(x+1)] = -\gamma x + \frac{1}{2\pi i} \int M(t) x^{-t} dt, \tag{17.50}$$

[2]See Whittaker and Watson, p. 269.

with the integration contour circling the entire left half plane. To reproduce the sum we need poles at negative integers with residue equal to $(-1)^t\zeta(-t)/t$, giving

$$M(t) = -\frac{\pi\zeta(-t)}{tsin(\pi t)}. \tag{17.51}$$

Now consider convergence of the integral. Writing $t = u + iv$, $sin(\pi t)$ gives for large v a factor $exp(-\pi|v|)$. Writing $x = re^{i\phi}$ and using $\zeta(-t) \sim 1$ we have $\zeta(-t)x^{-t}/(tsin(\pi t)) \sim exp(-2\pi|v|)/vr^t$, determining convergence.

But note that this contour gives terms for s from zero to ∞ rather than from 2 to ∞, so it is necessary to subtract the contributions from the poles at $t = 0, -1$. The Mellin representation becomes

$$ln[\Gamma(x+1)] = -\gamma x - \frac{1}{2\pi i}\int \frac{\pi\zeta(-t)x^{-t}dt}{tsin(\pi t)} - R_0 - R_{-1}, \tag{17.52}$$

with R_0, R_{-1} the residues at $t = 0, -1$. Expanding the integrand about $t = 0$ and using the fact that $\zeta(s)$ is analytic at zero we have

$$\frac{\pi\zeta(-t)x^{-t}}{tsin(\pi t)} \simeq \frac{1}{t^2}[\zeta(0) - t\zeta'(0) + \ldots][1 - tln(x) + \ldots], \tag{17.53}$$

giving

$$R_0 = \zeta'(0) + \zeta(0)ln(x). \tag{17.54}$$

To find the contribution from the singularity at $t = -1$, use $t = -1 + \epsilon$ and make use of the fact that ζ has a first order pole at $s = 1$ to write $\zeta(1 - \epsilon) = -1/\epsilon + K$, with K an unknown constant. We then have

$$\frac{\pi\zeta(-t)x^{-t}}{tsin(\pi t)} \simeq \frac{x(1 - \epsilon K)(1 - \epsilon ln(x))}{\epsilon^2(1 - \epsilon)} \tag{17.55}$$

giving

$$R_{-1} = x[1 - K - ln(x)]. \tag{17.56}$$

Comparing Eq. 17.52 with Eq. 8.15 we then find three important facts about the zeta function

$$K = \gamma, \quad \zeta(0) = -\frac{1}{2}, \quad \zeta'(0) = -\frac{ln(2\pi)}{2}. \tag{17.57}$$

The Mellin representation then becomes

$$ln[\Gamma(x+1)] = \frac{ln(2\pi x)}{2} + x[ln(x) - 1]$$
$$-\frac{\pi}{2\pi i}\int \frac{\zeta(-t)}{t sin(\pi t)} x^{-t} dt. \tag{17.58}$$

This form is inconvenient because $\zeta(-t)$ is not known for large t. We can replace the negative argument by a positive one by using Eq. 17.13 in the form

$$\zeta(-t) = -\frac{\Gamma(1+t)\zeta(1+t) sin(\pi t/2)}{\pi(2\pi)^t}. \tag{17.59}$$

Then move the contour to the right of the singularities, adding a new term to the expression for each singularity passed, giving the full Stirling series

$$ln[\Gamma(x+1)] = \frac{ln(2\pi x)}{2} + x(ln(x) - 1)$$
$$+\frac{1}{\pi}\sum_{m=odd}(-1)^{(m-1)/2}\frac{\Gamma(m)\zeta(m+1)}{(2\pi x)^m}, \tag{17.60}$$

and since $\zeta(m) \to 1$ for large m, this series diverges as $\Gamma(m)/x^m$.

Note that this series contains only odd powers of $1/x$. But using for small a the expansion $ln(1+a) \sim a - a^2/2 + ...$ we have, upon subsitituting Eq. 8.15

$$ln\left(1 + \frac{1}{12x} + \frac{1}{288x^2} + ...\right) \sim \frac{1}{12x} + O\left(\frac{1}{x^3}\right) \tag{17.61}$$

so terms of $1/x^2$ do not appear. From the order $1/x$ term we also obtain the result $\zeta(2) = \pi^2/6$.

17.7 Problems

1. Evaluate

$$\zeta(2) = \sum_1^\infty \frac{1}{n^2}.$$

Hint: Consider

$$\oint \frac{dz}{z^2} \frac{cos(\pi z)}{sin(\pi z)}$$

with the contour chosen to be a large square passing through the x values $(-M - 1/2, M + 1/2)$. Evaluate the integral in two ways. First bound the integrand for large M. Second use the Cauchy residue theorem. Then let $M \to \infty$.

2. Find $\Sigma_1^\infty(1/n^4)$ exactly. Check your result to three places using a few terms.

3. Use Eq. 17.13 and $\zeta(2)$ to find $\zeta(-1)$.

4. Use Eq. 17.13 and $\zeta(4)$ to find $\zeta(-3)$.

5. Find the number of terms to be kept in the asymptotic expansion of the Gamma function, Eq. 17.60, for large positive x.

6. Find from Eq. 17.13 that the zeros of $\zeta(s)$ at $-2, -4, -6, \ldots$ are first order zeros.

7. Prove the Mellin representation

$$e^{-x} = \frac{1}{2\pi i} \int_{\gamma - i\infty}^{\gamma + i\infty} \Gamma(u) x^{-u} du$$

with $0 < \gamma < 1$. What is the restriction on x?

Chapter 18

Boundary Layer Problems

Ludwig Prandtl was born in 1875 in Freising, Germany. He became a professor of mechanics at the University of Hanover in 1901. In 1904 he delivered an important paper entitled *Fluid Flow in Very Low Friction* in which he solved a paradox which had existed for some time concerning the motion of solid bodies through fluids. Newton in the *Principia* calculated resistance to the motion of a solid body through a gas by finding the momentum transfer caused by the collisions of the body with the small particles of the gas. Leonhard Euler, however, introduced the concept of a fluid, with pressure transmitting fluid elements rather than particles, ruining Newton's calculation by making the particles no longer independent. Exact mathematical solutions of the flow of the fluid around a solid body using potential theory gave the result that the drag should be zero, in clear contradiction with experiment. Prandtl recognized that in a very narrow layer next to the solid body, the viscosity was important, no matter how small it was, and resulted in a narrow boundary layer in which the flow was highly turbulent, and not described by potential theory. Kinetic effects, due to collisions of the individual particles, cause drag and heat dissipation to occur in this layer, and the understanding of the layer made clear the process of aerodynamic flow over a wing, and permitted the engineering design of more efficient airfoils.

From 1904 to 1953 he was professor of applied mathematics at the University of Göttingen, where he established a school of aerodynamics. In 1908 he developed the first theory of supersonic shock waves and created the first supersonic wind tunnels. Theodore von Karman, a student of his at Göttingen, developed the full theory of supersonic flow. In 1918 he published the Lanchaster–Prandtl wing theory, which simplified the design of airplane wings. The analysis of induced drag and wingtip vortices was included. The crater Prandtl on the far side of the moon is named for him.

18.1 Introduction

Many problems in physics involve a very small parameter ϵ such as viscosity or resistivity which appears in the equations of interest multiplying a derivative of higher order than that which appears in any other terms. In this case the solutions often exhibit boundary layer phenomena, i.e. the existence of narrow layers which scale as some power of ϵ, in which the solution is rapidly varying. It is desirable to find an analytic approximation to the solution, so that the scaling of the layer, and the behavior of the solution, is known as an analytic function of ϵ. This is especially true because in practice the layers may be extremely narrow, requiring particular care to treat numerically, and a numerical determination of the scaling would require a very large number of calculations.

For equations of this type, it is possible to construct an analytic solution order by order in the parameter ϵ. The process is simplified by using a Kruskal–Newton diagram constructed by finding the scaling of each term in the independent variable x and in ϵ, with terms scaling as $x^p \epsilon^q$, giving a point in the p, q plane.[1] To find all possible combinations of dominant balance for a given equation, find all possible placements of a line so that it includes two or more terms of the equation, with all other points lying above the line. Each such placement represents a potential solution to the equation whereby the dominant balance of the solution is given by the points on the line, and all points above the line are associated with terms which are small corrections to this solution. Graphically this may be understood as bringing the line up from below until it makes contact with a point, and then rotating it one way or the other until it makes contact with a second point. A line making contact with two or more points, with all remaining points of the diagram above the line, is called a support line. The scaling of x is quickly determined by the slope of the line in the diagram. If the terms defining the line are given by $x^p \epsilon^q$ and $x^r \epsilon^s$ then balancing them gives the scaling $x \sim \epsilon^{(s-q)/(p-r)}$, the power of ϵ being minus the slope of the line. Such a plot is known as a Kruskal–Newton diagram. It was first used by Newton and subsequently further developed by Kruskal

[1] The use of these diagrams for bounday layer problems was developed with Thomas Fischaleck.

[1963]. The use of such diagrams for simple algebraic equations is discussed in Chapter 1.

A given support line, by the neglect of all terms lying above the line, produces a simplified differential equation for y,

$$Dy(x) = 0. \tag{18.1}$$

There are two means of proceeding. Points lying above the line constitute small corrections to the solution. The lowest order solution $y_0(x)$ is given by neglecting them. Approximating the neglected terms using $y_0(x)$ one can solve the resulting inhomogeneous equation for y. A simple iteraton equation of the form

$$Dy_{n+1}(x) = f(y_n, x) \tag{18.2}$$

results. This iteration is rapidly convergent according to the smallness of ϵ. Alternatively one can perform a perturbation expansion in ϵ.

Each dominant balance gives either the exterior solution or a layer width $\delta(\epsilon)$. One then finds the solutions inside and outside the layers, and chooses constants of integration to match the solutions to each other and to the boundary conditions. The matching is done in an asymptotic sense. First assume that the solution and all derivatives are of order one, and solve for the outside solution $y_{out}(x)$, order by order in the small parameter ϵ. Inside the layer choose a new independent variable through $x = \delta X$, with $\delta(\epsilon)$ given by the Kruskal–Newton diagram. Using dominant balance inside the layer gives a solution with derivatives in the internal variable X of order one. Solve the resulting equation order by order in ϵ, giving the internal solution y_{layer}. Then match asymptotically to the outside solution by requiring the interior and outside solutions to approach the same function y_{match} in the region which is asymptotic to both domains, $\delta << x << 1$. If the layer is very narrow, it is often sufficient to find the solution to lowest order in ϵ. Even if ϵ is not particularly small, the technique can be useful by providing approximate analytic solutions in cases in which no analytic solution is available for the original equation. Finally, a global uniform solution is obtained by writing $y_{unif} = y_{out} + y_{layer} - y_{match}$. Inside the layer $y_{out} = y_{match}$, and outside the layer $y_{layer} = y_{match}$, this solution thus providing a continuous uniform approximate solution to the differential equation over the whole domain. The method will become clear with a few examples, for which ϵ is taken fairly large, to make visible the difference between the numerical solutions and the boundary layer treatment.

18.2 Layer Location

Consider a second order differential equation with the highest order derivative multiplied by a small parameter

$$\epsilon \frac{d^2y}{dx^2} + f(x)\frac{dy}{dx} + g(x)y = 0, \tag{18.3}$$

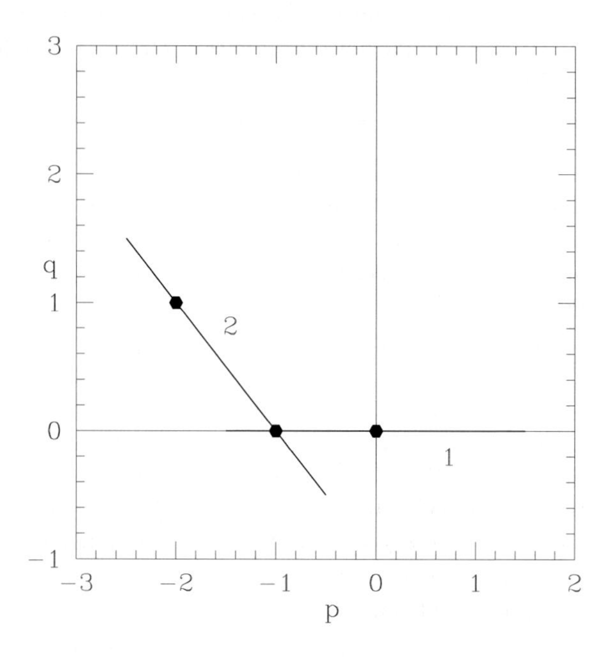

Fig. 18.1 Kruskal–Newton diagram for Eq. 18.3.

with boundary conditions $y(0) = A$, $y(1) = B$. Construct a Kruskal–Newton diagram for this equation, with terms given by $x^p\epsilon^q$. The functions y, f, g are all assumed to be of order one. The first term is of order $x^{-2}\epsilon$, the second of order x^{-1} and the third of order one. The Kruskal–Newton diagram for this is shown in Fig. 18.1. Line 1 gives the exterior solution, lowest order given by the neglect of terms of order ϵ, giving the external (external to the boundary layers) solution

$$y_0(x) = ce^{-\int dx g(x)/f(x)}. \tag{18.4}$$

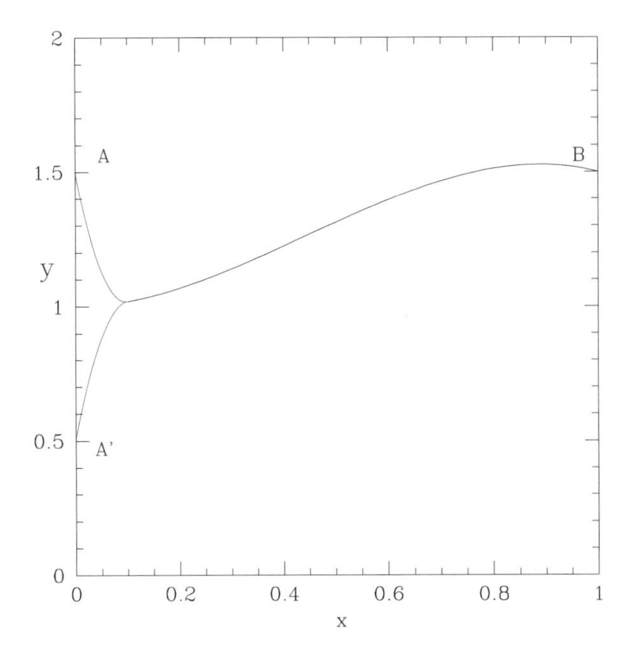

Fig. 18.2 Layer on the left.

If the integration constant c can be chosen so that the boundary conditions are matched, the problem is solved and there is no boundary layer phenomenon. If the boundary conditions cannot be matched then there must exist a layer in which the derivatives are not order one. Line 2 in the diagram gives the dominant balance of the layer,

$$\epsilon \frac{d^2 y}{dx^2} + f(x) \frac{dy}{dx} \simeq 0, \tag{18.5}$$

and thus $dx \sim \epsilon$. In general the scaling of dx is quickly determined by the slope of the line in the diagram. If the terms defining the line are given by $x^p \epsilon^q$ and $x^r \epsilon^s$ then balancing them gives the scaling $dx \sim \epsilon^{(s-q)/(p-r)}$, the power of ϵ being minus the slope of the line.

This determines the width of the layer, but not its location. The location of the layer is determined by the function $f(x)$. Figure 18.2 shows a solution with a layer to the left. The boundary condition has been matched at the right, and the external solution continued to the left until it is close to the boundary. Two possibilities are shown, with the external solution too low

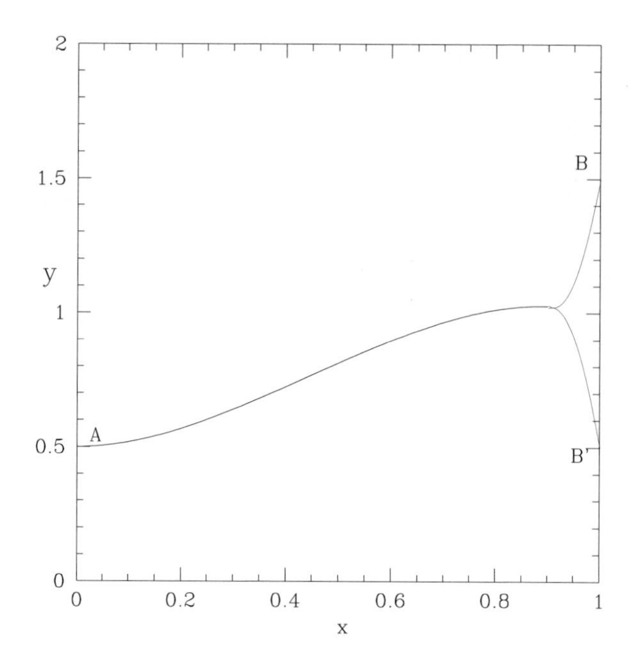

Fig. 18.3 Layer on the right.

to match (A), and two high to match (A′). It is clear that in either case, to match the external solution which has a derivative of order 1, the solution inside the layer must have dy/dx of opposite sign from d^2y/dx^2. Thus to satisfy Eq. 18.5 the function $f(x)$ must be positive.

Figure 18.3 shows the analogous plot for a layer at the right boundary. The boundary condition has been matched at the left, and the external solution continued to the right until it is close to the boundary. Now it is clear that in the layer dy/dx must have the same sign as d^2y/dx^2. Thus to satisfy Eq. 18.5 the function $f(x)$ must be negative.

Finally suppose that a layer exists inside the integration domain, as shown in Fig. 18.4. From this figure it is clear that to the right of the center line dy/dx has the opposite sign from d^2y/dx^2, and to the left of the center line dy/dx has the same sign as d^2y/dx^2. Thus there must be a sign change in $f(x)$ at the center of the layer with $df/dx > 0$. In the figure the case with $A < B$ is shown, but the same result holds if $A > B$. It is easy to see that no internal layer solution is possible with $df/dx < 0$.

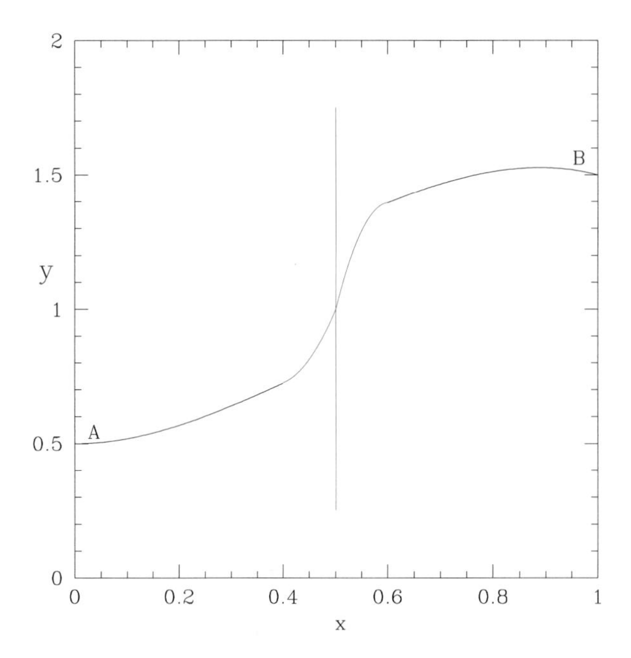

Fig. 18.4 Internal layer.

18.3 Layer at Left Boundary

Consider the equation

$$\epsilon \frac{d^2y}{dx^2} + (1+x)\frac{dy}{dx} + y = 0 \tag{18.6}$$

in the domain $(0, 1)$, with $0 < \epsilon \ll 1$ and $y(0) = 1$, $y(1) = 2$. From the previous analysis the only possible layer is at $x = 0$. To determine the possible scalings of the layer, construct a Kruskal–Newton diagram for the variables x, ϵ. The first term is of order ϵx^{-2}, dy/dx is of order x^{-1} and xdy/dx and y are of order 1. The resulting diagram is shown in Fig. 18.5 with terms given by $x^p \epsilon^q$.

Line 1 of the diagram has slope zero, so x is of order 1. This gives the exterior solution, with leading order balance given by

$$(1+x)\frac{dy}{dx} + y = 0. \tag{18.7}$$

Perform a perturbation expansion of the solution $y(x)$ in ϵ, $y = y_0(x) + \epsilon y_1(x) + ...$ Matching the boundary condition at $x = 1$ gives

$$y(x) = \frac{4}{1+x} + \epsilon \left[\frac{4}{(1+x)^3} - \frac{1}{1+x} \right] + \dots . \tag{18.8}$$

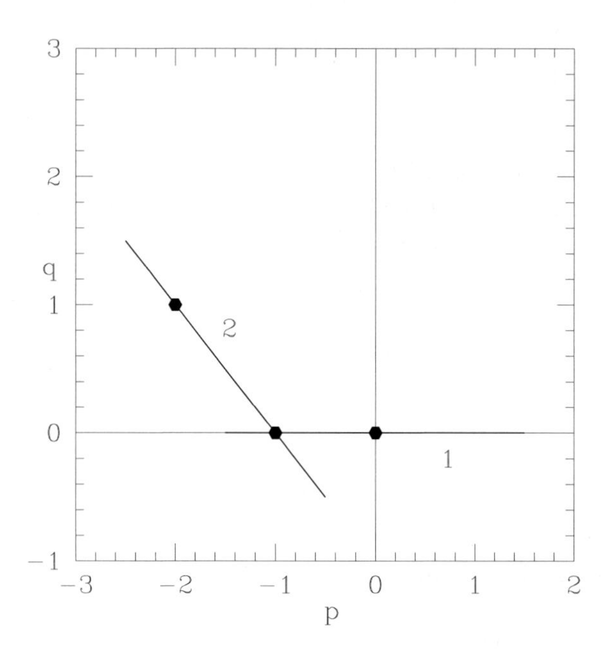

Fig. 18.5 Kruskal–Newton diagram for Eq. 18.6.

There is only one other possible dominant balance. Line 2 has slope -1, so $x \sim \epsilon$. Setting $x = \epsilon X$ and $y(x) = Y(X)$ gives to leading order

$$Y'' + Y' = 0, \tag{18.9}$$

with solution $Y_0 = a + be^{-X}$. Next order gives $Y_1 = (-bX^2/2 + c)e^{-X} - aX + d$, and the boundary condition at $X = 0$ gives $a + b = 1$ and $c + d - a = 0$.

The matching solution is obtained by the small x limit of the exterior solution, or the large X limit of the interior solution. To perform the matching take $\epsilon << x << 1$ and expand the interior and exterior solutions. The exterior solution becomes

$$y \to 4 - 4x + 3\epsilon + O(x^2, \epsilon^2, \epsilon x). \tag{18.10}$$

The interior solution becomes

$$Y \to a - ax + d\epsilon + O(x^2, \epsilon^2, \epsilon x), \tag{18.11}$$

giving $a = 4$, $d = 3$, $b = -3$, and also $c = -3$.

The uniform solution to this order is constructed by adding the outside and interior solutions and subtracting the matching solution

$$y_{unif} = \frac{4}{1+x} + \epsilon\left[\frac{4}{(1+x)^3} - \frac{1}{1+x}\right] - 3e^{-x/\epsilon} + [3x^2/(2\epsilon) - 3\epsilon]e^{-x/\epsilon}. \tag{18.12}$$

This procedure gives a solution which is uniformly accurate over the whole domain, and matches the given boundary conditions. Note that the dependence on ϵ is non-analytic; the solution does not have the form of a simple perturbation expansion.

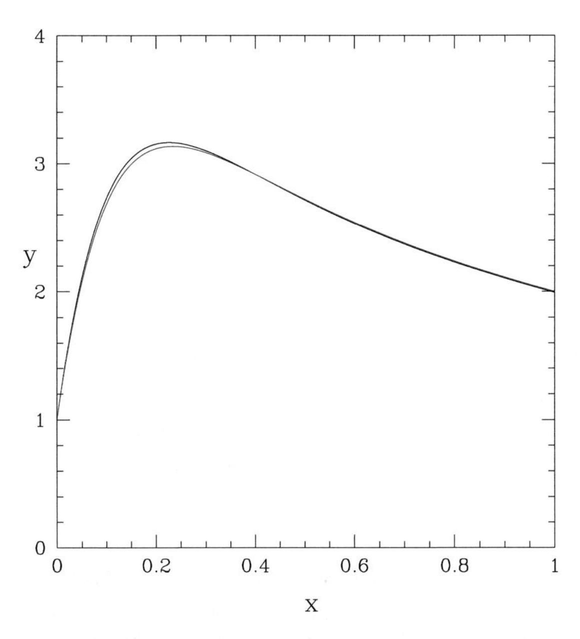

Fig. 18.6 Lowest order uniform solution (red) and numerical integration for Eq. 18.6 with $\epsilon = .1$.

Figure 18.6 shows the first order uniform solution obtained with $y(0) = 1$, $y(1) = 2$, and $\epsilon = .1$. The derivative is seen to be large only in the domain $x < \epsilon$. Shown is the uniform solution (red) and a numerical integration (black). In physical problems ϵ is much smaller than this, we choose it large to illustrate the difference between layer and numerical solutions. With very small values of ϵ the functions could not be distinguished even using lowest order solutions.

18.4 Layer in Domain Center

Consider the equation

$$\epsilon \frac{d^2y}{dx^2} + x\frac{dy}{dx} = x sinx \tag{18.13}$$

in the domain $(-\pi, \pi)$, with $0 < \epsilon \ll 1$ and the boundary conditions $y(\pm\pi) = \pm 1$. From the previous analysis the only possible layer is at $x = 0$.

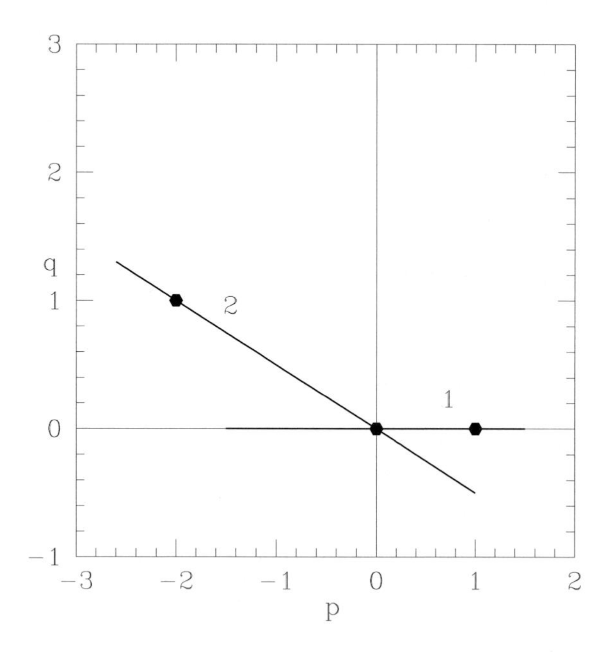

Fig. 18.7 Kruskal–Newton diagram for Eq. 18.13.

To determine the possible scalings of the layer, construct a Kruskal–Newton diagram for the variables x, ϵ. The first term is of order ϵx^{-2}, xdy/dx is of order 1, and $xsinx$ is of order x^p with $1 < p < 2$. The resulting diagram is shown in Fig. 18.7, with terms given by $x^p \epsilon^q$.

Line 1 of the diagram has slope zero, so x is of order 1. This gives the exterior solution, with lowest order balance given by

$$x\frac{dy}{dx} = xsinx \tag{18.14}$$

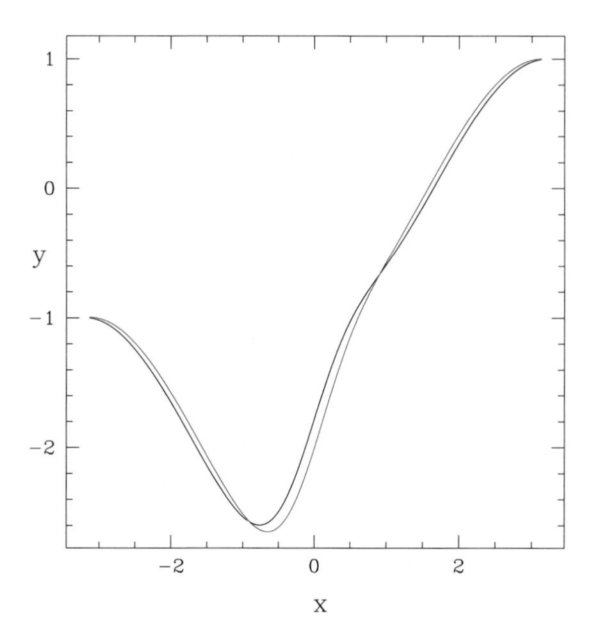

Fig. 18.8 Lowest order uniform solution (red) and numerical integration for Eq. 18.13 with $\epsilon = 0.2$.

This gives upon matching to the boundary conditions at $\pm\pi$ the solutions to the left and to the right of the internal layer

$$y(x) = \begin{cases} -cosx - 2 & x < 0 \\ -cosx & x > 0. \end{cases} \tag{18.15}$$

From the left the outside solution approaches -3 as $x \to 0$ and from the right it approaches -1.

Line 2 of the diagram has slope $-1/2$, so $x = \sqrt{\epsilon}X$, with $y(x) = Y(X)$, giving to lowest order

$$\frac{d^2Y}{dX^2} + X\frac{dY}{dX} = 0, \tag{18.16}$$

with solution $Y = a\int_0^X e^{-s^2/2}ds + b$, with a, b integration constants. To the right we have $Y \to a\sqrt{\pi/2} + b$ and to the left $Y \to -a\sqrt{\pi/2} + b$. Matching to the exterior solutions gives $a\sqrt{\pi/2} + b = -1$, and $-a\sqrt{\pi/2} + b = -3$, or $b = -3/2$, $a = 3/\sqrt{2\pi}$. The lowest order uniform solution is then obtained by adding the internal and external solutions and subtracting the matching solution for $x < 0$ and that for $x > 0$ we find

$$y(x) = -1 - cosx + \sqrt{2/\pi}) \int_0^{x/\sqrt{\epsilon}} e^{-s^2/2}ds. \tag{18.17}$$

Figure 18.8 shows the lowest order uniform solution obtained with $\epsilon = .2$. Shown is the uniform solution (red) and a numerical integration (black).

18.5 Layer at Right Boundary

Consider the equation

$$\epsilon\frac{d^2y}{dx^2} - \frac{dy}{dx} + (x-1)y = 0 \tag{18.18}$$

in the domain $(0, 1)$, with $0 < \epsilon \ll 1$ and the boundary conditions $y(0) = y(1) = 1$. The only possible layer is at $x = 1$, so to discover possible scalings we change variables to $z = 1 - x$, giving

$$\epsilon\frac{d^2y}{dz^2} + \frac{dy}{dz} - zy = 0 \tag{18.19}$$

The Kruskal–Newton diagram for this equation is shown in Fig. 18.9. Line 1, with zero slope, gives the lowest order exterior equation,

$$\frac{dy}{dz} - zy = 0. \tag{18.20}$$

Using this balance, solving Eq. 18.19 and matching to the boundary condition at $x = 0$ gives to first order in ϵ

$$y(x) = e^{x^2/2-x} + \epsilon(x^3/3 - x^2 + 2x)e^{x^2/2-x} + \dots. \tag{18.21}$$

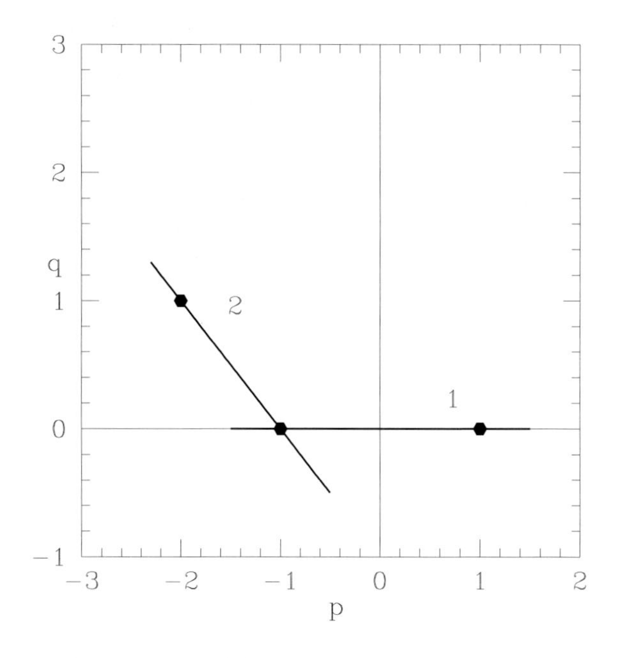

Fig. 18.9 Kruskal–Newton diagram for Eq. 18.19.

The layer is given by line 2, with slope -1, so z is of order ϵ. Transforming to the internal variable through $z = \epsilon Z$, with $y(z) = Y(Z)$, gives

$$Y'' + Y' - \epsilon^2 ZY = 0, \tag{18.22}$$

with solution $Y = Y_0 + \epsilon Y_1$, $Y_0 = ae^{-Z} + b$, with a, b integration constants, and $Y_1 = ce^{-Z} + d$ with c, d integration constants. Matching to the boundary condition at $x = 1$ gives

$$Y = ae^{-Z} + 1 - a + \epsilon c[e^{-Z} - 1]. \tag{18.23}$$

For $1 << Z << 1/\epsilon$ the asymptotic limit is

$$Y \to 1 - a - c\epsilon. \tag{18.24}$$

The external solution becomes, substituting $x = 1 - \epsilon Z$ and using $\epsilon Z << 1$

$$y \to e^{-1/2}(1 + 4\epsilon/3) + \ldots, \tag{18.25}$$

giving $a = 1 - e^{-1/2}$, $c = -4e^{-1/2}/3$. The lowest order uniform solution is then obtained by adding the internal and external solutions and subtracting the matching solution

$$y(x) = e^{x^2/2-x} + \epsilon(x^3/3 - x^2 + 2x)e^{x^2/2-x} \\ +(1 - e^{-1/2} - 4\epsilon e^{-1/2}/3)e^{(x-1)/\epsilon}. \tag{18.26}$$

Figure 18.10 shows the first order uniform solution obtained with $\epsilon = .1$. Shown is the uniform solution (red) and a numerical integration (black).

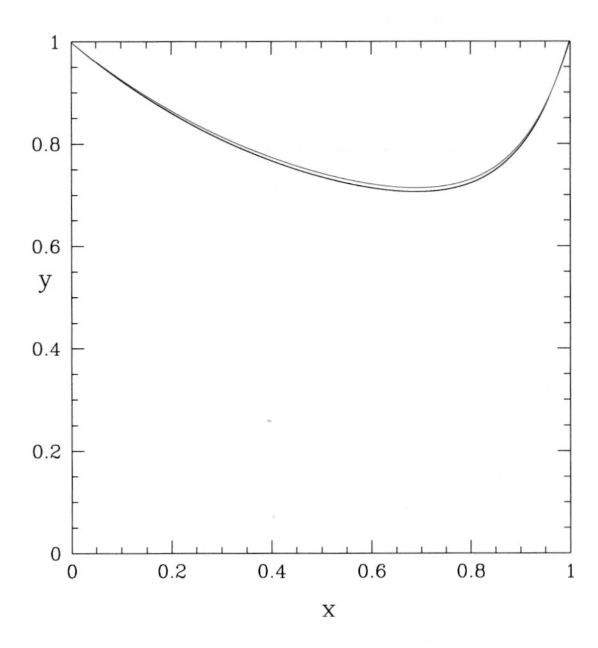

Fig. 18.10 Lowest order uniform solution (red) and numerical integration for Eq. 18.19 with $\epsilon = .1$.

18.6 Nested Boundary Layers

It can happen that an equation possesses two different possible dominant balances, giving rise to two layers with different scalings, one inside the other. An example of this behavior is given by Bender–Orszag [1978] by the equation

$$\epsilon^3 x y'' + x^2 y' - y(x^3 + \epsilon) = 0, \tag{18.27}$$

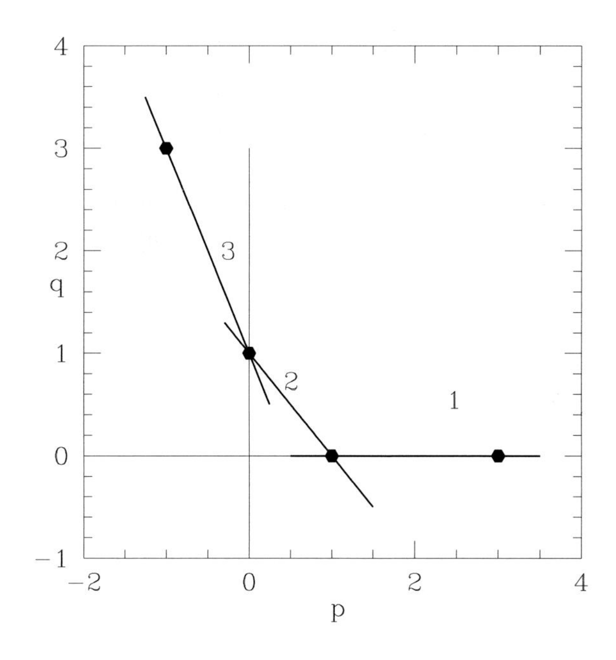

Fig. 18.11 Kruskal–Newton diagram for Eq. 18.19.

with boundary conditions $y(0) = 1, y(1) = \sqrt{e}$. The point $x = 0$ is a regular singular point, and the indicial equation gives $y \simeq x^\alpha$ with $\alpha = 0, 1$. The coefficient of y' is positive, so there can be no layer at $x = 1$. The Kruskal–Newton diagram for Eq. 18.27 is given in Fig. 18.11. As usual, the outer solution is given by the terms of order zero in ϵ, but there are seen to be two other possible balances. Line 2 in the graph has slope -1, giving $x \sim \epsilon$. Line 3 has slope -2, giving $x \sim \epsilon^2$. Thus line 3 gives a layer contained within the layer given by line 2. Begin with the outer solution. Write

$$y = y_0 + \epsilon y_1 + \ldots, \tag{18.28}$$

giving

$$x^2 y_0' - x^3 y_0 = 0, \qquad x^2 y_1' - x^3 y_1 = y_0, \tag{18.29}$$

and upon matching the boundary condition at $x = 1$ we have the solutions

$$y_0 = e^{x^2/2}, \qquad y_1 = (1 - 1/x)e^{x^2/2}. \tag{18.30}$$

Boundary conditions are not met at $x = 0$ by this solution because $y_1 \to \infty$, so there must be a layer there. First examine the outer possible layer, given by line 2 in the Kruskal–Newton diagram. Let $x = \epsilon X$, with $y(x) = Y(X)$, giving

$$X^2 \frac{dY}{dX} - Y = -\epsilon X \frac{d^2Y}{dX^2} + \epsilon^2 X^3 Y. \tag{18.31}$$

Expanding $Y = Y_0 + \epsilon Y_1$ we find

$$Y_0 = \alpha_0 e^{-1/X}, \quad Y_1 = \alpha_1 e^{-1/X} + \alpha_0 \left(\frac{2}{3X^3} - \frac{1}{4X^4} \right) e^{-1/X}. \tag{18.32}$$

Matching to the outer solution we find $\alpha_0 = \alpha_1 = 1$, with the layer solution given by

$$Y = e^{-1/X} \left[1 + \epsilon \left(1 + \frac{2}{3X^3} - \frac{1}{4X^4} \right) \right] + O(\epsilon^2), \tag{18.33}$$

but it is still not possible to match the condition $y(0) = 1$. The matching solution in the domain $\epsilon << x << 1$ is given by $y_{match} = 1 + \epsilon - \epsilon/x + O(\epsilon x, x^2, \epsilon^2)$.

The second layer is given by line 3 in the Kruskal–Newton graph, with $x \sim \epsilon^2$. Writing $x = \epsilon^2 Z$, $y(x) = \mathcal{Y}(Z)$ we then have

$$Z \frac{d^2\mathcal{Y}}{dZ^2} - \mathcal{Y} = -\epsilon Z^2 \frac{d\mathcal{Y}}{dZ} + \epsilon^5 Z \mathcal{Y}. \tag{18.34}$$

Writing $\mathcal{Y} = \mathcal{Y}_0 + \epsilon \mathcal{Y}_1$ we find

$$Z \frac{d^2\mathcal{Y}_0}{dZ^2} - \mathcal{Y}_0 = 0. \tag{18.35}$$

Let $\mathcal{Y}_0 = \sqrt{Z} f(2\sqrt{Z})$ giving

$$f'' + \frac{f'}{2\sqrt{Z}} - f \left(1 + \frac{1}{4Z} \right) = 0, \tag{18.36}$$

which is the modified Bessel equation (section 13.6), giving the solution

$$\mathcal{Y}_0 = \beta_0 \sqrt{Z} I_1(2\sqrt{Z}) + \beta_1 \sqrt{Z} K_1(2\sqrt{Z}). \tag{18.37}$$

Integral representations are readily found for these two solutions,

$$K_\nu(w) = \int_0^\infty e^{-w cosh(t)} cosh(\nu t) dt, \tag{18.38}$$

$$I_n(w) = \frac{1}{\pi}\int_0^{\pi} e^{wcos(t)}cos(nt)dt. \tag{18.39}$$

To carry out the matching we need the asymptotic behavior of these two solutions for small and large argument. Note that expanding the integrand of K_ν for small w results in infinity for each term, this function not being analytic at $w = 0$. However, the expression $wK_1(w)$, upon a change of variable to $u = wcosh(t)$, becomes

$$wK_1(w) = \int_w^{\infty} \frac{e^{-u}udu}{\sqrt{u^2 - w^2}}, \tag{18.40}$$

and thus for small w we have $K_1(w) \simeq 1/w$. For the behavior at infinity let $u = w + z$, giving

$$wK_1(w) = \int_0^{\infty} \frac{e^{-w}e^{-z}(w+z)dz}{\sqrt{z(z+2w)}}. \tag{18.41}$$

Letting w tend to infinity we then have

$$K_1(w) \simeq \sqrt{\frac{\pi}{2}}\frac{e^{-w}}{\sqrt{w}}. \tag{18.42}$$

For small w the function $I_1(x)$ can be developed in a Taylor series, giving

$$I_1(w) = \sum_1^{\infty} \frac{w^{2k-1}}{(2k-1)!}\frac{(2k-1)!!}{(2k)!!}. \tag{18.43}$$

For large argument, the dominant contribution comes from the end point $t = 0$. Expanding the cosine function we have

$$I_1(w) \simeq \frac{1}{\pi}\int_0^{\pi} e^{w(1-t^2/2)}(1 - t^2/2)dt, \tag{18.44}$$

which reduces to

$$I_1(w) \simeq \frac{e^w}{\pi}\frac{1}{\sqrt{2w}}\left[\Gamma(1/2) - \frac{2\Gamma(3/2)}{w} + \ldots\right], \tag{18.45}$$

giving for large w $I_1(w) \simeq e^w(1 - 1/w)/\sqrt{2\pi w}$.

Now carry out the matching. Since Y tends to zero for small X we must require that $\mathcal{Y}$ tend to zero for large Z. Thus we find $\beta_0 = 0$. Matching the solution to 1 at zero we find $\beta_1 = 2$ and

$$\mathcal{Y}_0 = 2\sqrt{Z}K_1(2\sqrt{Z}). \tag{18.46}$$

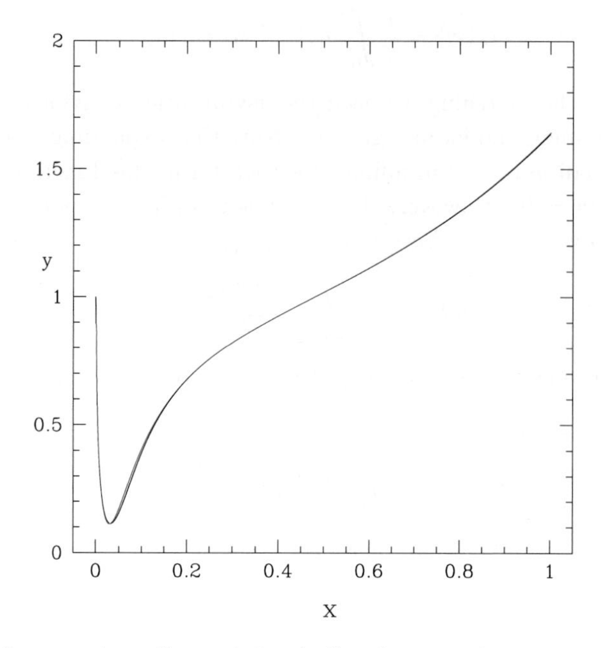

Fig. 18.12 Lowest order uniform solution (red) and numerical integration for Eq. 18.27 with $\epsilon = .1$.

Finally the uniform solution to lowest order is given by

$$y_u = y_0(x) + Y_0(X) + \mathcal{Y}_0(Z) - y_{match}, \tag{18.47}$$

there being no matching term coming from the domain between the two layer solutions, since they each tend to zero exponentially in the matching region. Figure 18.12 shows a numerically integrated solution and the first order uniform solution for $\epsilon = 0.1$. Notice that with this large value of ϵ the two layer solutions do not come close to zero, not even falling below $y = 0.1$. Nevertheless the asymptotic matching gives a very good approximation to the numerical integration.

18.7 Inhomogeneous Equations

Some problems involving inhomogeneous equations simplify under a rescaling of the dependent variable. Consider a differential equation of the form $\sum_{q,m,n} C_{q,m,n}\, \epsilon^q x^m y^{(n)} = x^r$. Rescale y through $y = \epsilon^{-\beta} Y$ and multiply the equation by ϵ^β, giving

$$\sum_{q,m,n} C_{q,m,n}\, \epsilon^q x^m Y^{(n)} = x^r \epsilon^\beta. \tag{18.48}$$

Terms involving $Y^{(n)}(x)$ are represented in a Kruskal–Newton diagram at coordinates $(p, q) = (m - n, q)$ as in the homogeneous case. The position of the inhomogeneity is at coordinates $(p, q) = (r, \beta)$, and has shifted by β. Alternatively, without multiplying by ϵ^β it is the homogeneous terms that shift; only the relative displacement is relevant.

Draw the Kruskal–Newton diagram as in the homogeneous case. This leads to one or more support lines. Represent the inhomogeneity by a vertical line corresponding to the set of points $\{(r, \beta),\ -\infty < \beta < \infty\}$. Any vertical line must intersect all support lines, so the inhomogeneity can be included in the corresponding balances, since rescaling y corresponds to a vertical shift of all homogeneous points with respect to all inhomogeneous points. The value of β for this balance is given by the height of the intersection point.

Consider the equation

$$\epsilon^3 y'' - \epsilon y + x^3 y' = x, \tag{18.49}$$

with boundary condition $y(0) = A$ and $y(1) = B$. Figure 18.13 shows the Kruskal–Newton plot for this equation. The solid dots correspond to the

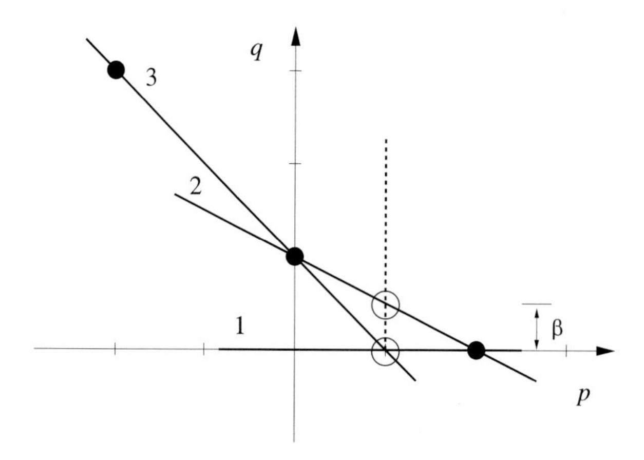

Fig. 18.13 The Kruskal–Newton plot for Eq. 18.49. The inhomogeneity is represented by the dotted vertical line at $p = 1$. The empty circles mark intersection points with support lines 1, 2 and 3.

homogeneous part and the support lines 1 to 3 are drawn neglecting the inhomogeneous terms. The dotted line represents the inhomogeneity.

We now use Kruskal's asymptotic principle of maximal complexity [1963] according to which "the most informative ordering is that which simplifies the least, maintaining a maximal set of comparable terms". The intersection point with line 2 lies at height 1/2 and the dependent variable must be rescaled according to $y = \epsilon^{-\frac{1}{2}}Y$ in order to allow this balance. Otherwise the inhomogeneity would lie below line 2 and it would be a term of lower order. The other intersections lie on the p-axis giving $\beta = 0$, and there is no need to rescale. Line 1 has slope zero, the dotted line intersects it, so include the inhomogeneity. The intersection lies on the p-axis, so $\beta = 0$, giving the exterior solution, with leading order balance given by $x^3 y_0'(x) = x$.

Matching the boundary condition at $x = 1$ gives

$$y \sim y_0 = -\frac{1}{x} + B + 1. \tag{18.50}$$

Line 2 has slope $-1/2$, so it intersects the dotted line at height $\beta = 1/2$, and the rescaled variables in the middle region are given by $y(x) = \epsilon^{-1/2}Y(X)$, and $x = \sqrt{\epsilon}X$. The leading order balance is given by $-Y_0(X)+X^3Y_0'(X) = X$. Change variable to $Z = 1/X$ giving

$$ZY(Z) + dY/dZ = -1. \tag{18.51}$$

Attempt an integral solution of the form

$$Y(Z) = \int e^{Zt} f(t)dt \tag{18.52}$$

giving

$$f(t) = Ae^{t^2/2}, \qquad f(t)e^{Zt}|_a^b = -1 \tag{18.53}$$

where a, b are the end points of the integration contour. We thus find $a = 0$, $b = i\infty$, and $A = 1$. Changing integration variable through $t \to it$ we have

$$Y_0(X) = Ce^{-Z^2/2} + i\int_0^\infty e^{iZt - t^2/2}dt. \tag{18.54}$$

Note that if Y is a solution to the inhomogeneous equation, then ReY is also a solution, and ImY is a solution to the homogeneous equation. Thus the imaginary part of Y_0 can be absorbed into the homogeneous part, giving

$$Y_0(X) = Ce^{-Z^2/2} - \int_0^\infty sin(Zt)e^{-t^2/2}dt. \tag{18.55}$$

We determine the constant C by matching to the exterior solution $y_0 = -1/x + B + 1$. The limit of large X is the limit of small Z, giving $Y_0(X) \to C - Z$, giving

$$y(x) \to \frac{C}{\sqrt{\epsilon}} - \frac{1}{x}. \tag{18.56}$$

Matching these solutions we find $C = \sqrt{\epsilon}(B+1)$ and determine the matching solution in this domain $y_{match_{12}} = -1/x + B + 1$.

Line 3 has slope -1 and it intersects the dotted line at zero height. The inner variables are therefore $y(x) = \mathcal{Y}(\mathcal{X})$, and $x = \epsilon\mathcal{X}$. We find to leading order $\mathcal{Y}_0''(\mathcal{X}) - \mathcal{Y}_0(\mathcal{X}) = \mathcal{X}$, with solution

$$\mathcal{Y}_0 = ce^{-\mathcal{X}} + de^{\mathcal{X}} - \mathcal{X}. \tag{18.57}$$

Finally, we match the inner solution with the middle solution $y \sim \epsilon^{-\frac{1}{2}} Y_0 = -\epsilon^{-\frac{1}{2}} \int_0^\infty \sin\left(\frac{t}{X}\right) e^{-\frac{t^2}{2}} dt$. Expanded in the intermediate domain $\epsilon \ll x \ll \sqrt{\epsilon}$ or equivalently $\sqrt{\epsilon} \ll X \ll 1$ the contribution for $X \ll 1$ comes from the end point, giving $y \sim -\epsilon x = -\mathcal{X}$. We then find $d = 0$, and matching to the boundary condition at $x = 0$ gives $c = A$. We then find for the intermediate domain $\mathcal{X} \gg 1$, $\mathcal{Y}_0 \to y_{match_{23}} = -\mathcal{X}$.

The uniform solution is obtained by adding the outside solution to the two layer solutions and subtracting the two matching solutions,

$$y_u(x) = (B+1)(1 + e^{-\epsilon^2/x^2}) - \int_0^\infty \sin\left(\frac{\sqrt{\epsilon}t}{x}\right) e^{-t^2/2}dt + Ae^{-x/\epsilon}. \tag{18.58}$$

Note that without the use of Kruskal's principle of maximum complexity and the rescaling of $y = Y/\sqrt{\epsilon}$ there is no layer of width $\sqrt{\epsilon}$, only the layer of width ϵ. In this case it is impossible to match the external solution $y = -1/x + B + 1$ to $\mathcal{Y}_0$ and a solution cannot be found.

18.8 Simplification through Expansion

In the case of the presence of transcendental functions it can be convenient to expand these functions in infinite series, producing Kruskal–Newton diagrams with an infinity of points. In carrying out this expansion for an internal layer, where the support line is not horizontal, only the leading terms remain on the support line, all higher order terms move off to the right. Thus the layer equations are greatly simplified.

18.8.1 *A corner layer*

Consider the equation

$$\epsilon y'' + 2tan(x)y' - 3y = 0, \tag{18.59}$$

with $y(0) = 1$, $y(\pi/2) = 1$. Expand $tan(x) = x + x^3/3 + 2x^5/15 + ...$ The Kruskal–Newton diagram is given in Fig. 18.14, where the solid dots

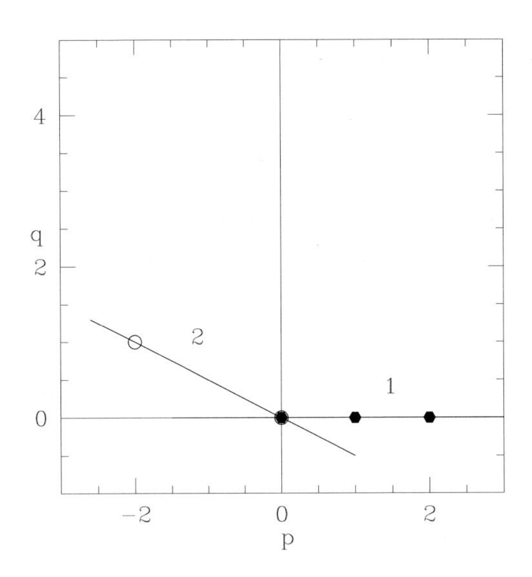

Fig. 18.14 Kruskal–Newton diagram for Eq. 18.59.

correspond to the second and third terms and the empty circle corresponds to the first term in Eq. 18.59. Although for the outer solution all dots are present and there is no simplification, for the layer solution only the leftmost dots contribute to the support line. The dots on the p-axis constitute the outer balance $2tan(x)y' \sim 3y$ with solution $y = sin(x)^{3/2}$. Line 2 gives the layer equation with $x = \sqrt{\epsilon}X$

$$Y'' + 2XY' - 3Y = 0. \tag{18.60}$$

Solutions are given by

$$Y_C = \frac{1}{i}\int_C e^{Xt+t^2/4}t^{-5/2}dt \tag{18.61}$$

where the contour must end at $\pm i\infty$ and can pass on either side of the cut originating at $t = 0$. If the contour is taken from $-i\infty$ to $i\infty$ the large X behavior is given by a saddle point at $t = -2X$, producing asymptotic behavior of $X^{-5/2}e^{-X^2}$. A contour beginning and ending at $i\infty$ and circling the cut has asymptotic behavior of $X^{3/2}$ given by the end points of the contour. Thus the interior solution is given by

$$Y = \frac{a}{i}\int_{-i\infty}^{i\infty} e^{Xt+t^2/4}t^{-5/2}dt + \frac{b}{i}\int_{i\infty}^{i\infty} e^{Xt+t^2/4}t^{-5/2}dt \tag{18.62}$$

with the second contour circling the cut in a clockwise manner. Evaluating the limits for $X \to 0$ we find

$$Y \sim (a+b)2^{-5/4}\Gamma(-3/4). \tag{18.63}$$

Evaluating the limits for $X \to \infty$ we find

$$Y \sim a2^{-3/2}\sqrt{\pi}X^{-5/2}e^{-X^2} + bX^{3/2}\Gamma(-3/2). \tag{18.64}$$

Matching to the outside solution we find $b = \epsilon^{3/2}/\Gamma(-3/2)$ and $a = 2^{5/4}/\Gamma(-3/4) - \epsilon^{3/2}/\Gamma(-3/2)$.

18.8.2 *Nested layers*

Consider the equation

$$\epsilon^3 y'' + tan(x)y' - [xsin(x) + \epsilon cot(x)]y = 0, \tag{18.65}$$

with $y(0) = 1$ and $y(\pi/2) = e^{\pi/2}$. The lowest order outside solution is

$$y = Ae^{cosx+xsinx} \tag{18.66}$$

and the layer is on the left, so the boundary condition on the right gives $A = 1$. Also for $x \to 0$ we have $y \to e$.

The Kruskal–Newton diagram leads to very complicated layer equations. However, since the support lines giving the layer are not horizontal, changing the power of x moves a point off the line. Thus expand $tan(x) = x + x^3/3 + 2x^5/15 + ...$, $cot(x) = 1/x - x/3 + ...$

The Kruskal–Newton diagram then takes the form given in Fig. 18.15. The points at $q = 0$ correspond to the expansion of $tan(x)y'$ and the points at $q = 1$ correspond to the expansion of $\epsilon cot(x)y$. For line 2 $x = \epsilon X$ and

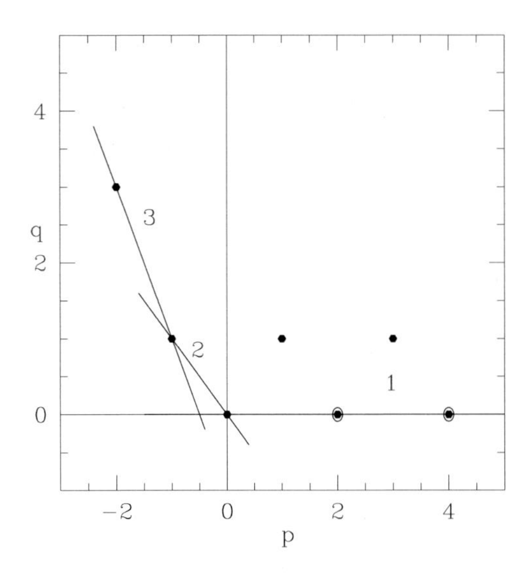

Fig. 18.15 Kruskal–Newton diagram for Eq. 18.65.

to leading order $XY' - Y/X = 0$, giving $Y = Ce^{-1/X}$. Matching to the outside soloution we find $C = 1$. But $X \to 0$ gives $Y \to 0$, so it is still not possible to satisfy the boundary conditions.

The second layer is given by line 3, with slope -2 so we take $x = \epsilon^2 Z$ giving

$$\mathcal{Y}'' - \frac{\mathcal{Y}}{Z} = 0. \qquad (18.67)$$

An integral solution tending to zero for large Z is given by

$$\mathcal{Y} = \int_0^\infty \frac{e^{-zt-1/t}}{t^2} dt \qquad (18.68)$$

and $\mathcal{Y}(0) = 1$, satisfying the boundary condition at the left. Saddle points for large z are located at $t = \pm 1/\sqrt{z}$. The integration contour for the exponentially decreasing solution with a saddle point at positive t, as well as that for the exponentially increasing solution, is shown in Fig. 18.16. The contour passing through the saddle at negative t cannot be deformed to give zero because of the essential singularity at $t = 0$. The uniform solution is then given by

$$y_u = e^{cosx+xsinx} - e + ee^{-\epsilon/x} + \int_0^\infty \frac{e^{-xt/\epsilon^2-1/t}}{t^2} dt. \tag{18.69}$$

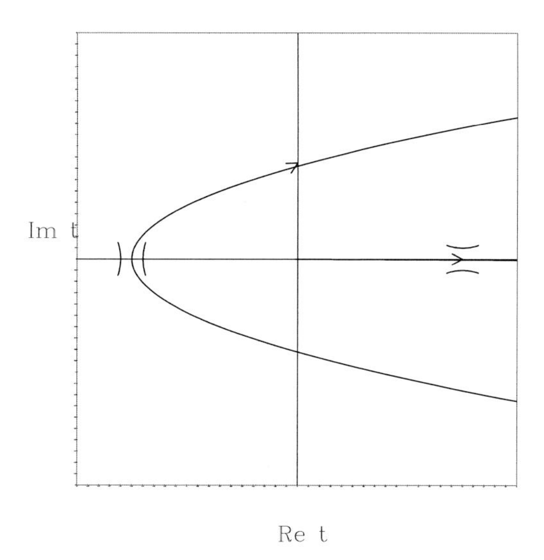

Fig. 18.16 Integration contours for Eq. 18.68.

18.9 Coupled Equations

The tearing mode in a toroidally confined plasma gives an example of the use of Kruskal–Newton diagrams for two coupled second order equations. The linear evolution equations for a single helicity mode (somewhat simplified for this presentation) are given by White [2001] by

$$\psi - x\xi = \frac{1}{\gamma\tau_R}\nabla_\perp^2\psi, \tag{18.70}$$

$$(\gamma\tau_A)^2\nabla_\perp^2\xi = x\left(F\psi - \nabla_\perp^2\psi\right) \tag{18.71}$$

where ψ is the helical magnetic field flux, ξ is the plasma displacement, and $\nabla_\perp^2 = \frac{1}{r}\frac{d}{dr}r\frac{d}{dr} - \frac{m^2}{r^2}$ and $x = r - 1$. The point $r = 1$ is the plasma radius at which the magnetic perturbation of the form $sin(n\phi - m\theta)$ is resonant, with ϕ the toroidal angle, θ the poloidal angle, and the equilibrium field helicity $q(r)$ equal to m/n at $r = 1$. The small parameters are $\gamma\tau_A$ and $1/(\gamma\tau_R)$ and F is order one. A boundary layer occurs at $x = 0$. The outside solutions are immediately given by $\nabla_\perp^2\psi = F\psi$ and $\psi = x\xi$ and the boundary conditions then require that for small x in the case $m = 1$, ξ is order one, constant to the left of the layer, and zero to the right, whereas for $m \neq 1$ instead ψ is continuous but there is a jump in its derivative across the layer $\psi'(0_+) - \psi'(0_-) = \psi\Delta'$.

First examine $m \neq 1$. Expanding $\nabla_\perp^2\xi = \xi'' + x\xi'' + \xi' + R_1$, with R_1 of order 1, substituting $\nabla_\perp^2\psi$, and making use of the fact that for small x the function ψ is continuous at $x = 0$ to treat it as a constant for small x, we thus obtain a single second order differential equation for ξ

$$\psi - x\xi(x) = \frac{1}{\gamma\tau_R}\left[\frac{\gamma^2\tau_A^2(\xi''(x) + x\xi''(x) + \xi'(x) + R_1)}{x} - F\psi\right]. \quad (18.72)$$

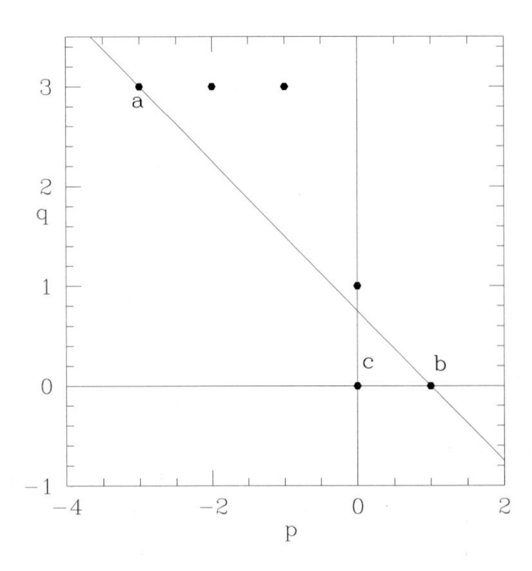

Fig. 18.17 Kruskal–Newton diagram for the $m \geq 2$ tearing mode.

The Kruskal–Newton diagram is shown in Fig. 18.17. The dominant terms are point a, the ξ'' term, point b, the $x\xi$ term, and point c, the constant

ψ appearing on the left side of the equation. Other terms lie above the line connecting points a and b. We have plotted $\gamma^2\tau_A^2/(\gamma\tau_R) \sim \epsilon^3$, but any small ordering would produce the same graph.

However, because there are two dependent variables ψ and ξ in this equation, the graph can be changed by a relative normalization of them. We now use Kruskal's asymptotic principle of maximal complexity. We renormalize ξ through $\xi = \psi(\gamma\tau_R/(\gamma^2\tau_A^2))^{1/4}\chi(z)$ to bring points a, b to lie on the same line as point c. The slope of the line gives the scaling of the layer as $x \sim (\gamma^2\tau_A^2/\gamma\tau_R)^{1/4}$, and we introduce the independent variable z through $x = (\gamma^2\tau_A^2/\gamma\tau_R)^{1/4}z$, and keep only the dominant terms of the Kruskal–Newton diagram, giving

$$\chi'' - z^2\chi = z. \tag{18.73}$$

To solve this equation let $\chi = zf(w)$, with $w = z^2/2$, giving

$$2wf'' + 3f' - 2wf = 1. \tag{18.74}$$

Use a Fourier–Laplace representation $f = \int e^{ws}g(s)ds$ giving upon integration by parts

$$g = A(1 - s^2)^{-1/4}, \qquad 2e^{ws}(1 - s^2)g(s)|_a^b = 1 \tag{18.75}$$

giving for the integration contour $a = 0, b = 1, A = 1/2$ and

$$\chi = -\frac{z}{2}\int_0^1 d\mu e^{-z^2\mu/2}(1 - \mu^2)^{-1/4}. \tag{18.76}$$

For the necessary matching to the exterior solution we have $\int \psi'' dx = \Delta'\psi(0)$ or

$$\gamma^{5/4}\tau_A^{5/4}S^{3/4}\int_{-\infty}^{\infty}\frac{dz}{z}\chi'' = \Delta' \tag{18.77}$$

with $S = \tau_R/\tau_A \gg 1$ and $\gamma\tau_R \gg 1$ and $\gamma\tau_A \ll 1$, as assumed.

To evaluate the integral $I = \int_{-\infty}^{\infty}(dz/z)\chi''$, substitute χ and integrate over z, giving an Euler beta function (see section 8.10)

$$I = \sqrt{\pi/2}\int_0^1 d\mu\frac{\mu^{1/2}}{(1 - \mu^2)^{1/4}} = \pi\Gamma(3/4)/\Gamma(1/4). \tag{18.78}$$

Thus

$$\gamma\tau_A = \left(\frac{\Gamma(1/4)\Delta'}{\pi\Gamma(3/4)}\right)^{4/5} S^{-3/5}. \tag{18.79}$$

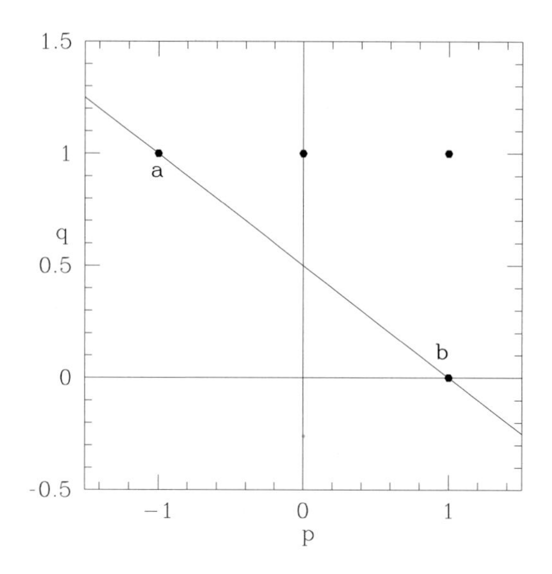

Fig. 18.18 Kruskal–Newton diagram for the $m = 1$ tearing mode.

For the $m = 1$ mode substitute $\psi = xG$ into Eqs. 18.70 and 18.71 giving equations in the order one quantities ξ and G

$$xG - x\xi(x) = \frac{(2G' + xG'' + R_1 + xR_2)}{\gamma\tau_R},$$

$$\gamma^2\tau_A^2\xi''(x) = Fx^2G - x^2G'' - 2xG' - xG - x^2G' - xG \qquad (18.80)$$

where R_1 and R_2 are terms of order one.

The Kruskal–Newton diagrams for these equations are shown in Figs. 18.18 and 18.19. In Fig. 18.18 $\epsilon = 1/(\gamma\tau_R)$ and point a is the derivative terms, point b the xG and $x\xi$ terms. This plot gives a layer scaling of $x \sim 1/\sqrt{\gamma\tau_R}$. In Fig. 18.19 point a is the ξ'' term, and point b the xG' and x^2G'' terms. This plot gives a layer scaling of $x \sim \gamma\tau_A$. In both diagrams there are other smaller terms given by points above these lines. But of course these equations are coupled and there is only one layer, and equating these scalings we find $\gamma \sim \tau_A^{-2/3}\tau_R^{-1/3}$.

Introducing the layer variable $z = (\gamma\tau_R)^{1/2}x$, and keeping only the dominant terms of the Kruskal–Newton diagram and letting $\psi = f/\sqrt{\gamma\tau_R}$ we find the equations

$$f - z\xi = f', \qquad \xi'' = -zf''. \qquad (18.81)$$

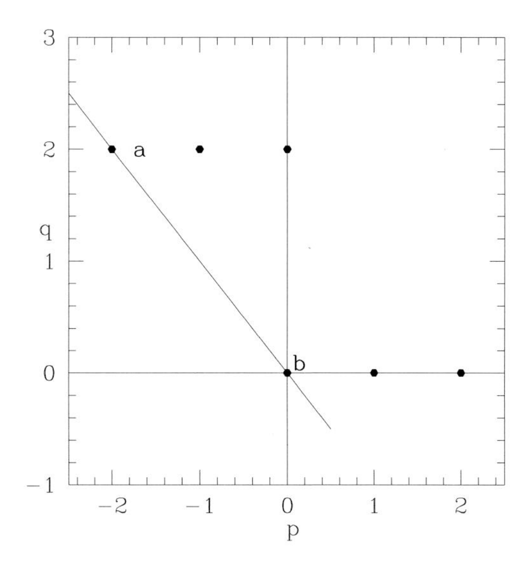

Fig. 18.19 Kruskal–Newton diagram for the $m = 1$ tearing mode.

The second equation is immediately integrated to give $\xi' = -zf' + f$ plus a constant of integration, which must be zero since ξ', f, and f' all tend to zero at infinity. Since ξ must tend to zero for large positive z, but to a constant for large negative z, try the form

$$\xi = \int_z^\infty w(s)ds. \tag{18.82}$$

We then have $\xi' = -w = -zf' + f$. But this suggests the form

$$f = z\int_z^\infty w(s)ds - w(s). \tag{18.83}$$

Substituting we find $w'(z) = -zw$, or $w(z) = e^{-z^2/2}$.

We then have an exact solution which satisfies the boundary conditions $\xi \to \xi_0$, $x \to -\infty$, and $\xi \to 0$, $x \to \infty$ and $\psi \to 0$, $x \to +\infty$ and $\psi \to -\xi_0 x$, $x \to -\infty$ given by

$$\xi = \frac{\xi_0}{\sqrt{2\pi}}\int_z^\infty e^{-z^2/2}dz,$$

$$\psi = \frac{\xi_0}{\sqrt{2\pi}(\gamma\tau_R)^{1/2}}\left[e^{-z^2/2} - z\int_z^\infty e^{-z^2/2}dz\right] \tag{18.84}$$

and $\gamma = \tau_R^{-1/3}\tau_A^{-2/3}$.

18.10 The Nonlinear Cole Equation

Consider the nonlinear equation

$$\epsilon y'' + yy' - y = 0, \tag{18.85}$$

which has been studied by Cole [1968]. Take boundary conditions $y(0) = A$, $y(1) = B$. The Kruskal–Newton diagram is given in Fig. 18.20 where the nonlinear term is shown with an open circle. This diagram indicates an outside solution given by line 1 with $y = x + constant$ or $y = 0$. Note that all successive iterations either using perturbation theory or Newtonian iteration give zero there are no higher order corrections.

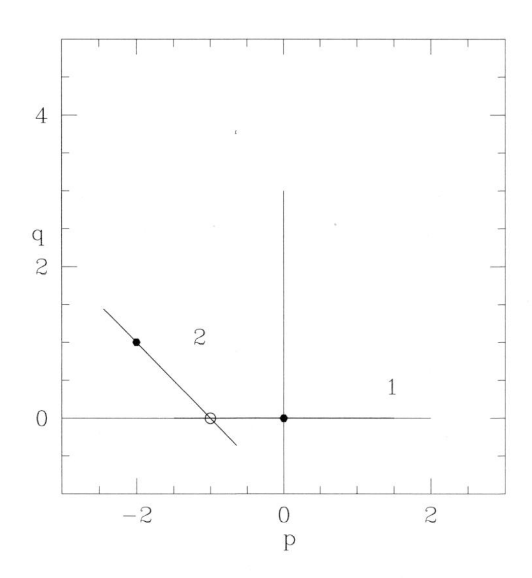

Fig. 18.20 Kruskal–Newton diagram for Eq. 18.85.

The layer location is determined by the sign of the function multiplying the derivative term. We will see that for $A + B \geq 1$ and $B > 0$ there is a layer on the left, and we first consider this case. The outside solution is given by $y = x + B - 1$, or $y = 0$. The layer is given by line 2 of Fig. 18.20 with $x = \epsilon X$. The lowest order equation is

$$Y'' + YY' = 0, \tag{18.86}$$

immediately integrating to $Y' + Y^2/2 = c$. Furthermore since $Y' > 0$ if $Y < \sqrt{2c}$ and $Y' < 0$ if $Y > \sqrt{2c}$ we find for large X that $Y \rightarrow \sqrt{2c}$. Matching to the outside solution we then have $c = (B-1)^2/2$. Consider the large X expansion of the solution $Y = B - 1 + f$, with $f \ll 1$ giving $f' + (B-1)f + f^2/2 = 0$. For $f \ll 1$ we then have $f \simeq e^{-\alpha X}$ with $\alpha = B - 1$. But iteration using the Newton method will introduce terms of the form $e^{-2\alpha X}$ etc., so we take the ansatz $Y_0 = \sum_0^\infty c_n e^{-n\alpha X}$. Substituting into the differential equation we find

$$c_n = \frac{1}{2}\sum_0^n \frac{c_m c_{n-m}}{n c_0} \tag{18.87}$$

with the solution $c_n = 2c_0\beta^n$, for $n \neq 0$ with $\beta = c_1/(2c_0)$ undetermined. The boundary condition at the left edge fixes β through $\sum_0^\infty c_n = A$, giving the solution

$$Y_0 = \frac{2c_0}{1 - \beta e^{-\alpha X}} - c_0, \tag{18.88}$$

and for $\alpha > 0$ $c_0 = B - 1$, and $\beta = (A - B + 1)(A + B - 1)$.

Note that $B \rightarrow 1$ is a singular limit, and we have in this case the solution $Y = 2/(c + X)$ which for $Y(0) = A$ gives

$$Y_0 = \frac{2A}{2 + AX}. \tag{18.89}$$

Now consider next order. Note that for any nonlinear equation the perturbation expansion always leads to linear equations for the higher order corrections. Substituting $Y = Y_0 + \epsilon Y_1$ we find the linear inhomogeneous equation

$$Y_1'' + Y_0 Y_1' + Y_0' Y_1 = Y_0. \tag{18.90}$$

For the singular case Eq. 18.89 we have the general solution

$$Y_1 = c_1 Z + c_2 Z^{-2} + (4/3) Z ln(Z) \tag{18.91}$$

with $Z = (2 + AX)/(2A)$. But $Y_0 + \epsilon Y_1$ cannot be matched to the outside solution $y = const + x$ because of the $Zln(Z)$ term, and to go to higher order an additional term must be added, making the solution have the form $y = Y(Z) + \epsilon ln\epsilon Y_{11}(Z) + \epsilon Y_{11}(Z)$.

Now consider the general case, Eq. 18.88. Since the Y_0 appearing in the equation is of the form $\sum_n e^{-n\alpha X}$ we try the ansatz $Y_{1,h} = \sum_0^\infty f_n e^{-n\alpha X}$. Substituting into the differential equation we find for the homogeneous equation

$$n(n-1)f_n = n\sum_0^{n-1} c_{n-m}f_m. \tag{18.92}$$

Substituting the value for c_n and letting $\alpha^2 f_n = F_n\beta^n$ we find the solution $F_n = nF_1$ with F_1 free. We then find the homogeneous solution

$$Y_{1,h} = \frac{F_1}{\alpha}\sum_1^\infty n\beta^n e^{-n\alpha X} = \frac{F_1\beta e^{-\alpha X}}{\alpha(1-\beta e^{-\alpha X})^2}. \tag{18.93}$$

Now look for a particular solution of the inhomogeneous equation. Also note that to match the exterior solution in the asymptotic domain $\epsilon \ll x \ll 1$ to order ϵ we need to take $Y_1 \to X$, so that $\epsilon Y_1 \to x$. Thus take as ansatz

$$Y_{1,i} = \sum_0^\infty d_n e^{-n\alpha X} + X. \tag{18.94}$$

Also note that

$$\begin{aligned}\sum_1^\infty \alpha n c_n e^{-n\alpha X}X &= \sum_1^\infty nc_n e^{-n\alpha X}(1-e^{-\alpha X}) \\ &= \sum_1^\infty [nc_n - (n-1)c_{n-1}]e^{-n\alpha X}\end{aligned} \tag{18.95}$$

where we have neglected terms of order $c_n X^2 e^{-n\alpha X}$. Substituting into the differential equation we find

$$n(n-1)\alpha d_n = n\sum_0^{n-1} c_{n-m}d_m + \frac{nc_n}{\alpha} - \frac{(n-1)c_{n-1}}{\alpha}. \tag{18.96}$$

This equation has a solution with $\alpha d_n = D_n\beta^n$, $D_0 = -1$, and

$$D_n = n\left(D_1 - \frac{\gamma_n}{2\beta}\right) \qquad n \geq 1 \tag{18.97}$$

with

$$\gamma_n = \frac{2}{n(n-1)}\sum_1^{n-1} m\gamma_m + \frac{4}{n^2}, \qquad \gamma_1 = 0. \tag{18.98}$$

To evaluate $\gamma = lim_{n\to\infty}\gamma_n$ note that $\gamma_{n+1} - \gamma_n = \frac{4}{n^2(n+1)^2}$. Using $\gamma_1 = 0$ we have

$$\gamma_n = \sum_1^{n-1}(\gamma_{m+1} - \gamma_m) = 4\sum_1^{n-1}\frac{1}{m^2(m+1)^2}, \tag{18.99}$$

and using $1/(m^2(m+1)^2) = (1-2m)/m^2 + 1/(m+1)^2 + 2/(m+1)$ we find

$$\gamma = 8\zeta(2) - 12 = 4\pi^2/3 - 12 = 1.15947326718\ldots. \tag{18.100}$$

To optimize convergence of the series for $Y_{1,i}$ we take $D_1 = \gamma/(2\beta)$. Finally we have

$$Y_1(X) = \frac{1}{2\beta\alpha}\sum_1^{\infty} n(\gamma - \gamma_n)\beta^n e^{-n\alpha X} + X + \frac{F_1\beta e^{-\alpha X}}{\alpha(1-\beta e^{-\alpha X})^2}. \tag{18.101}$$

It is easily seen that $(\gamma - \gamma_n) \simeq 4/3n^3$ so this series is convergent for all $X \geq 0$ in the range $-1 \leq \beta \leq 1$, which restricts the possible values of A to lie in the range $0 \leq A < \infty$. Thus there do not exist first order series solutions for arbitrary boundary conditions.

Table 18.1

The Sequence γ_n				
n	$\gamma_{n+1} - \gamma_n$	$\gamma - \gamma_n$	$n(\gamma - \gamma_n)$	$n^2(\gamma - \gamma_n)$
1	1	1.1594732672	1.1594732672	1.1594732672
2	$\frac{1}{9}$	.15947326718	.31894653435	.63789306871
3	$\frac{1}{36}$	.04836215606	.14508646820	.43525940459
4	$\frac{1}{100}$	.02058437829	.08233751315	.32935005261
5	$\frac{1}{225}$	.01058437829	.05292189144	.26460945721
6	$\frac{1}{441}$	.00613993384	.03683960306	.22103761838

The value of F_1 is fixed by the requirement that $Y_1(0) = 0$, giving

$$\beta F_1(\beta) = -\frac{(1-\beta)^2}{2}\sum_1^{\infty} n(\gamma - \gamma_n)\beta^{n-1}. \tag{18.102}$$

The series for $F_1(\beta)$ converges rapidly, so only a few terms are necessary to obtain an accurate value. Note that the derivative at $X = 0$

$$Y_1'(0) = -\frac{1}{2\beta}\sum_1^{\infty} n^2(\gamma - \gamma_n)\beta^n + 1 - \frac{F_1\beta}{(1-\beta)^2} + \frac{2F_1\beta}{(1-\beta)^3} \tag{18.103}$$

involves the sum $\sum_1^{\infty} n^2(\gamma - \gamma_n)\beta^n$ which can readily seen to be convergent in the range $-1 \leq \beta < 1$.

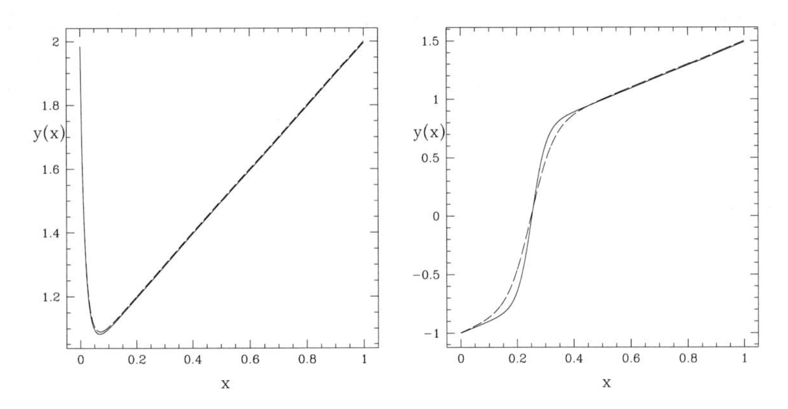

Fig. 18.21 Numerical and first order uniform solutions to the Cole equation, $\epsilon = 0.02$, $A = 2$, $B = 2$, $\beta = 1/3$ and $A = -1$, $B = 1.5$, $\beta = -1$.

The uniform solution for $B > 1$ is then

$$y(x) = \frac{\epsilon}{2\beta\alpha}\sum_{1}^{\infty} n(\gamma - \gamma_n)\beta^n e^{-n\alpha x/\epsilon} + x$$

$$+\frac{\epsilon F_1 \beta e^{-\alpha x/\epsilon}}{\alpha(1-\beta e^{-\alpha x/\epsilon})^2} + \frac{2(B-1)}{1-\beta e^{-\alpha x/\epsilon}} - B + 1. \tag{18.104}$$

There is also a large class of cases in which the layer is internal, when $-1 < A{+}B < 1$, and $B > 0$, $A < 0$. For all these solutions y crosses through zero with positive slope, so an internal layer solution is possible. Let the crossing point be x_0, and defining the variable Z through $x - x_0 = \epsilon Z$ we again find Eq. 18.86. Also noting that if $Y(Z)$ is a solution, so is $-Y(-Z)$, we find a solution very similar to the previous, but the boundary condition at $Z = 0$ is that $Y = 0$, giving $\beta = -1$ and

$$Y_0(Z) = \begin{cases} \frac{2(B-1+x_0)}{1+e^{-\alpha Z}} - B + 1 - x_0, & Z > 0 \\ \frac{2(A+x_0)}{1+e^{\alpha Z}} - A - x_0, & Z < 0 \end{cases} \tag{18.105}$$

with the layer location determined in the narrow layer width approximation to be $x_0 = (1 - A - B)/2$.

The next order solution Y_1 is also constructed using the solution from the first case, $B > 1$. For $Z > 0$ the solution is given by Eq. 18.101 with X replaced by Z, and for $Z < 0$ the solution is given by minus Eq. 18.101 with X replaced by $-Z$.

We also encounter another layer phenomenon, which happens if $B < 1$ and $A > 0$. Namely the layer solution matches on to the exterior solution

$y = 0$ rather than to the solution $x + constant$, at the point $x_0 = 1 - B$. Since the asymptotic value is zero, the layer solution is then given by Eq. 18.89. Nevertheless there is layer phenomenon near the point where the outside solution would cross through zero, and the integration constants are determined by matching both function and derivative at this point. In this case the lowest order uniform solution is

$$Y_0(Z) = \frac{2A}{2 + AX} + \begin{cases} \frac{2\sqrt{\epsilon}}{2+\sqrt{\epsilon}Z} + x - x_0, & Z > 0 \\ \frac{2\sqrt{\epsilon}}{2-\sqrt{\epsilon}Z}, & Z < 0 \end{cases} \tag{18.106}$$

with $x = \epsilon X$, $x - x_0 = \epsilon Z$, and $x_0 = 1 - B$, and the function is equal to $\sqrt{\epsilon}$ at $x = x_0$.

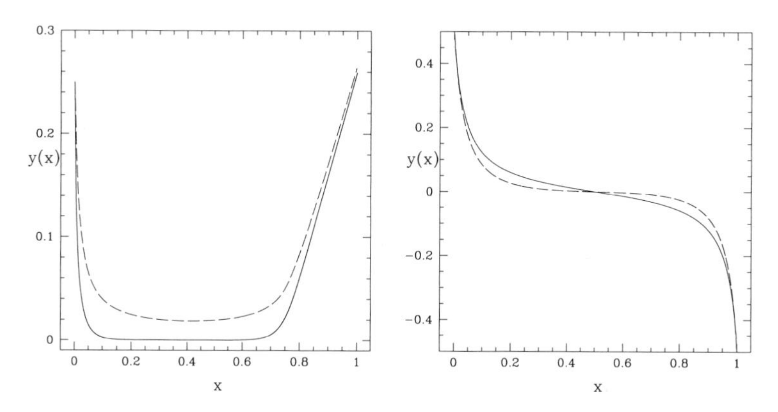

Fig. 18.22 Numerical and lowest order uniform solutions to the Cole equation, for $\epsilon = 0.002$, in domain c, $A = .25$, $B = .25$, and in the lower right quadrant, $\epsilon = 0.01$ $A = .5$, $B = -.5$.

In the lower right quadrant $A > 0, B < 0$, there cannot be an internal layer because the slope would be negative, so the solution forms two layers matching to the outside solution $y = 0$, as shown in Fig. 18.22. In this case the lowest order uniform solution is

$$Y_0(Z) = \frac{2A}{2 + AX} + \frac{2B}{2 + BZ} \tag{18.107}$$

with $x = \epsilon X$, $x - 1 = \epsilon Z$.

Finally we note that the original equation, along with the boundary conditions, is invariant under the substitutions

$$y \to -y, \qquad x \to 1 - x, \qquad A \to -B, \qquad B \to -A, \tag{18.108}$$

and this completes the space of solution. The space of solutions is shown in Fig. 18.24, with symmetry about the line $A + B = 0$. In the two upper right triangular sections with $-1 < \beta < 1$ the layer is on the left, with an example for $\beta = 1/3$ shown in Fig. 18.21. In the central part of the left upper quadrant the layer is internal with $\beta = -1$, also with an example shown in Fig. 18.21. The numerical integrations were done using an initial guess for $y'(0)$ given by $Y_0'(0) + \epsilon Y_1'(0)$, which is already quite accurate, and iterating until the boundary conditions were matched. If the layer is not close to $x = 0$ then $y'(0)$ is equal to 1 plus an exponentially small correction. In all these regions the series solutions converge, giving analytic expressions for the first order uniform solutions.

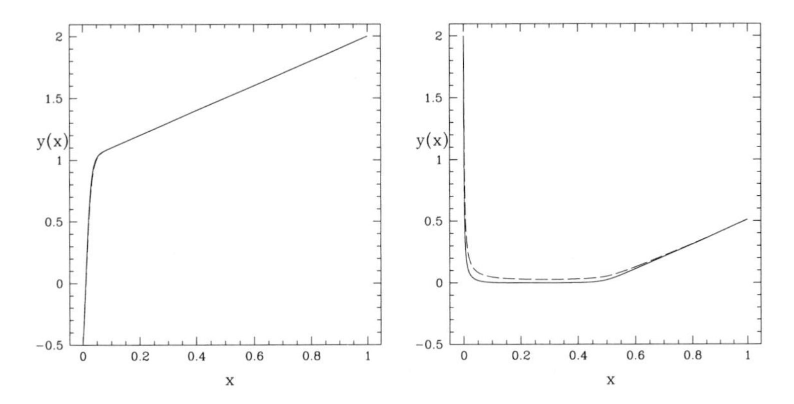

Fig. 18.23 Numerical and lowest order uniform solutions to the Cole equation, for $\epsilon = 0.002$, in domain a, $A = -.5$, $B = 2$, $\beta = -3$, and in domain b, $A = 2$, $B = .5$.

In section a we have $\beta < -1$, and the series expression for Y_1 is divergent. In sections b and c Y_0 is given by the singular limit Eq. 18.89. Examples of solutions in domains a, b, and c are shown in Fig. 18.23. In all these domains there is a layer on the left, but in domains b and c there is a second layer at the point where the two external solutions $y = 0$ and $y = x + constant$ meet.

In the lower left quadrant with $-1 < \beta < 1$ solutions are obtained with the substitution given by Eq. 18.108 and the layer is on the right. Similarly solutions exist in the domains obtained by reflecting a, b, c across the line $A = -B$.

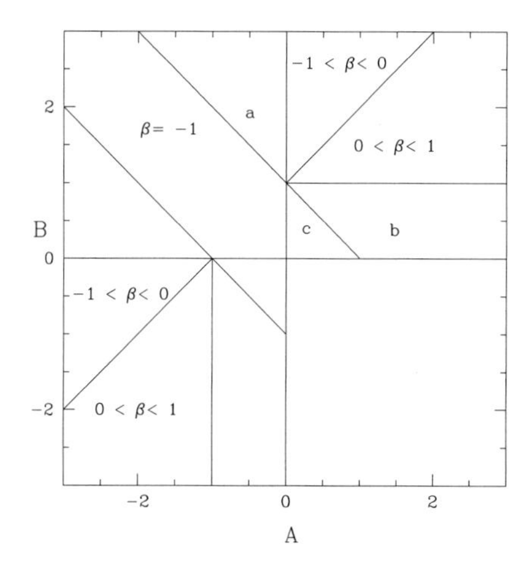

Fig. 18.24 Solution domains for Eq. 18.85. Series solutions converge in all domains but a, b and c.

18.11 Failure of Asymptotic Matching

Asymptotic matching of inner and outer solutions can fail, and a solution still exist. Consider the equation[2]

$$\epsilon y'' + x^2 y' + y = 0 \tag{18.109}$$

with boundary conditions given at $x = 0$ and $x = 1$. Choose $y(0) = 1$, $y(1) = e$, although the analysis can be completed for any values. The Kruskal–Newton diagram is shown in Fig. 18.25.

The outside solution, given by line 1 is $y = Ce^{1/x}$ which can be matched to the boundary condition at $x = 1$ giving $C = 1$. The outside equation has an essential singularity at $x = 0$, and the solution is large for small x. Support line 2 gives the layer as $x = \sqrt{\epsilon}X$, giving the layer solution

$$y = AcosX + BsinX \tag{18.110}$$

and matching to the left boundary condition gives $A = 1$. But the outside solution is singular as $x \to 0$ and asymptotic matching of these two solutions fails, there existing no common matching domain.

Neglect of the $x^2 y'$ term for the inner solution with $y \sim cos(x/\sqrt{\epsilon})$ requires $x \ll \epsilon^{1/4}$. But neglect of the $\epsilon y''$ term for the outer solution with

[2] I am indebted to Vasily Geyko for bringing this equation to my attention.

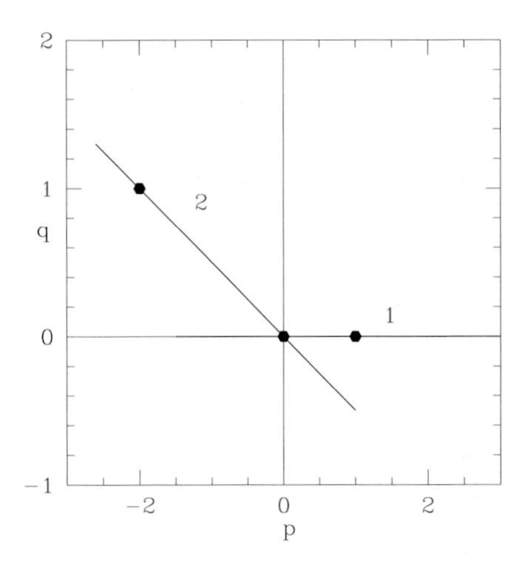

Fig. 18.25 Kruskal–Newton diagram for Eq. 18.109.

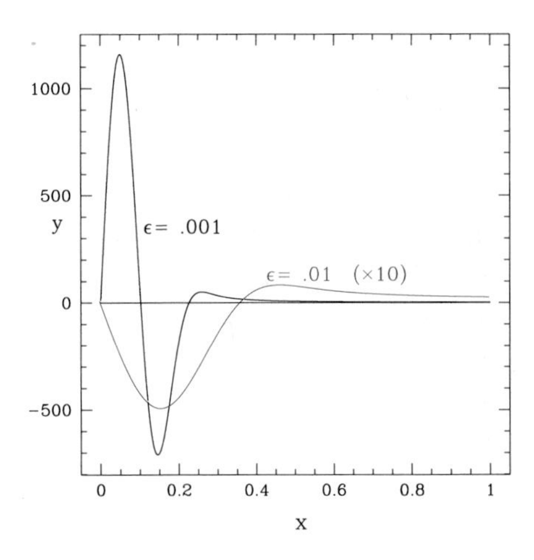

Fig. 18.26 Numerical solutions of Eq. 18.109 with $y(0) = 1$, $y(1) = e$, $\epsilon = 0.001$ and $\epsilon = 0.01$.

$y \sim exp(1/x)$ requires $x \gg \epsilon^{1/4}$, and thus there is no point at which these solutions can be matched. Numerical integration indicates that the solution is indeed sinusoidal for $x \ll \epsilon^{1/4}$ and exponential for $x \gg \epsilon^{1/4}$, as shown in Fig. 18.26 for $\epsilon = .001$ and for $\epsilon = .01$, but these solutions cannot be obtained using asymptotic matching.

18.12 Boundary Layers in the Complex Plane

Asymptotic matching in the complex plane is a strategy for calculating exponentially small terms that has been developed by Kruskal and Segur.

To apply the method determine the singularities of the leading order term of the naive asymptotic expansion and find an inner equation in the vicinity. Find the exponentially small terms in the far field of the inner solution and match them with the possible exponentially small corrections to the naive expansion. Because the inner problem exhibits Stokes phenomenon in its far field [1962; 1979], the exponentially small corrections are only present in certain sectors. If the naive perturbation expansion contains only trivial information, the exponentially small corrections can play an important role when they control qualitative new physical phenomena or if they prevent the existence of solutions.

18.12.1 *Crystal growth*

We consider the equation

$$\epsilon\psi''(s) + \psi(s) = \frac{1}{1+s^2}, \tag{18.111}$$

with boundary conditions $\psi \to 0$ as $s \to \pm\infty$. This problem is the steady state version of a problem considered by Chapman and Mortimer. It is motivated by the geometrical model for crystal growth. Chapman and Mortimer used Stokes smoothing to show that no steady state solution exists although a naive perturbation expansion indicates the contrary. We reproduce this result using the Kruskal–Segur method.

Expanding $\psi = \psi_0 + \epsilon\psi_1 \ldots$, we find at leading order

$$\psi_0(s) = \frac{1}{1+s^2}. \tag{18.112}$$

The result satisfies the boundary conditions and it can be verified that all higher order terms do also. The leading order term has singularities at $s = \pm i$. We first study the vicinity of $s = -i$. It is useful to shift the singularity to the origin by a change of the independent variable, $s = -i+it$, $\psi(-i+it) = \phi(t)$. Equation 18.111 transforms to

$$-2\epsilon t\ddot{\phi}(t) + \epsilon t^2\ddot{\phi}(t) + 2t\phi(t) - t^2\phi(t) = 1. \tag{18.113}$$

The Kruskal–Newton diagram of this equation is shown in Fig. 18.27. Support line 1 marks the outer balance giving at leading order $\phi_0(t) = 1/[t(2-t)]$ in agreement with Eq. 18.112. Possible exponentially small corrections to the outer balance involve superpositions of the WKBJ solutions of the homogeneous version of Eq. 18.113

$$\phi_{\pm} = e^{\pm t/\sqrt{\epsilon}}. \tag{18.114}$$

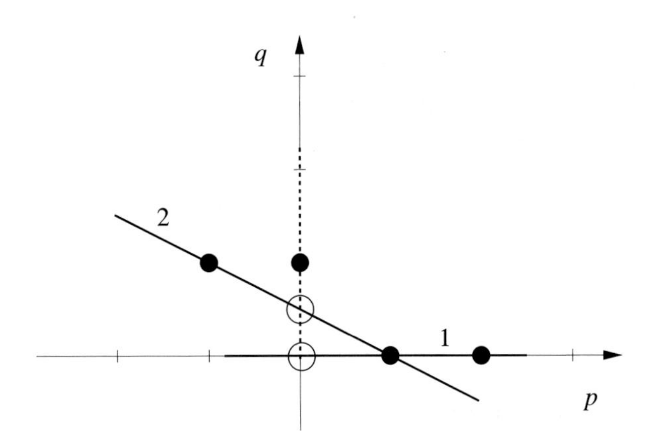

Fig. 18.27 The Kruskal–Newton diagram for Eq. 18.113. The dotted line represents the inhomogeneity and empty circles mark intersection points with the support lines.

The inner equation is marked by support line 2. The scaling of the inner variables can be read off from the diagram: $t = \epsilon^{1/2}\tau$ and $\phi = \epsilon^{-1/2}\Phi$ and the inner equation reads

$$-2\tau\frac{d^2\Phi}{d\tau^2} + 2\tau\Phi = 1.$$

A solution of this equation is given by

$$\Phi(\tau) = \frac{1}{2}\int_C \frac{e^{-\tau z}}{1-z^2}dz. \tag{18.115}$$

The path C starts at $z = 0$ and goes to $+\infty$ in the right half plane. We choose the singularity $z = 1$ to lie below the path. We can deform this contour to the steepest descent contour $arg(\tau z) = 0$ without passing through $z = 1$ provided $Im(\tau) < 0$. Continuing to $Im(\tau) > 0$ the contour must be indented to include a clockwise circuit around the pole, Fig. 18.28. Thus, the line $Im(\tau) = 0$ is a Stokes line for $\Phi(\tau)$ and we pick up an extra pole contribution

$$\Phi_{pole} = -\pi i Res(e^{-\tau z}/(1-z^2), z=1) = -\frac{\pi}{2} i e^{-\tau}$$

when crossing it. The end point contribution $\Phi_{ep} \sim 1/2\tau$ is present everywhere.

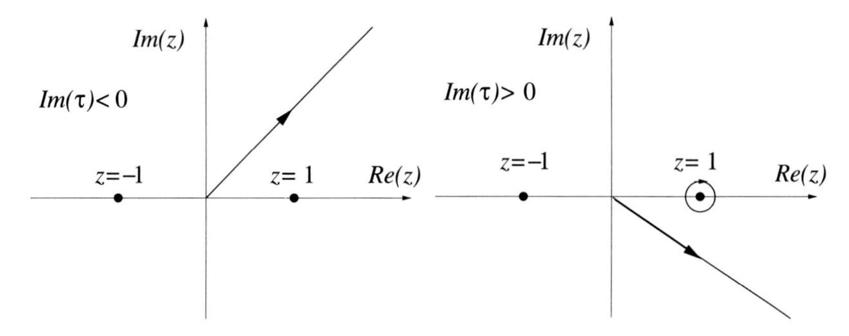

Fig. 18.28 The steepest decent path for the integral Eq. 18.115 (left), showing the contribution from the pole (right).

Thus,

$$\Phi(\tau) \sim \begin{cases} 1/2\tau & Im(\tau) > 0 \\ 1/2\tau - \frac{\pi}{4} i e^{-\tau} & Im(\tau) = 0 \\ 1/2\tau - \frac{\pi}{2} i e^{-\tau} & Im(\tau) < 0 \end{cases} .$$

The outer solution $\phi_0(t)$ must be generalized by adding the WKBJ solutions $\phi_\pm(t)$, Eq. 18.114, as needed:

$$\phi(t) \sim \begin{cases} \phi_0(t) & Im(t) > 0 \\ \phi_0(t) + \frac{a}{2}\phi_-(t) & Im(t) = 0 \\ \phi_0(t) + a\phi_-(t) & Im(t) < 0 \end{cases} .$$

The constant a can be found by matching the inner and outer expansion along the ray $Im(t) = 0$. We find

$$a = -\frac{\pi}{2} i \epsilon^{-1/2}.$$

Extrapolating back to the real s-axis we also have to take into account the contributions from the singularity at $s = i$. Since the solution $\psi(s)$ must be real on the real s-axis, this other contribution is the complex conjugate. We arrive at the uniformly valid solution for real s satisfying the boundary conditions for $s \to -\infty$:

$$\psi(s) \sim \begin{cases} 1/(1+s^2) & s < 0 \\ 1/(1+s^2) + \pi\epsilon^{-1/2}e^{-1/\sqrt{\epsilon}}sin(s/\sqrt{\epsilon}) & s > 0 \end{cases}.$$

The result shows that the exact solution is not well approximated by $1/(1+s^2)$ for $s >> 1$. At large s, the exponentially small (in ϵ) correction becomes visible. The solution fails to satisfy the boundary condition as $s \to +\infty$.

To confirm the Kruskal–Segur method we obtain an exact solution. Perform a Fourier transformation of Eq. 18.111

$$\int_{-\infty}^{\infty} dze^{-kz}[\epsilon\psi''(z) + \psi(z)] = \int_{-\infty}^{\infty} dze^{-kz}\frac{1}{1+z^2} \tag{18.116}$$

and write $\psi(z) = \int_{-\infty}^{\infty} dze^{ikz}\phi(k)$. Then using

$$\int_{-\infty}^{\infty} e^{-ikz}\frac{dz}{1+z^2} = \pi e^{-|k|}, \qquad \int_{-\infty}^{\infty} e^{ikz}e^{-|k|}dk = \frac{2}{1+z^2} \tag{18.117}$$

we find $\phi(k) = e^{-|k|}/[2(1-\epsilon k^2)]$ and an exact solution to the inhomogeneous differential equation

$$\psi(z) = \frac{1}{2}\int_{-\infty}^{\infty} e^{ikz}\frac{e^{-|k|}}{1-\epsilon k^2} = Re\int_{0}^{\infty} e^{ikz}\frac{e^{-k}}{1-\epsilon k^2}. \tag{18.118}$$

Now look at $z \to +\infty$. Writing the integrand as $exp(ikz - k - ln(1 - \epsilon k^2))$. we find a saddle point for large $|z|$ at $k = 1/\sqrt{\epsilon} - i/z$ giving a contribution of $\psi_s(z) = cos(z/\sqrt{\epsilon})/\sqrt{2\epsilon}$. In addition there is an end point contribution from $k \simeq 0$, obtained to lowest order by dropping the ϵk^2 term in the denominator and using Eq. 18.117, giving $\psi_e(z) = 1/(1+z^2)$ so for large positive z the integral solution is asymptotic to $\psi(z) \simeq 1/(1+z^2) + cos(z/\sqrt{\epsilon})/(\sqrt{2\epsilon})$. We recognize the second term as a solution to the homogeneous equation, and thus the solution

$$\psi(z) = Re\int_{0}^{\infty} e^{ikz}\frac{e^{-k}}{1-\epsilon k^2} - \frac{cos(z/\sqrt{\epsilon})}{\sqrt{2\epsilon}}. \tag{18.119}$$

is asymptotic to $1/(1+z^2)$ for large positive z, and thus satisfies the requisite boundary condition.

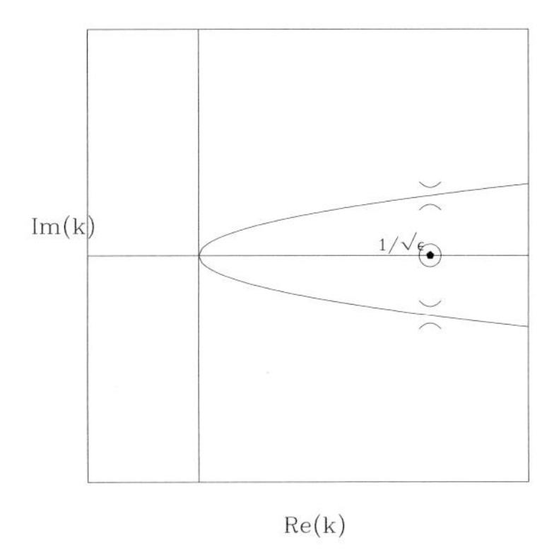

Fig. 18.29 Integration paths for $z > 0$ (lower) and $z < 0$ (upper) showing saddle points.

Now continue to negative z. The integration path must be moved to the new saddle point above the real axis, as shown in Fig. 18.29, producing an additional circular path around the pole at $k = 1/\sqrt{\epsilon}$. Thus for negative z we have

$$\psi(z) = Re \int_0^{\infty} e^{ikz} \frac{e^{-k}}{1-\epsilon k^2} - \frac{cos(z/\sqrt{\epsilon})}{\sqrt{2\epsilon}} + \frac{\pi e^{-1/\sqrt{\epsilon}}}{\sqrt{\epsilon}} sin(z/\sqrt{\epsilon}) \quad (18.120)$$

with the integration path now passing through the upper saddle point. Evaluating again the end point and saddle point contributions we have for large negative z $\psi(z) \simeq 1/(1+z^2)+\pi e^{-1/\sqrt{\epsilon}} sin(z/\sqrt{\epsilon})/\sqrt{\epsilon}$ so it is impossible to require $\psi(z) \to 0$ for $z \to \pm\infty$.

18.12.2 *Viscous fingering*

We consider the equation

$$\epsilon\psi''(s) + (1+s^2)^2\psi(s) = 1, \quad (18.121)$$

with boundary condition $s\psi(s) \to 0$ as $s \to \pm\infty$. Since we expect this problem to have no solution, we also discuss the following modifications

$$\epsilon\psi''(s) + [(1+s^2)^2 + a]\psi(s) = 1, \quad (18.122)$$

and

$$\epsilon\psi''(s) + [(1+s^2)^2 + a + b/(1+s^2)^2]\psi(s) = 1, \tag{18.123}$$

where both a and b are small parameters. These equations lead to inner equations that have been discussed in the context of viscous fingering by Tanveer [2000].

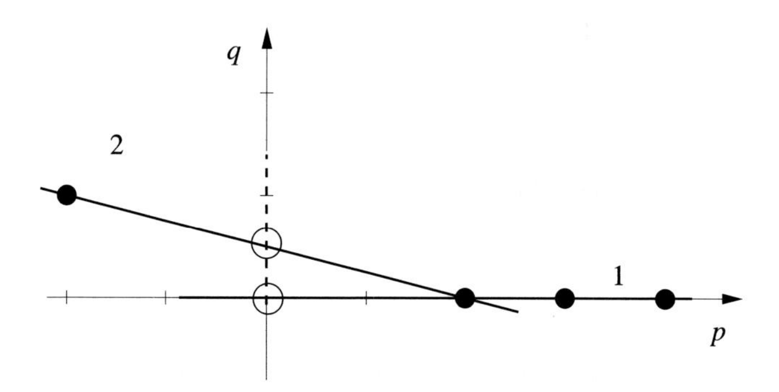

Fig. 18.30 The Kruskal–Newton diagram for Eq. 18.124.

18.12.3 *Model equation with $a = b = 0$*

We first consider Eq. 18.121. Expanding $\psi(s) = \psi_0(s) + \epsilon\psi_1(s)\ldots$ we find at leading order $\psi_0(s) = 1/((1+s^2)^2)$. This term satisfies the boundary conditions at infinity and the higher order terms do also. The leading order term is singular at $s = \pm i$ and we study the vicinity of $s = -i$. Using shifted coordinates $s = -i + it$, $\psi(-i+it) = \phi(t)$, Eq. 18.121 transforms to

$$-\epsilon\ddot{\phi}(t) + (4t^2 - 4t^3 + t^4)\phi(t) = 1. \tag{18.124}$$

The Kruskal–Newton diagram for this equation is shown in Fig. 18.30. Line 1 gives the outer balance and we get at leading order $\phi_0(t) = 1/(t^2(2-t)^2)$. The possible exponentially small corrections are the WKBJ solutions of the homogeneous version of Eq. 18.124:

$$\phi_\pm(t) = \frac{1}{\sqrt{t(2-t)}}e^{\pm t^2(3-t)/3\sqrt{\epsilon}}. \tag{18.125}$$

Line 2 gives the inner equation. We find from the diagram $t = \epsilon^{1/4}\tau$, $\phi = \epsilon^{-1/2}\Phi$, and

$$-\frac{d^2\Phi}{d\tau^2} + 4\tau^2\Phi = 1. \tag{18.126}$$

Seeking a solution that vanishes for large τ in the sector $0 < \arg(\tau) < \pi/2$ a WKBJ analysis indicates the following behavior:

$$\Phi(\tau) \sim \begin{cases} 1/4\tau^2 & 0 < arg(\tau) < \frac{\pi}{2} \\ 1/4\tau^2 + \frac{A}{2}\frac{1}{\sqrt{\tau}}e^{-\tau^2} & arg(\tau) = 0 \\ 1/4\tau^2 + A\frac{1}{\sqrt{\tau}}e^{-\tau^2} & -\frac{\pi}{2} < arg(\tau) < 0 \end{cases}. \tag{18.127}$$

The constant A has to be determined numerically or from an exact solution of Eq. 18.126. We have to add the WKBJ solutions $\phi_{\pm}(t)$, Eq. 18.125, to the naive regular expansion as needed:

$$\phi(t) \sim \begin{cases} 1/t^2(2-t)^2 & 0 < arg(t) < \frac{\pi}{2} \\ 1/t^2(2-t)^2 + \frac{B}{2}\phi_-(t) & arg(t) = 0 \\ 1/t^2(2-t)^2 + B\phi_-(t) & -\frac{\pi}{2} < arg(t) < 0 \end{cases}.$$

The constant B can be related to A by matching along the ray $arg(t) = 0$: $B = \sqrt{2}\epsilon^{-3/8}A$.

Extrapolating back to the real s-axis and adding the complex conjugate, which is the contribution from $s = i$, we find the uniformly valid solution satisfying the boundary condition at $s = -\infty$:

$$\psi(s) \sim \begin{cases} 1/(1+s^2)^2 & s < 0 \\ 1/(1+s^2)^2 + 2^{-1/2}\epsilon^{-3/8}e^{-2/3\sqrt{\epsilon}}Acos((s^3/3+s)/\sqrt{\epsilon})/\sqrt{1+s^2} & s > 0 \end{cases}.$$

Due to the presence of the exponentially small correction the solution does not satisfy the boundary condition as $s \to +\infty$.

18.12.4 Model equation with $a \neq 0$ and $b = 0$

Seeking a model that possesses solutions we may discuss the modified Eq. 18.122. For a small, the leading order outer solution is not modified. It is singular at $s = -i$ and we use shifted coordinates, $s = -i + it$, $\psi(-i + it) = \phi(t)$. Equation 18.122 transforms to

$$-\epsilon\ddot{\phi}(t) + (4t^2 - 4t^3 + t^4 + a)\phi(t) = 1. \tag{18.128}$$

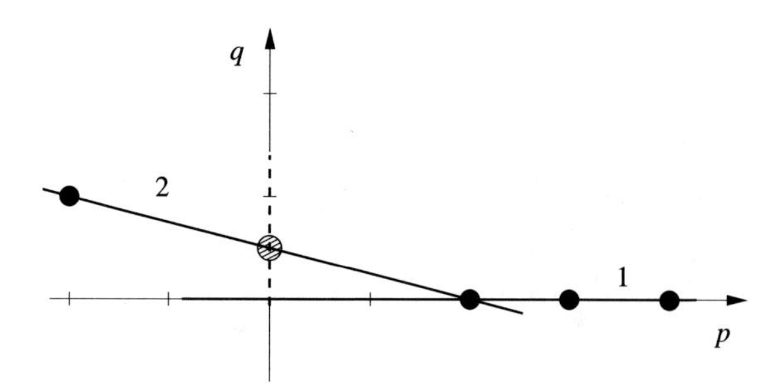

Fig. 18.31 Kruskal–Newton diagram for Eq. 18.128. The shaded dot represents the term $a\phi$ with a scaled to lie on support line 2.

The Kruskal–Newton diagram for this equation is shown in Fig. 18.31. The parameter a has to be scaled according to $a = \alpha\epsilon^{1/2}$, with α of order unity, for the corresponding dot to lie on support line 2 and not below. The modified inner equation reads

$$-\frac{d^2\Phi}{d\tau^2} + (4\tau^2 + \alpha)\Phi = 1. \tag{18.129}$$

The solution of this equation for $\tau >> 1$ is still of the form in Eq. 18.127 but with the parameter A being a function of α, $A = A(\alpha)$. If α is chosen such that $A(\alpha) = 0$, we do not pick up an exponentially small contribution when crossing the Stokes line $Im(t) = 0$ and the boundary conditions can be fulfilled. It follows from Eq. B.5 that this happens for $\alpha = \alpha_n = 8n + 6$, $n = 0, 1, 2, \ldots$. This has also been shown by Tanveer [2000] using a different approach.

18.12.5 *Model equation with $a \neq 0$ and $b \neq 0$*

It is interesting to study how this result is modified if the term multiplied by b in model Eq. 18.123 is present. Instead of Eq. 18.128 we get, using shifted coordinates

$$-\epsilon\ddot{\phi}(t) + \left[4t^2 - 4t^3 + t^4 + a + \frac{b}{t(2-t)}\right]\phi(t) = 1. \tag{18.130}$$

Expanding $1/t(2-t) = 1/4t^2 + 1/4t + 3/16 + \ldots$, we can draw the Kruskal–Newton diagram, see Fig. 18.32. No dot lies below support line 2 if we scale b according to

$$b = \beta\epsilon,$$

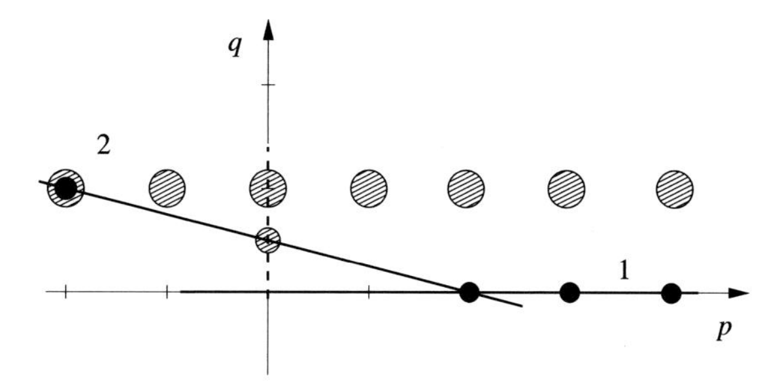

Fig. 18.32 Kruskal–Newton diagram for Eq. 18.130. The term multiplied by b is represented by a series of shaded dots. The parameter b is scaled, so no dot lies below support line 2.

with β of order unity. The inner equation reads

$$-\frac{d^2\Phi}{d\tau^2} + \left(4\tau^2 + \alpha + \frac{\beta}{4\tau^2}\right)\Phi = 1. \tag{18.131}$$

Tanveer also studied Eq. 18.131. He found solutions for a discrete sets of eigenvalues α_n that are functions of β. This shows that the term multiplied by b in model Eq. 18.123 has an effect on the value of a even though it is numerically irrelevant on the real s-axis.

18.13 Problems

1. Draw the Kruskal–Newton diagram and find a leading order (in ϵ) uniform approximation to

$$\epsilon y'' + (1 + x^2)y' - xy = 0,$$

with $\epsilon \ll 1$ and $y(0) = y(1) = 1$.

2. Draw the Kruskal–Newton diagram and find a leading order (in ϵ) uniform approximation to

$$\epsilon y'' - 2xy' - (5 + x)y = 0,$$

with $\epsilon \ll 1$ and $y(\pm 1) = 1$.

3. Draw the Kruskal–Newton diagram and find a leading order (in ϵ) uniform approximation to

$$\epsilon y'' + sin(x)y' - sin(2x)y = 0,$$

with $\epsilon \ll 1$ and $y(0) = 1$, $y(\pi) = 0$.

4. Draw the Kruskal–Newton diagram and find a leading order (in ϵ) uniform approximation to

$$\epsilon y'' + 2xy' - 8x^2 y = 0,$$

with $\epsilon \ll 1$ and $y(-1) = 0$, $y(1) = 2$.

5. Draw the Kruskal–Newton diagram and find a leading order (in ϵ) uniform approximation to

$$\epsilon y'' + 2xy' = xcos(2x),$$

with $\epsilon \ll 1$ and $y(\pm 1) = 3$.

6. Draw the Kruskal–Newton diagram and find a leading order (in ϵ) uniform approximation to

$$\epsilon y'' + (1 + 2x^2)y' + 2y = 0,$$

with $\epsilon \ll 1$ and $y(0) = y(1) = 1$.

7. Draw the Kruskal–Newton diagram and find a leading order (in ϵ) uniform approximation to

$$\epsilon y'' + y' + x cos(x^2) y = 0,$$

with $\epsilon \ll 1$ and $y(0) = 1$, $y(1) = 2$.

8. Draw the Kruskal–Newton diagram for the layer and find a leading order (in ϵ) uniform approximation to

$$\epsilon \frac{d^2y}{dx^2} + 2tan(x)\frac{dy}{dx} - 3y = 0$$

with the boundary conditions $y(0) = 1$, $y(\pi/2) = 1$, $\epsilon > 0$.

9. Draw the Kruskal–Newton diagram for the layer and find a leading order (in ϵ) uniform approximation to

$$\epsilon^2 \frac{d^2y}{dx^2} + tan(x)\frac{dy}{dx} - [x sin(x) + \epsilon cot(x)]y = 0$$

with $y(0) = 1$, $y(\pi/2) = e^{\pi/2}$, $\epsilon > 0$.

10. Draw the Kruskal–Newton diagram for the layer and find a uniform approximation accurate to order ϵ to

$$\epsilon \frac{d^2y}{dx^2} + x\frac{dy}{dx} = x$$

with $y(0) = y(1) = 1$, $\epsilon > 0$.

11. Draw the Kruskal–Newton diagram for the layer and find a leading order (in ϵ) uniform approximation to

$$(\epsilon^2 + x^2)\frac{d^2y}{dx^2} + (\epsilon + x^2)\frac{dy}{dx} = 0$$

with $y(0) = 0$, $y(1) = 1$, $\epsilon > 0$.

Appendix A

Lagrange's Theorem

Let $f(z)$ and $g(z)$ be analytic inside the contour C, a a point inside the contour, and t a complex number such that $|tg(z)| < |z - a|$ for all points z on C. Then the equation $\zeta = a + tg(\zeta)$ regarded as an equation in ζ has one root inside C, and

$$f(\zeta) = f(a) + \sum_{1}^{\infty} \frac{t^n}{n!} \frac{d^{n-1}}{da^{n-1}} [f'(a) g^n(a)]. \tag{A.1}$$

To prove this suppose $\theta(z)$ has a single zero inside C at $z = a$, and for all points z of C and all x inside C we have $|\theta(x)| < |\theta(z)|$. Then $\theta(z) - \theta(\zeta)$ has a single zero inside C at $z = \zeta$. Write the Cauchy integral expression for f

$$f(\zeta) = \frac{1}{2\pi i} \int_C \frac{f(z)\theta'(z)}{\theta(z) - \theta(\zeta)} dz, \tag{A.2}$$

since the contour can be contracted to circle the single root $z = \zeta$ where $\theta(z) \simeq \theta(\zeta) + \theta'(\zeta)(z - \zeta)$. Now make a Laurent expansion

$$\frac{1}{\theta(z) - \theta(\zeta)} = \sum_n \frac{\theta(\zeta)^n}{\theta(z)^{n+1}}, \tag{A.3}$$

giving

$$f(\zeta) = \frac{1}{2\pi i} \sum_n \theta(\zeta)^n \int_C \frac{f(z)\theta'(z)}{\theta(z)^{n+1}} dz. \tag{A.4}$$

Now integrate by parts in all terms except $n = 0$,

$$f(\zeta) = \frac{1}{2\pi i} \int_C \frac{f(z)\theta'(z)}{\theta(z)} dz + \frac{1}{2\pi i n} \sum_{1}^{\infty} \theta(\zeta)^n \int_C \frac{f'(z)}{\theta(z)^n} dz. \tag{A.5}$$

Substituting $\theta = (z-a)/g$ we find, with g analytic inside C

$$f(\zeta) = f(a) + \frac{1}{2\pi i n} \sum_{1}^{\infty} \frac{(\zeta - a)^n}{g(\zeta)^n} \int_C \frac{f'(z) g(z)^n}{(z-a)^n} dz, \tag{A.6}$$

or

$$f(\zeta) = f(a) + \sum_{1}^{\infty} \frac{t^n}{n!} \frac{d^{n-1}}{da^{n-1}} [f'(a) g(a)^n]. \tag{A.7}$$

Appendix B

Integral Solution for Eq. 18.129

In this appendix we give an integral representation of a particular solution of Eq. 18.129 and study its asymptotic behavior. Change variable to $w = \tau^2/2$ with $\Phi(\tau) = f(w)$, giving

$$2wf'' + f' - (8w + \alpha)f = -1, \tag{B.1}$$

and use a Fourier–Laplace representation $f = \int e^{wt} g(t)dt$, giving a solution with

$$g(t) = A(t-2)^{-(\alpha+6)/8}(t+2)^{(\alpha-6)/8}, \qquad 2(t^2-4)ge^{wt}|_a^b = -1 \tag{B.2}$$

with (a, b) the contour end points. We then have the solution

$$\Phi(\tau) = A \int_{-C} e^{wt}(t-2)^{-(\alpha+6)/8}(t+2)^{(\alpha-6)/8}dt \tag{B.3}$$

with $2\sqrt{2}A(-1)^{(-\alpha+2)/8} = -1$. Changing the integration variable $t \to -t$,

$$f(w) = A \int_C e^{-wt}(-t-2)^{-(\alpha+6)/8}(-t+2)^{(\alpha-6)/8}dt. \tag{B.4}$$

The contour C starts at $t = 0$ and goes to infinity in the sector $Re(wt) > 0$. Starting in the domain $Im(w) < 0$ we choose the singularities at $t = \pm 2$ to lie below the contour. We can deform the contour to the steepest descent contour $arg(wt) = 0$ without passing through a singularity provided $Im(w) < 0$ (Fig. B.1). Continuing to $Im(w) > 0$ the steepest descent contour must be deformed to go around the branch cut. Thus, the line $Im(w) = 0$ is a Stokes line for $f(w)$ and we pick up an extra contribution from the cut when passing it:

$$f_{bc} \simeq \frac{i}{2\sqrt{2}}e^{-2w}(4w)^{-(\alpha+2)/8} sin\left[\pi(\alpha/8 + 1/4)\right]\Gamma(\alpha/8 + 1/4). \tag{B.5}$$

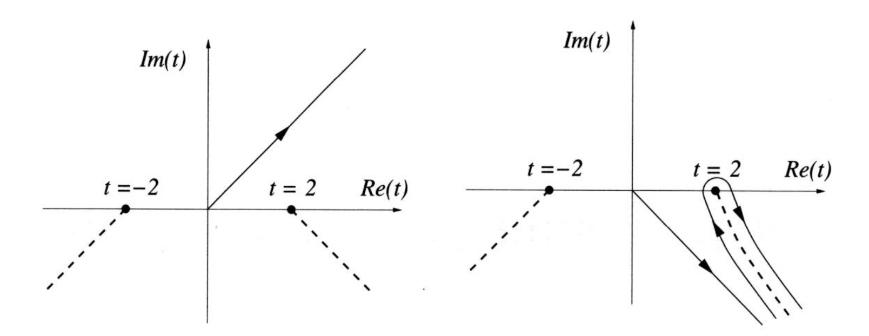

Fig. B.1 Left: The steepest descent contour for Eq. B.4 if $Im(w) < 0$. Right: Same but with $Im(w) > 0$.

The contribution $f_{ep} \simeq 1/(8w)$ from the end point $t \simeq 0$ is present everywhere. Thus,

$$\Phi(\tau) \simeq \begin{cases} 1/4\tau^2 & -\pi/2 < arg(\tau) < 0 \\ 1/4\tau^2 + f_{bc}/2 & arg(\tau) = 0 \\ 1/4\tau^2 + f_{bc} & 0 < arg(\tau) < \pi/2 \end{cases} . \tag{B.6}$$

Bibliography

Airy, G. B., *Autobiography of Sir George Airy*, Cambridge University Press (1896).

Airy, G. B., Trans. Camb. Phil. Soc. vi, 379 (1838).

Antione, J. P., *Two Dimensional Wavelets and their Relatives*, Cambridge University Press (2004).

Bell, E. T., *Men of Mathematics*, Simon and Schuster, New York (1965).

Bender, C. M. and Orszag, S. A., *Advanced Mathematical Methods for Scientists and Engineers*, McGraw-Hill, New York, NY (1978).

Berk, H. L., Nevins, W. M., and Roberts, K. V., J. of Math. Phys. 23, 988 (1982).

Berry, M. V., Proceedings of the Royal Society 427, 265 (1990).

Bohm, D., *Quantum Theory*, Prentice-Hall, Englewood Cliffs, NJ (1951) p. 41.

Bohr, N., Philosophical Magazine 26, 1 (1913).

Boyer, C. B., *The History of the Calculus and its Conceptual Development*, Dover, New York (1959).

Bombieri, E., A web search using "Riemann hypothesis" will lead to the Clay Institute article by Enrico Bombieri and to many others.

Brillouin, L., C. R. Acad. Sci. Paris 183, 24 (1926).

Budden, K. G., Phil. Trans. Royal Soc. London 290, 405 (1979).

Burrrus, C. S., Gopinath, A. R., and Haitao Guo, *Introduction to Wavelets and Wavelet Transforms*, Appendix A, Prentice-Hall, Upper Saddle River, NJ (1998).

Chapman, S. J. and Mortimer, D. B., Proc. R. Soc. A 461, 2385 (2005).

Chen, L., Kaw, P. K., Oberman, C. R., Guzdar, P., and White, R. B., Phys. Rev. Lett. 41, 649 (1978).

Cohen, L., *Time-Frequency Analysis*, Prentice-Hall, Englewood Cliffs, NJ (1995).

Cole, J. D., *Perturbation Methods in Applied Mathematics*, Blaisdell, Walthan, MA (1968).

Copson, E. T., *Introduction to the Theory of Function of a Complex Variable*, Oxford Press, London (1935) p. 118.

Daubechies, I., *Ten Lectures on Wavelets*, SIAM, Philadelphia, PA (1992).

Daubechies, I., Communications on Pure and Applied Mathematics 41, 909–996 (1988).

Davis, P. J., Amer. Math. Monthly 66, 849–869 (1959).
Dewar, R. L. and Davies, B., J. Plasma Phys. 32, 443 (1984).
Dingle, R. B., *Asymptotic Expansions: Their Derivation and Interpretation*, Academic Press, London (1973).
Eliot, T. S., *Old Possum's Book of Practical Cats*, Faber, London (1939).
Erdelyi, E. T., *Asymptotic Expansions*, Dover Publications (1955).
Flandrin, P., *Time-Scale Analysis*, Academic Press, London (1999).
Ford, L. R., *Differential Equations*, McGraw-Hill, New York, NY (1955).
Ford, Ann. Phys. NY 7, 287 (1959).
Franceschetti, D. R., *Biographical Encyclopedia of Mathematicians*, Marshall Cavendish, New York (1999).
Friedrichs, K. O., Bull. Amer. Math Soc. 61, 485 (1955).
Feingold, M., *The Newtonian Moment*, Oxford University Press, Oxford (2004).
Fuchs, L., Journal für Math., LXVI (1866) pp. 121–160.
Furry, W. H., Phys. Rev. 71, 360 (1947).
Gleick, James, *Isaac Newton*, Pantheon Books, New York, NY (2003).
Gleick, James, *Genius*, Pantheon Books, New York, NY (1992).
Gradstein, E. S. and Riezek, E. M., *Tables of Integrals, Sums, Sequences and Continuations*, Government Publication of Physical-Mathematical Literature, Moscow (1963).
Haar, A., Mathematische Annalen 69, 331–371 (1910).
Heading, J., *An Introduction to Phase Integral Methods*, Wiley, NY (1962).
Hubbard, Barbara, *The World According to Wavelets*, A. K. Peters, Wellesley, MA (1996).
Jaffard, S. and Meyer, Y., Wavelet methods for pointwise regularity and local oscillations of functions, Am. Math. Soc., Volume 587 (1996).
Jaffard, S., Meyer, Y., and Ryan, R., *Wavelets, Tools for Science and Technology*, SIAM, Philadelphia (2001).
Jeffries, H., Proc. Lond. Math. Soc. 23, 428 (1923).
Jeffries, H., *Asymptotic Approximations*, Oxford University Press (1962).
Jeffries, H., Philos. Mag. [7], 33, 451–456 (1942).
Kartashev, L. P. and Kartashev, S. I. (eds.), Supercomputers and the Riemann zeta function, *Proc. 4th Intern. Conf. on Supercomputing*, International Supercomputing Institute (1989) 348–352.
Knopp, K., *Elements of the Theory of Functions*, Dover Publications (1952).
Kramers, H. A., Zeit. f. Phys. 39, 828 (1926).
Kruskal, M., Asymptotology, in "Mathematical Models in Physical Sciences" (proceedings of conference of that name at University of Notre Dame, Apr 15–17, 1962), edited by S. Drobot and P. A. Viebrock, Prentice-Hall, Englewood Cliffs, NJ, 17–47, with discussion 47–48 (1963).
Kruskal, M. D. and Segur, H., Stud. Appl. Math. 85, 129 (1991).
Landau, L. D., J. Physics (U.S.S.R.) 10, 25 (1946).
Mallat, S. M., *A Wavelet Tour of Signal Processing*, Academic Press, London (1998).
Mallat, S. G., IEEE Trans. Pattern Anal. Mach. Intell. 11(7), 674 (1989).

Mehta, M. L., *Random Matrices and the Statistical Theory of Energy Levels*, Academic Press, New York (1967).

Morse, P. M. and Feshbach, H., *Methods of Theoretical Physics, I, II*, McGraw Hill (1953).

The MacTutor History of Mathematics Archive, http://www-history.mcs.st-andrews.ac.uk/history/index.html.

Montgomery, H. L., "The pair correlations of zeros of the zeta function", pp. 181–193 in "Analytic Number Theory", ed. H. G. Diamond, Proc. Symp. Pure Math. 24, [Providence, Am. Math. Soc.].

Montgomery, H. L. and Dyson, F., for full account see Computing Science 91 [4] 296 (2003).

Newton, Isaac, *Methods of series and fluxions*, in *The Mathematical Papers of Isaac Newton*, edited by D. T. Whiteside, Volume III, 1670-1673, pp. 50–71, Cambridge University Press (1969).

Olver, F. W. J., *Asymptotics and Special Functions*, Academic Press, New York (1974) pp. 1–2.

Olver, F. W. J., Phil. Tran. of the Royal Society London 278, 137 (1975).

Pearlstein, L. D. and Berk, H. L., Phys. Rev. Lett. 23, 220 (1969).

Poincaré, Acta Mathematica, viii. (1886) pp. 295–344.

Schiff, L. I., *Quantum Mechanics*, McGraw-Hill, New York (1955) pp. 155–158.

Stix, T., *Waves in Plasmas*, American Insititue of Physics, NY (1992).

Stokes, G. G., Trans. Camb. Phil. Soc. 10, 105 (1857).

Stokes, G. G., Proc. Camb. Phil. Soc. 6, 362 (1889).

Soop, M., Ark. Fys. 30, 217 (1965).

Tanveer, S., J. Fluid Mech. 409, 273 (2000).

Taylor, Brook, *Methodus incrementorum directa et inversa* (1715).

Titchmarsh, E. C., *Introduction to the Theory of Fourier Integrals*, Oxford University Press (1948).

Tsang, K. T., Catto, P. J., Whitson, J. C., and Smith J., Phys. Rev. Lett. 40, 327 (1978).

Turnbull, H. W., "The Discovery of Infinitesimal Calculus", Nature 167, no. 4261 (June 30, 1951) 1048–1050.

Turnbull, H. W., *The Great Mathematicians*, University Press, New York (1961).

Van den berg, J. C., editor, *Wavelets in Physics*, Cambridge University Press (1999).

Veneziano, G., Il Nuovo Cimento 57A, 190 (1968).

Watson G. N., *A Treatise on the Theory of Bessel Functions*, Cambridge University Press (1922).

WaveLab is a library of MatLab routines for wavelets and time–frequency transforms. MatLab is a product of the Mathworks company, based in Natick, Massachusetts. Wavelab can be retrieved at http://www-stat.stanford.edu/$\sim$ wavelab.

Welland, G. V., *Beyond Wavelets*, Academic Press (2003).

Wentzel, G., Zeit. f. Phys. 38, 518 (1926).

Whittaker, E. T. and Watson, G. N., *A Course of Modern Analysis*, Cambridge University Press (1962).

White, R. B. and Chen, F. F., Plasma Physcis 16, 565 (1974).
White, R. B., J. of Comput. Phys. 31, 409 (1979).
White, R. B., *The Theory of Toroidally Confined Plasmas*, Imperial College Press (2001).
White, R. B., (2000) The code WKB, written in fortran 99, is available by anonymous ftp. Simply type "ftp ftp.pppl.gov" and in reply to user type "anonymous", for password give your e-mail address. Then change directory through "cd /pub/white/Wkb" after which "get * " will retrieve all files.

Index

Abel's formula, 17
Adelman L., 336
Airy
  asymptotic, $z \to -\infty$, 204
  asymptotic expansion, 202
  asymptotic, $z \to +\infty$, 204
  differential equation, 196
  integral representation, 200
  matching local solutions, 208
  Taylor series, 206
  WKB analysis, 196
Airy G.B., 195
analytic continuation, 37
analytic function, 32
anti-Stokes, 75
anti-Stokes line, 75
asymptotic dominance, 49
asymptotic limits of integrals, 118
asymptotic sequence, 51
asymptotic series
  bound on error, 53
  definition, 51
  existence of sum, 52
  origin of divergence, 63
  oscillating, 61
  truncation, 60

Bernoulli, 133
Bernoulli Daniel, 15
Bernoulli equation, 23
Bernoulli Jakob, 15
Bernoulli Johann, 15
Bernoulli's law, 15
Bessel
  analytic continuation in ν, 249
  asymptotic expansion, 256
  divergent series, 250
  equation, 247
  Fourier–Laplace integral, 252
  modified Bessel equation, 45, 263
  Sommerfeld integral, 258
  Wronskian, 258
Bessel F.W., 247
Bessel's inequality, 177
Bohr–Sommerfeld, 87
Borel summation, 68
bound state, 86, 236
boundary layer, 344
boundary layer
  center, 352
  left, 349
  nested, 356
  right, 354
boundary layer location, 347
broken symmetry, 110
Brownian motion, 23
Budden, 93, 230
Budden problem, 93, 232

Cauchy A., 31
Cauchy integral formula, 33
Cauchy integral theorem, 33
Cauchy residue theorem, 35
Cauchy–Riemann conditions, 32

causality, 85, 90, 159, 307
code - WKB, 76
Cole equation, 372
confluent hypergeometric function, 285
connection formulae, 78
convergence test
  radical, 34
  ratio, 34

difference equation, 157
diffusion equation, 22
Dirac delta function, 18
Dirichlet, 327
divergent series, 49
dominant balance, 2, 50
dominant WKB, 77

eigenvalues of a Hermitian matrix, 109
eigenvalues of a random matrix, 110
eikonal, 77
end point contribution, 119
error function, 62, 96
Euclidean algorithm, 337
Euler beta function, 146
Euler kernel, 153, 167
Euler L., 133
Euler phi function, 133, 337
Euler product for ζ, 332
Euler summation, 67
Euler theorem, 337
Euler–Mascheroni constant, 137, 249
exponential tunneling, 110

fast Fourier transform, 186
fast wavelet transform, 186
Fermat's last theorem, 133
Feynman diagrams, 105
Feynman R., 105
fixed point convergence, 5
Fourier J., 151
Fourier–Laplace kernel, 153
Fredholm alternative theorem, 21
Fuchs classification, 45

Gamma function
  $\Gamma(2z)$, 146
  definition, 134
  digamma function, 148
  Euler product, 144
  Euler–Mascheroni constant, 137
  integral representation, 140, 144
  Stirling approximation, 135, 338
Gauss, 247, 271, 327, 334
generating function
  Bessel, 261, 262
  Legendre, 173
Gram–Schmidt, 176
Green's function, 18, 19, 159, 160, 308
group velocity, 85
grystal growth, 381

Hadamard, 334
Hankel functions, 252
Heading, 90
Heading J., 73
Heading's rules, 83
Heaviside function, 18
Heisenberg uncertainty principle, 190
Hermite
  definition, 278
  integral representation, 274, 278
Hermite C., 271

indicial equation, 45
indicial exponent, 45
inhomogeneous differential equation, 152
instability, 86
integral representation
  Airy function, 200, 207
  Bessel, 252, 257, 258, 261
  bound state, 238
  Budden problem, 232
  error function, 62
  Hankel functions, 252
  Hermite, 274, 278
  overdense barrier, 223
  underdense barrier, 228
  Weber, 274, 282
  Weber's equation, 277

Whittaker, 288, 291, 292
integral solutions, 152
inverse functions, 39
irregular singular point, 48

Jeffries H., 73

kernel, 152
kernel
Euler, 153
Fourier, 153
Mellin, 153
Sommerfeld, 153
Kruskal M., 2
Kruskal–Newton diagram, 3, 344, 349, 352
Kruskal–Segur, 381
Kummer E., 31

l'Hôpital, 15, 333
Lagrange, 247
Landau prescription for a pole, 94
Laplace integral, 118
Laplace P., 117
Laplace's equation, 32
leading asymptotic form, 50
Legendre
integral representation, 167
integral representation for P_ν, 169
integral representation for Q_ν, 169
Rodrigues' formula, 168
Legendre A.M., 165
Legendre equation, 166
Legendre polynomials, 167
level repulsion, 110
level spacing, 110
Levi-Civita tensor, 17
Littlewood, 334
local approximations, 45

Maxwell, 271
Mellin integral representation, 158
Mellin kernel, 153
Mellin representation, 157
Mellin representation
Airy, 207
Gamma function, 338
parabolic cylinder function, 281
Mittag-Lefler, 138
modified Bessel equation, 263
modified Bessel equation, 45

Napoleon B., 31, 117, 151
Neumann, 248
Newton I., 1, 343
Newton–Raphson, 8

ordinary differential equation, 16
ordinary point, 45
orthogonal polynomials
Chebychev, 177
Gegenbauer, 177
Hermite, 177
Jacobi, 177
Laguerre, 177
Legendre, 177
orthonormal, 176
overdense barrier, 87, 221
overestimated prime conjecture, 334

parabolic cylinder function, 272
parallel postulate, 165, 327
perturbation theory, 106
phase plane analysis, 24
phase plane barrier, 26
phase-integral methods, 74
Picard's great theorem, 48
Poincaré H., 43
Poussin, 334
Poynting flux, 80
Prandtl L., 343
prime number distribution, 332
principle of maximal complexity, 362, 369
propagator, 160
public key codes, 134, 336

quantum electrodynamics, 105
quantum tunneling, 113

radius of convergence, 35
reflection, 89

regular singular point, 45
residue theorem, 36
resistive tearing, 305
Riccati equation, 23
Riemann G., 327
Riemann hypothesis, 335
Riemann sheet, 38, 218
Riemann zeta
  $\zeta'(0)$, 339
  $\zeta(0)$, 339
  $\zeta(s)$ and $\zeta(1-s)$, 329
  definition, 328
  integral representation, 328
  zeros, 335
Rivest R., 336
Rosenbluth M.N., 299

saddle point
  and Stokes constant, 214
  contribution, 124
  form, 118
scattering, 87, 222
Schlafli's integral formula, 169
Schwartz inequality, 175
Schwinger J., 105
separation of variables, 23
Shamir A., 336
Shanks transformation, 67
signal processing, 180
singular perturbation theory, 11, 106
solubility conditions, 21
Sommerfeld contour, 258
Sommerfeld kernel, 153
steepest descent, 119
Stirling approximation, 135, 340
Stokes, 75
Stokes constant
  definition, 78
  derivation, 81
  half value, 81, 215, 220
  integral representation, 216, 221
Stokes diagram
  Airy, 196
  Bessel, 252
  bound state, 86
  Budden, 230
  Budden problem, 94
  complex eigenvalue, 97
  definition, 76
  scattering, 87
Stokes lines, 77
Stokes phenomenon
  Airy, 205
  Bessel, 257
  error function, 63
  integral representations, 152
  Mellin representation, 158
Stokes theorem, 33
Struve equation, 303
subdominant WKB, 77

Tanveer, 386
Taylor series, 34, 106
Te Riele, 334
theory of relativity, 43
time–frequency atoms, 190
Tomonaga S., 105
transcendental function, 134
transcendental number, 134, 137
transmission, 87
truncation of series, 57

underdense barrier, 227

viscous fingering, 385

Watson's lemma, 120
wave packet, 85
wavelet
  basis, 179
  filter, 180
  quadratic mirror filter, 180
  scaling function, 178
Weber W., 271
Weber's integral, 267
Weber–Hermite equation, 272, 301
Whittaker E.T., 285
Whittaker function, 285
Wigner–Ville transform, 191
Wilkinson polynomial, 13
WKB
  accuracy, 241

Airy, 196
Bessel, 250
bound state, 86
Budden problem, 93, 230
connection formulae, 78
continuation rules, 83
derivation, 74
description, 74
error function, 97
scattering, 87
Stokes constants, 81
Wronskian
Airy, 210
Bessel, 258
definition, 16
parabolic cylinder function, 280
second order equation, 20